Mode-Locked Lasers

Introduction to
Ultrafast Semiconductor and
Fiber Lasers

Other World Scientific Titles in Laser Physics

Fundamentals of Laser Optoelectronics
Second Edition
ISBN: 978-981-12-5475-8
ISBN: 978-981-12-5498-7 (pbk)

Fundamentals of Laser Physics
by Kyungwon An
ISBN: 978-981-12-6527-3
ISBN: 978-981-12-9321-4 (pbk)

Mode-Locked Lasers

Introduction to Ultrafast Semiconductor and Fiber Lasers

Niloy K Dutta
University of Connecticut, USA

World Scientific

NEW JERSEY • LONDON • SINGAPORE • BEIJING • SHANGHAI • HONG KONG • TAIPEI • CHENNAI • TOKYO

Published by

World Scientific Publishing Co. Pte. Ltd.
5 Toh Tuck Link, Singapore 596224
USA office: 27 Warren Street, Suite 401-402, Hackensack, NJ 07601
UK office: 57 Shelton Street, Covent Garden, London WC2H 9HE

British Library Cataloguing-in-Publication Data
A catalogue record for this book is available from the British Library.

MODE-LOCKED LASERS
Introduction to Ultrafast Semiconductor and Fiber Lasers

ISBN 978-981-12-9016-9 (hardcover)
ISBN 978-981-12-9017-6 (ebook for institutions)
ISBN 978-981-12-9018-3 (ebook for individuals)

For any available supplementary material, please visit
https://www.worldscientific.com/worldscibooks/10.1142/13758#t=suppl

Desk Editor: Joseph Ang

Typeset by Stallion Press
Email: enquiries@stallionpress.com

Preface

The concept of the laser dates to 1958. The successful demonstration of a solid-state ruby laser and He–Ne gas laser was reported in 1960. Soon after the development of the laser, mode-locked laser action was demonstrated. Mode locking using a saturable absorber, which even today is widely used, was demonstrated in the mid-1960s. The mode-locked pulses are generated essentially by phase locking of many lasing longitudinal modes of the laser cavity. Simple theory predicted that to get very short pulses, many lasing longitudinal modes need to be locked in phase. Thus, it was quickly realized that a wide gain spectrum of the laser medium will produce short mode-locked pulses. Dye lasers (where the laser medium are organic dye molecules dissolved in a solvent) were demonstrated also in the mid-1960s. The dye has a large fluorescence spectral width (~50–100 nm) and a large gain spectrum width when pumped (excited to higher energy levels) by an external laser. This led to the investigation of mode locking of dye lasers where very short pulses (~100 fs) were produced. Dye lasers continued to be the laser of choice for many scientific investigations of materials and physical phenomena and chemistry for several decades because of the wide range of wavelengths and short pulses that a dye laser could produce.

Semiconductor lasers, invented in the early 1960s and developed in the 1970s and 1980s, became the laser source of choice for fiber optics communication systems because of their ease of operation. Fiber optics communication systems are installed worldwide today. It was recognized that mode-locked semiconductor lasers are important for very high-speed optical time division multiplexed systems, which led to the development of short pulse generation and mode locking of semiconductor lasers at precise repetition rates.

The market demand for higher-capacity transmission was helped by the fact that computers continued to become more powerful and needed to be interconnected. This is one of the key reasons why the explosive growth of optical fiber technology parallels that of computer processing and other key information technologies. The need for higher capacity is continuing to encourage research in wavelength division multiplexed (WDM)-based transmission and optical time division multiplexed transmission.

The laser emission in various rare-earth (Nd, Pr, Er, Ho, Tm, and Yb)-doped glasses has been studied since the 1960s. However, the potential for rare-earth-doped fiber amplifiers and fiber lasers was not recognized until the late 1980s and it began with a demonstration of a very low threshold fiber amplifier. This pioneering demonstration conveyed an important message that using well-established low-loss fiber fabrication technology and semiconductor laser fabrication technology a new class of rare-earth-doped lasers and amplifiers could be produced for new applications. This message was particularly well received immediately by optical communication technologists, and they went on to develop Er-doped fiber amplifiers for various fiber communication applications.

Rare-earth-doped amplifiers have a wide gain spectrum. Thus mode-locked operation of fiber laser began to be investigated for a wide range of applications. The wavelength of operation can be varied by varying the rare-earth dopant and the pump laser. High-power fiber lasers producing multi-kWs of power using double-clad fiber geometries are expected to become increasingly important in future industrial applications. The development of short-pulse fiber lasers is expected to become important in many applications, such as tomography and sensors.

Fiber-based Raman lasers and amplifiers represent an important area for lasers and amplifiers in wavelength regions where a rare-earth dopant may not be efficient. Using the cascaded Raman process and multiple Stokes shifts, very efficient fiber Raman lasers and amplifiers have been demonstrated.

Applications of mode-locked and -pulsed lasers in various domains such as harmonic generation, surgery, microscopy, communication, and materials processing are continuing to increase. Pump–probe experiments using pulsed laser systems allow starting and tracing of chemical reactions and studies of ultrafast relaxation processes in materials. High-power fiber lasers are candidates for metals processing applications. Very high-power pulses are being utilized for nuclear fusion studies. Although significant developments have occurred in high-power optical fiber lasers, the need for higher power for commercial (metal cutting and shaping) applications will continue to drive the research on higher-power fiber lasers using Nd- or Yb-based systems and possibly others.

This book is aimed at researchers already engaged in or wishing to enter the field of mode-locked lasers. It is also useful for graduate students, scientists, and engineers interested in fiber optics communication. An attempt has been made to make this book self-contained and each chapter has a set of references which can be consulted for further study.

N. K. Dutta

Contents

Chapter 1

Introduction

1.1 Historical Developments

The concept of the laser dates to 1958 [1]. The successful demonstration of a solid-state ruby laser and He–Ne gas laser was reported in 1960 [2, 3]. Soon after the development of the laser, mode-locked laser action was demonstrated [4, 5]. Mode locking using a saturable absorber which even today is used very widely was demonstrated in the mid-1960s [5]. The mode-locked pulses are generated essentially by phase locking of many lasing longitudinal modes of the laser cavity. Simple theory (Section 1.3) predicted that to get very short pulses, many lasing longitudinal modes need to be locked in phase. Thus, it was quickly realized that a wide gain spectrum of the laser medium will produce short mode-locked pulses. Dye lasers (where the laser medium are organic dye molecules dissolved in a solvent) were demonstrated also in the mid-1960s [6, 7]. The dye has a large fluorescence spectral width (~50–100 nm) and a large gain spectrum width when pumped (excited to higher energy levels) by an external laser. This led to the investigation of mode locking of dye lasers where very short pulses (~100 fs) were produced. This is discussed in detail in Section 1.5. Dye lasers continued to be the laser of choice for many scientific investigations of materials and physical phenomena and chemistry for several decades because of the wide range of wavelengths and short pulses that a dye laser could produce.

The semiconductor injection laser was demonstrated in 1962 [8–11]. CW operation of a semiconductor laser was demonstrated in 1970 [12, 13] using the concept of double heterostructure [14]. Since then, the semiconductor injection laser has emerged as an important device in many optoelectronic systems, such as optical recording, high-speed data transmission through fibers, optical sensors, high-speed printing, and guided wave signal processing.

The laser has a gain medium and an optical feedback mechanism. The latter is generally provided by the cleaved facets of the semiconductor itself or by a grating of suitable periodicity etched close to the gain medium. Semiconductor laser has the unique ability to be modulated at gigahertz speeds simply by modulation of the injection current. Mode-locked semiconductor lasers were initially demonstrated using an external cavity. Semiconductor laser chip designs exist today which produce mode-locked pulses. Although not as versatile as dye lasers, the compactness and ease of operation are advantages of semiconductor mode-locked lasers for practical systems. Semiconductor laser properties are discussed in several books [15–18].

The optical fiber used in optical transmission systems is based on silica (SiO_2) material. Low-loss silica fibers were first proposed in 1966 [19], and by the early 1970s, low-loss fibers had an attenuation of 20 dB/km. Rare-earth-doped silica fibers can provide optical gain and laser action when the rare earth is excited by a suitable pump laser. These devices are called fiber lasers. Mode-locked fiber lasers have been extensively studied. They can be made to operate at different wavelengths using different rare-earth doping (Nd, Er, Yb, Ho, Sm, etc.). This book provides a comprehensive account of design, fabrication, and performance of mode-locked semiconductor and fiber lasers.

1.2 Principles of Laser

A typical laser (shown in Figure 1.1) has a medium which provides optical gain and a set of mirrors on each side of the gain medium which form a Fabry–Perot resonator and provide optical feedback. The two together form the optical cavity of the laser. The light generated by spontaneous emission in the gain medium is amplified by the gain medium (the atoms emit by stimulated emission normal to the mirrors) and it bounces between the mirrors to form the laser light. A fraction of

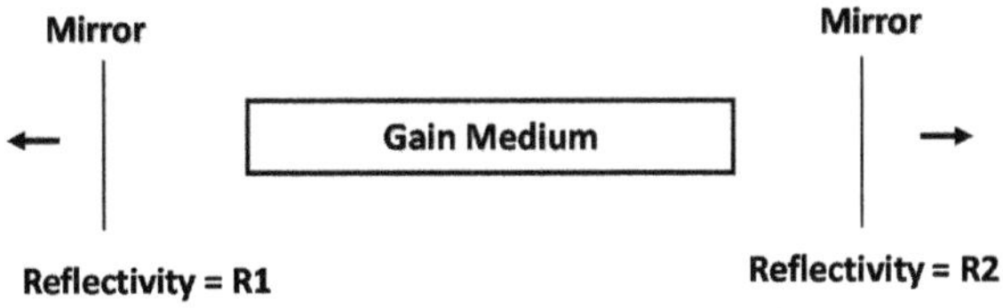

Figure 1.1. Schematic of a typical laser. R_1 and R_2 are reflectivity of the mirrors which provide optical feedback. The laser emission is normal to the optical cavity along the z direction. The separation between the mirrors is the laser cavity length, L.

this light is transmitted by the mirrors and is observed in the form of laser emission. If g is the net optical gain per unit length of the laser medium of length L and R_1 and R_2 are the reflectivity of the mirrors, the condition for laser action (based on net round-trip optical loss = optical gain) is

$$R_1R_2 \exp(2gL) = 1. \tag{1.1}$$

In addition, after one round trip, the light must have a phase change which is a multiple of 2π. Assuming λ_m is the wavelength of laser action (in air), this implies $E(z) = E(z + 2L)$, where $E(z)$ is the electric field. For a plane wave, $E(z) = E_0\exp(\text{i}nk_0z)$, where $k_0 = 2\pi/\lambda_0$, and λ_0 is the wavelength in air, and n is the refractive index of the laser medium. Thus, the condition $E(z) = E(z + 2L)$ implies

$$2nLk_0 = 2m\pi, \tag{1.2}$$

where m is an integer. The above equation states only certain discrete values of wavelength λ_0 can undergo laser action. Writing $k_0 = 2\pi/\lambda_m$, the wavelengths at which the laser can operate are given by

$$2nL = m\lambda_m. \tag{1.3}$$

The above describes the wavelengths of operation of the laser and are called longitudinal modes (for different m values) of the laser. For $L =$ ~0.5 m for gas lasers ($n \sim 1$), and $\lambda \sim 0.5$ μm, m is ~2×10^6, and, for a semiconductor laser with $L = 250$ μm, $n = 3.0$, and $\lambda = 1$ μm, $m \sim 1500$.

Thus, m is a large number. The separation between the laser modes for large m is given by

$$\Delta\lambda = \lambda^2/2nL. \tag{1.4}$$

For the above examples of gas laser and the semiconductor laser, the $\Delta\lambda$ values are 2.5×10^{-7} μm and 7×10^{-4} μm, respectively. The gain spectrum of a laser is typically much larger than the mode spacing. For 250 μm long semiconductor lasers emitting near 1.3 μm, the mode spacing ($\Delta\lambda$) is ~1 nm and the gain spectrum is ~50 nm wide. This allows multiple longitudinal mode operation of a typical semiconductor laser. The gain spectral width of most other types of lasers is much larger than the mode spacing. Generally, these laser modes are emitted with no phase correlations between them. A mode-locked laser pulse is obtained when all lasing modes within the gain spectrum have the same phase (i.e., locked in phase). This is discussed in the following section.

1.3 Principle of Mode Locking

Mode locking is a technique to obtain ultra-short pulses of light, which involves the "locking" together of the phases of many lasing longitudinal modes [20]. In this technique, the phases of the longitudinal modes are locked when the gain or loss of the laser is modulated at the inverse of the round-trip time of the optical cavity.

The electric field of the mth longitudinal mode may be written as

$$E(z,t) = \hat{\varepsilon}_m E_m(z)\sin(\omega_m t + \phi_m) = \hat{\varepsilon}_m E_m \sin k_m z \sin(\omega_m t + \phi_m), \tag{1.5}$$

where

$$k_m = m\frac{\pi}{L} \tag{1.6}$$

and

$$\omega_m = k_m c = m\frac{\pi c}{L}, \tag{1.7}$$

$m = 1, 2, 3,\ldots$. For simplicity, it is assumed that the mode fields all have the same magnitude and polarization. Furthermore, consider the case when the phases of all modes are equal, i.e., consider the simple case $\phi_m = 0$. Then the total electric field in the cavity is

$$E(z,t) = \sum_m E_m(t) = E_0 \sum_m \sin(k_m z) \sin \omega_m t, \tag{1.8}$$

where the summation is over all oscillating modes. The integer m typically is a very large number. Consider, $m \rightarrow M + n$ and write Equations (1.6) and (1.7) as

$$k_m = \frac{(M+n)\pi}{L}, \tag{1.9}$$

$$\omega_m = \frac{(M+n)\pi c}{L}, \tag{1.10}$$

where M is a very large positive integer ($M \approx 10^{4-6}$) and n runs from $-(N-1)/2$ to $(N-1)/2$, corresponding to a total of N modes (few hundred) centered at the frequency $M\pi c/L$ which is near the peak of the gain spectrum. Then Equation (1.8) becomes

$$\begin{aligned} E(z,t) &= E_0 \sum_{-(N-1)/2}^{(N-1)/2} \sin \frac{(M+n)\pi z}{L} \sin \frac{(M+n)\pi ct}{L} \\ &= \frac{1}{2} E_0 \sum_n \left[\cos \frac{(M+n)\pi(z-ct)}{L} - \cos \frac{(M+n)\pi(z+ct)}{L} \right] \end{aligned} \tag{1.11}$$

for the total electric field in the laser cavity. Using the general identity

$$\sum_{-(N-1)/2}^{(N-1)/2} e^{inx} = \frac{\sin(Nx/2)}{\sin(x/2)}, \tag{1.12}$$

the sum

$$\begin{aligned} \sum_{-(N-1)/2}^{(N-1)/2} \cos \frac{(M+n)\pi(z-ct)}{L} &= \mathrm{Re} \sum_{-(N-1)/2}^{(N-1)/2} e^{i(M+N)\pi(z-ct)/L} \\ &= \mathrm{Re} \left\{ e^{iM\pi(z-ct)/L} \frac{\sin[\pi N(z-ct)/2L]}{\sin[\pi(z-ct)/2L]} \right\} \\ &= \left[\cos \frac{M\pi(z-ct)}{L} \right] \frac{\sin[\pi N(z-ct)/2L]}{\sin[\pi(z-ct)/2L]}. \end{aligned} \tag{1.13}$$

From Equation (1.13), then

$$E(z,t) = \frac{E_0}{2}\left\{\begin{array}{l} \cos k_0(z-ct)\dfrac{\sin[\pi N(z-ct)/2L]}{\sin[\pi(z-ct)/2L]} \\ -\cos k_0(z+ct)\dfrac{\sin[\pi N(z+ct)/2L]}{\sin[\pi(z+ct)/2L} \end{array}\right\}, \tag{1.14}$$

where $k_0 = \pi M/L$.

The functions

$$A_N^{(\pm)}(z,t) = \frac{\sin[\pi N(z \pm ct)/2L]}{\sin[\pi(z \pm ct)/2L]}. \tag{1.15}$$

appearing in Equation (1.14) have maxima occurring at

$$z \pm ct = m(2L), \quad m = 0, \pm 1, \pm 2, \dots . \tag{1.16}$$

Consider now a fixed value of z inside the cavity, for instance, then there are pulses of peak amplitude $NE_0/2$ appearing at time intervals of $2L/c$, each pulse having a duration T/N, as indicated in Figure 1.2. Consider again the scalar electric field in the exponential form

$$E(z,t) = E_m e^{i(\omega_m t+\phi_m)} e^{ik_m z}. \tag{1.17}$$

suppose that the amplitude E_m is not constant but rather is modulated periodically in time according to the formula

$$E_m = E_0\,(1 - \varepsilon \cos \Omega t), \tag{1.18}$$

where Ω is the modulation frequency and E_0 and ε are constants.

It creates an amplitude-modulated field:

$$E_m(z,t) = E_0(1-\varepsilon\cos\Omega t)e^{i(\omega_m t+\phi_m)}e^{ik_m z}. \tag{1.19}$$

Since

$$\cos\Omega t e^{i\omega_m t} = \frac{1}{2}e^{i(\omega_m+\Omega)t} + \frac{1}{2}e^{i(\omega_m-\Omega)t}, \tag{1.20}$$

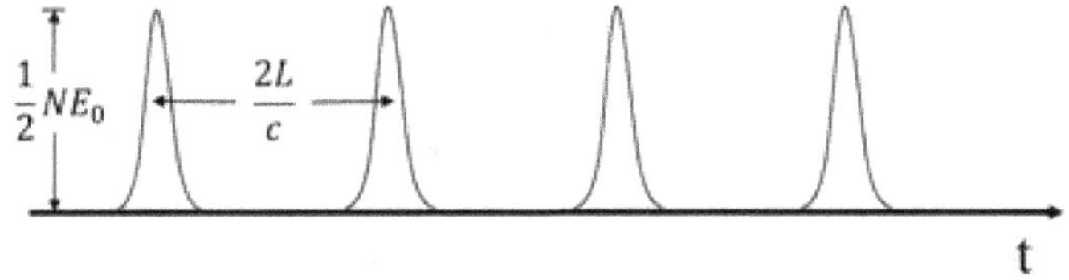

Figure 1.2. A mode-locked pulse train as a function of time, observed at a fixed position z.

the electric field may be written as a sum of harmonically varying parts:

$$E_m(z,t) = -\frac{1}{2}E_0[e^{i(\omega_m+\Omega)t} - 2e^{i\omega_m t} + e^{i(\omega_m-\Omega)t}]e^{i\phi_m}e^{ik_m z}. \tag{1.21}$$

Equation (1.21) shows that the amplitude modulation of the field of frequency ω_m has generated sidebands of frequency $\omega_m \pm \Omega$. If Ω equals $\pi c/L$ i.e., longitudinal mode spacing, the side bands produced reinforce the adjacent modes and make them oscillate in the same phase. This creates a situation that the highest gain is obtained if all the modes within the gain spectrum oscillate in the same phase; this is equivalent to a pulsed operation as shown in Figure 1.2. This is the principle of amplitude-modulated mode locking. Often a saturable absorber is used (in the optical cavity) in addition to current modulation for mode locking.

1.4 Various Types of Mode-Locked Lasers

Several types of mode-locked lasers have been reported beginning in the 1960s. These lasers operate at different regions of the spectrum and demonstrate pulsed operation using the mode-locking principle [4, 5]. Mode-locked operation using a saturable absorber was demonstrated in 1966 [5]. From Section 1.3, it is clear that a wide spectrum would lead to shorter pulse operation. Among the laser systems known in the 1960s, the dye laser had a very wide spectrum. Thus, a dye laser was extensively investigated for mode locking.

The fiber laser has been extensively studied for mode locking, using both a ring laser cavity and a straight optical cavity. Both active and passive mode-locking techniques are suitable for fiber lasers [21] (see later chapters).

The semiconductor laser, which subsequently became the laser of choice for many applications because of its small size and ease of operation, has also been studied for mode-locked operation. A typical laser chip is about 250 μm long with cleaved ends (facets), thus the mode spacing is large (~0.5–1 nm), and hence the number of modes within the gain spectrum is much smaller than that using an external cavity (~10 cm long, mode spacing ~0.005–0.01 nm). The semiconductor laser was initially investigated for mode-locked operation using an external cavity and later compact mode-locked lasers were demonstrated [22–25].

1.5 Mode-Locked Dye Lasers

A dye laser uses an organic dye (which fluoresces in the visible region of the spectrum) as the gain medium. The dye is dissolved into a liquid solution which mixes the dye molecules evenly into the liquid. The dye emits over a wide range of wavelengths, spanning over 50–100 nm. The wide bandwidth makes them particularly suitable for laser emission which can be tuned over a wide range as mode-locked pulsed lasers. A typical dye is Rhodamine 6G which can be tuned from 560 nm to 635 nm.

The dye solution is generally circulated through a dye cell at high speed to reduce triplet state absorption and decrease gradual degradation of the dye. The schematic of the energy levels of a dye molecule is shown in Figure 1.3. A high energy source of light is needed to "pump" the dye liquid beyond its lasing threshold. An external laser (typically Ar laser, N_2 laser, or Ti-Sapphire laser) is usually used in later-generation dye lasers. Mirrors are also needed to produce laser action. The light generated by the dye is amplified with each pass through the liquid. The typical reflectivity of the output mirror is 80% and all other mirrors have ~99% or higher reflectivity. For a tunable laser, a filter or a prism or a diffraction grating is usually configured in the beam path. The schematic of a tunable dye laser is shown in Figure 1.4 [26].

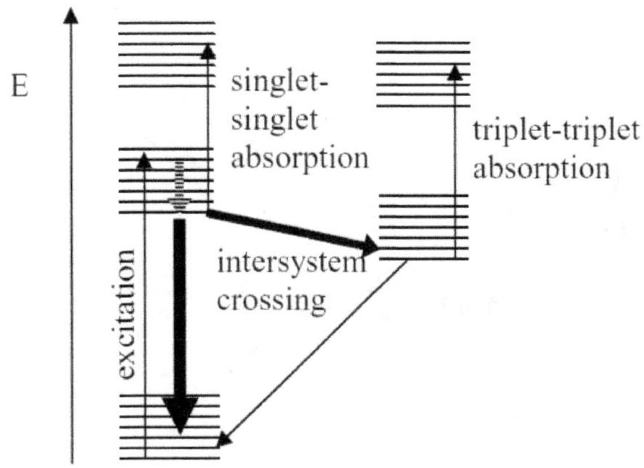

Figure 1.3. Energy level and mechanism for dye laser operation and the processes that compete with laser operation are shown. The solid bold vertical line shows the laser transition [26].

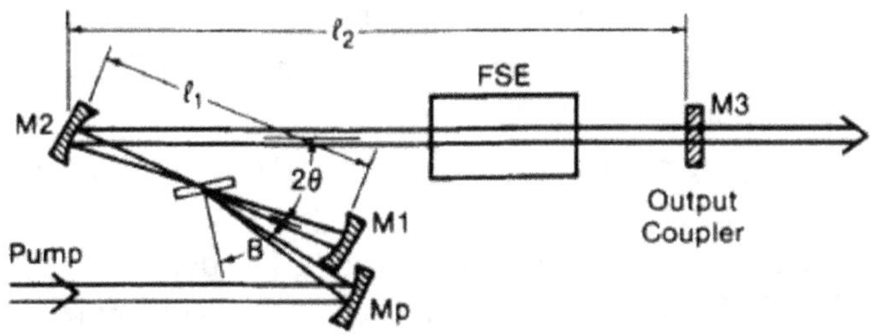

Figure 1.4. Typical schematic of a dye laser setup. The pump light from a laser is focused onto the dye. The mirrors *M*1 and *M*2 focus the internally propagating laser light into the dye. The output mirror is *M*3 and FSE is the frequency-selective element for varying the laser emission wavelength [26].

Passive mode locking of a dye laser has been reported using a saturable absorber in the optical cavity [27]. The dye laser cavity has a cell containing a 2×10^{-5} M solution of DODCI (diethyloxadicarbocyanine iodide) in methanol. This second dye acts as a saturable absorber for the Rhodamine 6G dye which is pumped with Ar laser to produce gain. The experimental arrangement is shown in Figure 1.5. A stable continuous output of mode-locked pulses is obtained over the range of

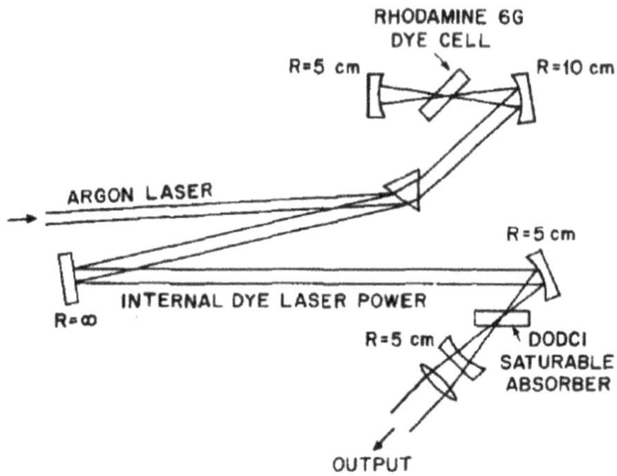

Figure 1.5. Schematic of a mode-locked dye laser [27].

590–610 nm. Autocorrelation measurements using second-harmonic generation (SHG) show the pulses to be as short as 1.5 ps. The pulses are transform-limited.

Using a colliding pulse mode-locked (CPM) dye laser system (similar to Figure 1.5), Fork *et al.* [28] were able to generate a continuous stream of mode-locked pulses with 0.1 ps pulse width. The dye laser was pumped with a CW argon laser at 5145 A. In this case, the synchronized counter-propagating pulses in a ring dye laser interact simultaneously in a thin saturable absorber. The requirement for a CPM laser structure is that the difference in the arrival time of the two interfering pulses is small compared to the pulse duration. In addition, it is necessary that the saturable absorber be sufficiently thin that the optical path in the absorber is of the order of or less than the desired pulse width.

The saturable absorber used is 3,3′-diethyloxadicarbocyanine iodide dissolved in ethylene glycol with a small signal loss of ~20%. A special nozzle was used to generate a dye stream of approximately 10 μm. thick.

Active mode locking has also been reported using dye lasers. Using an acoustic modulator as an active mode-locking element in a CW Rhodamine 6G dye laser, pulses as short as 55 psec have been obtained [29]. A detailed review of the physics of dye lasers is given in Ref. [30].

1.6 Book Overview

Mode-locked lasers have both scientific and industrial applications. These include scientific studies, such as harmonic generation, measurement of material characteristics (including chemical processes), surgery, and microscopy. High-power pulsed lasers have applications in material processing and metal cutting. Very high-power laser pulses are being studied for controlled nuclear fusion.

Mode-locked semiconductor and fiber lasers and pulse generation using such lasers are the main content areas in this book. These lasers are relatively more compact and easier to use compared to previous types of short-pulse lasers, such as dye lasers. Semiconductor and fiber lasers are discussed in several books (for references, see Chapters 2 and 4). These books include a short discussion on mode locking.

The fundamental operating principles of semiconductor lasers used for the generation of light are described in Chapter 2. The principles of optical gain and optical waveguide are described. The epitaxial growth of semiconductor materials is described. The growth, fabrication method, design, and performance of lasers based on the InGaAsP/InP material system are described. Single-wavelength lasers based on distributed feedback mechanism are described. The fabrication and performance of lasers with quantum well gain region are described. These materials emit in the wavelength range of 1300–1600 nm which is a region of low loss of optical fibers. Thus, a major commercial application of lasers using this material is in fiber optic communication systems.

Chapter 3 describes the pulse generation and mode-locking using semiconductor lasers. The mechanism of gain switching for pulsed operation and compression of gain-switched pulses are described. Optical fiber-based pulse compression is described here. The equations for pulse generation and pulse compression are developed in Chapter 3. Mode-locked semiconductor laser designs and performance are described. A semiconductor laser is generally used in a transmission system as an optical pulse-code generator i.e., the laser generates pulses that follow the pattern of the modulation current. The rate equations for this modulation are described. Relaxation oscillation and transient effects are discussed.

Chapter 4 discusses the principle of fiber lasers, fiber fabrication, and rare doped fiber fabrication. Pump lasers needed for fiber laser operation are described. These pump lasers are generally different for different rare-earth dopants. The performances of CW fiber lasers, fiber Raman lasers, and microstructure (photonic crystal) fiber are described.

Both active and passive mode locking of fiber lasers are discussed in Chapter 5. Actively mode-locked fiber lasers are an important source for high-speed fiber optic communication systems because the pulses can appear at a specific repetition rate and short pulses (~1 ps) can be generated. The stability of mode-locked operation has been improved using a phase-locked loop. Multiple wavelength mode-locked operation is also described.

Pulse compression techniques using gratings, optical fibers of various types, and saturable absorbers (including saturable absorber mirrors) are described in Chapter 6. The results in this chapter are for Er-doped fibers. Pulse propagation through an optical fiber and soliton formation is studied. Optical fiber-based pulse compression result using different fiber types is described. Passive and active mode-locking results using nonlinear fiber are described. The principle of a saturable absorber and its use in mode locking of fiber lasers are described. A special case is a saturable absorber mirror whose properties and experimental results are described.

Mode-locking results for different types of rare-earth dopants are described in Chapter 7. These dopants (Yb, Nd, Tm, Ho, etc.) also produce fiber lasers at wavelengths different from that of Er. A suitable modulator needed for active mode locking is not available at all of these emission wavelengths, thus most of the descriptions in this chapter use passive mode-locking techniques, such as a saturable absorber or nonlinear fiber.

High-power pulse generation is described in Chapter 8. Double-clad fibers are used for high-power pump input and thus high-power coupling in the fiber is described. Double-clad amplifier is suitable for amplification of pulses to very high power. Chirped pulse amplification (first studied in the mid-1980s) is described. This process enables the generation of very high peak power pulses. Supercontinuum generation (i.e., generation of a very wide wavelength range) from high peak power pulses is described.

The application of mode-locked lasers for high-speed optical time division multiplexing (OTDM) transmission systems and optical logic studies are described in Chapter 9. Mode-locked lasers are important for the creation of the pulse train, for clock recovery, and for demultiplexing of the pulse train so that the error rates can be measured. Optical logic studies involve the study of Boolean logic elements optically and the study of encryption and decryption circuits. Mode-locked lasers are needed for generating the optical pulses, for clock recovery, and for demultiplexing.

Optical fiber has a circular waveguide. Planar waveguide-based rare-earth-doped materials can be fabricated. The planar waveguide-based pulse generation and mode locking are described in Chapter 10. Pump laser is also needed to produce gain in the planar waveguide. Er-, Nd-, and Yb-doped planar waveguide results are described. Supercontinuum generation in planar waveguides is also discussed. For compact fiber amplifiers and lasers, the planar waveguide geometry is an important candidate. Progress in waveguide design and packaging which includes optimum pump coupling and heat removal can make these devices competitive with fiber-based amplifiers with considerable reduction in size.

The application of mode-locked lasers in various domains such as harmonic generation, surgery, microscopy, and materials processing is described in Chapter 11. Improving Nd- or Yb-doped fiber amplifier performance for space communication applications is important. These materials emit at a shorter wavelength and thus the diffraction-limited spot size is smaller for these amplifiers. Along the same line of thought, the development of shorter wavelength optical fiber amplifiers is important. Although significant developments have occurred in high-power optical fiber lasers, the need for higher power for commercial (metal cutting and shaping) applications will continue to drive the research on higher-power fiber lasers using Nd- or Yb-based systems and possibly others.

Tremendous advances in mode-locked laser technologies have occurred over the last several decades. The advances in research and many technological innovations have led to improved designs of semiconductor and fiber-based mode-locked lasers. These lasers are relatively compact and can be fabricated at many different wavelengths. Many mode-locked laser systems are available commercially now and they are often designed for specific applications. As the application platform develops, many more advances in power output, pulse widths, and repetition rates are very likely.

Finally, much of the advances in mode-locked laser development would not have been possible without the advances in materials and processing technology. The challenges of much of the current mode-locked laser research are intimately linked with the challenges in material fabrication which include not only the investigation of new material systems but also the improvements in existing technologies, to make them more reproducible and predictable.

References

1. A. L. Schawlow and C. H. Townes, *Phys. Rev.*, **112**, (1958) 1940.
2. T. H. Maiman, *Nature*, **187**, (1960) 493.
3. A. Javan, W. R. Bennett, Jr., and D. R. Herriot, *Phys. Rev. Lett.*, **6**, (1961) 106.
4. L. E. Hargrove, R. L. Fork, and M. A. Pollack, *Appl. Phys. Lett.*, **5**, (1964) 4.
5. A. J. DeMaria, D. A. Stetser, and H. Heynau, *Appl. Phys. Lett.*, **8**, (1966) 174.
6. F. P. Schafer, W. Schmidt, and J. Volze, *Appl. Phys Lett.*, **9**, (1966) 306.
7. P. P. Sorokin and J. R. Lankard, *IBM J. Res. Dev.*, **10**, (1966) 162.
8. R. N. Hall, G. E. Fenner, J. D. Kingley, T. J. Soltys, and R. O. Carlson, *Phys. Rev. Lett.*, **9**, (1962) 366.
9. M. I. Nathan, W. P. Dumke, G. Burns, F. H. Dill, Jr., and G. Lasher, *Appl. Phys. Lett.*, **1**, (1962) 63.
10. T. M. Quist, R. H. Rediker, R. J. Keyes, W. E. Krag, B. Lax, A. L. McWhorter, and H. J. Ziegler, *Appl. Phys. Lett.*, **1**, (1962) 91.
11. N. Holonyak Jr. and S. F. Bevacqua, *Appl. Phys. Lett.*, **1**, (1962) 82.
12. Zh. I. Alferov, V. M. Andreev, E. L. Portnoi, and M. K. Trukan Sov, *Phys. Semiconductors*, **3**, (1970) 1107.
13. I. Hayashi, M. B. Panish, P. W. Foy, and S. Sumski, *Appl. Phys. Lett.*, **17**, (1970) 109.
14. H. Kroemer, *Proc. IEEE*, **51**, (1963) 1782.
15. H. Kressel and J. K. Butler, *Semiconductor Lasers and Heterojunction LEDs*. Academic Press, New York, 1977.
16. H. C. Casey, Jr. and M. B. Panish, *Heterostructure Lasers*. Academic Press, New York, 1978.

17. G. P. Agrawal and N. K. Dutta, *Semiconductor Lasers* (Van Nostrand Reinhold, New York, 1986, Second Ed. 1992).
18. L. A. Coldren and S. W. Corzine, *Diode Lasers and Photonic Integrated Circuits* (John Wiley, New York, 1995).
19. C. K. Kao and G. A. Hockmann, *Proc. IEEE*, **133**, (1966) 1151.
20. E. M. Garmire and A. Yariv, *IEEE JQE*, **QE-3**, (1967) 222.
21. N. K. Dutta, *Fiber Amplifiers and Fiber Lasers* (World Scientific, Singapore, 2015).
22. P. B. Hansen, G. Raybon, U. Koren, B. I. Miller, M. G. Young, M. Newkirk, M. D. Chien, B. Tell, and C. A. Burrus, *Appl. Phys. Lett.*, **62**, (1993) 1445.
23. Y. K. Chen, M. C. Wu, T. Tanbun-Ek, R. A. Logan, and M. A. Chin, *Appl. Phys. Lett.*, **58**, (1991) 1253, *IEEE Photon. Tech. Lett.*, **3**, (1991) 824.
24. P. A. Morton, V. Mizrahi, and T. Tanbun-Ek, *Electronics Lett.*, **28**, (1992) 2156.
25. H. Fan, C. Wu, M. El-Aasser, N. K. Dutta, U. Koren, and A. B. Piccirilli, *IEEE Photon. Tech. Lett.*, **12**, (2000) 972.
26. F. P. Schafer (ed.), *"Dye Lasers" Topics in Applied Physics*, Vol. 1 (Springer-Verlag, NY, 1977).
27. E.P. Ippen, C.V. Shank, and A. Dienes, *Appl. Phys. Lett.*, **21**, (1972) 348.
28. R. L. Fork, B. I. Greene, and C. V. Shank, *Appl. Phys. Lett.*, **38**, (1981) 671.
29. A. Dienes, E. P. Ippen, and C. V. Shank, 1971, *Appl. Phys. Lett.*, **19**, (1971) 258.
30. C. V. Shank, *Rev. Mod. Phys.*, **47**, (1975) 64.

Chapter 2

Semiconductor Lasers

2.1 Introduction

Significant advances in research results, development, and application of semiconductor lasers have occurred over the last three decades. The fiber optic revolution in telecommunication which provided several orders of magnitude improvement in transmission capacity at low cost would not have been possible without the development of reliable semiconductor lasers. Semiconductor injection lasers continue to be the laser of choice for various system applications primarily because of their small size, simplicity of operation, and reliable performance. For most transmission system applications, the laser output is encoded with data by modulating the current. However, for many high data rate applications, which require long-distance transmission, external modulators are used to encode the data. For early work and thorough discussion of semiconductor lasers, see Refs. [1–4].

The optical fiber has zero dispersion near 1300 nm wavelength and has lowest loss near 1550 nm wavelength. Thus, semiconductor lasers emitting near 1300 nm and 1550 nm are of interest for fiber optic transmission applications. Lasers emitting at these wavelengths are fabricated using the InGaAsP/InP material system and were first fabricated in 1976 [5]. Much of the fiber optic transmission systems around the world that are in use or are currently being deployed utilize lasers emitting near 1300 nm or 1550 nm. These commercial lasers are generally fabricated now using the Metal-Organic Chemical Vapor Deposition (MOCVD) growth technique.

2.2 Semiconductor Laser Fundamentals

The semiconductor injection laser was invented in 1962 [6–8]. With the development of epitaxial growth techniques and the subsequent fabrication of double heterojunction, the laser technology advanced rapidly in the 1970s and 1980s [1–4]. The demonstration of CW operation of the semiconductor laser in the early 1970s [9] (using a double heterostructure [10]) was followed by an increase in development activity in several industrial laboratories.

A typical laser (shown in Figure 2.1) has a medium which provides optical gain and a set of mirrors on each side of the gain medium which form a Fabry–Perot resonator and provide optical feedback. The two together form the optical cavity of the laser. The light generated by spontaneous emission in the gain medium is amplified by the gain medium (the atoms emit by stimulated emission normal to the mirrors) and it bounces between the mirrors to form the laser light. A fraction of this light is transmitted by the mirrors and is observed in the form of laser emission. Let the optical cavity be in the z direction. Then, in the plane wave approximation, the (assuming linear polarization) electric field of the laser emission along $+z$ is given by

$$E\,(z,t) = E_0 \exp[i(\omega t - nk_0 z)] = E_0 \exp[i\,(\omega t - n_r k_0 z)] \exp(n_i k_0 z), \qquad (2.1)$$

where ω is the frequency ($\omega = 2\pi\nu$) of the light and $k_0 = 2\pi/\lambda_0$ where λ_0 is the wavelength in free space and $n = n_r + in_i$ where n_r and n_i are the real

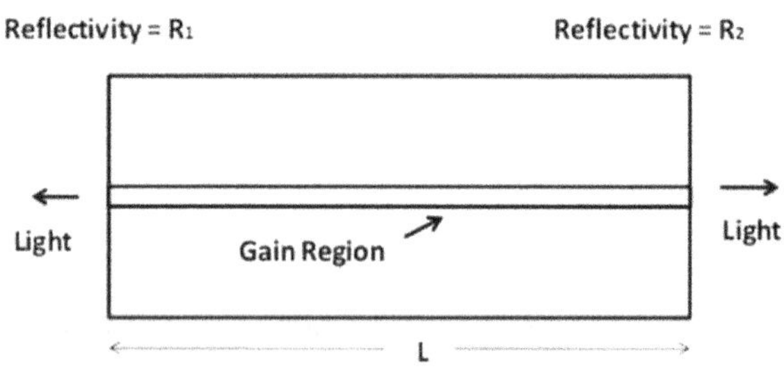

Figure 2.1. Schematic of a typical laser. R_1 and R_2 are reflectivity of the mirrors which provide optical feedback. The laser emission is normal to the optical cavity along the z direction.

and imaginary parts of the refractive index of the gain medium. The change in intensity, $I(z) = |E(z)|^2$ as a function of travel distance (δz) in the gain medium is,

$$I(z + \delta z) = I(z) \exp (2n_i k_0 \delta z). \tag{2.2}$$

The quantity $2n_i k_0 = g_n$, where g_n is the net gain as the wave travels in the gain medium. The net gain is the difference between the gain created by optical transitions (g) and optical losses (absorption due to other materials in the gain medium) and the scattering of laser light in the gain medium. This combined optical loss is noted as α. A fraction of the optical mode (known as confinement factor, Γ[11,12]) of the laser is confined in the active region of the laser [1–4]. Thus, the optical gain of the laser mode is Γg, where g is the material gain. Thus, net gain $g_n = \Gamma g - \alpha$. In steady state, round-trip gain must equal round-trip loss. The gain when the laser begins laser action is known as gain at threshold (gth). The threshold gain (gth) is given by

$$R_1 R_2 \exp [2(\Gamma g\text{th} - \alpha) L] = 1, \tag{2.3}$$

where R_1 and R_2 are the reflectivity of the mirrors and L is the cavity length. Equation (2.3) may be rewritten as

$$\Gamma g\text{th} = \alpha + (1/2L) \ln (1/(R_1 R_2)) = \alpha + \alpha_m, \tag{2.4}$$

where $\alpha_m = (1/2L) \ln (1/(R_1 R_2))$ is also known as "mirror losses". It essentially reflects the optical loss due to laser output. In addition, in steady state, after one round trip in the cavity, the electric field must be identical to the initial electric field i.e., $E(z) = E(z + 2L)$, where L is the length of the optical cavity. This leads to another condition for laser action i.e.,

$$2n_r k_0 L = 2m\pi, \tag{2.5}$$

where m is an integer. The above equation says that the round-trip change in phase must be an integer multiple of 2π which is the condition for Fabry–Perot resonance.

Thus, the Fabry–Perot modes of the optical cavity are the laser wavelengths, λ_0 ($k_0 = 2\pi/\lambda_0$). The laser operates at specific discrete wavelengths determined by Equation (2.5). They are called the longitudinal modes of the laser. From Equation (2.6), the longitudinal mode spacing (for large *m*) is given by

$$\Delta\lambda = \lambda^2/(2n_r L). \tag{2.6}$$

For a typical gain medium, the optical gain occurs over a small range of wavelengths. The gain as a function of wavelength is called a gain spectrum. A typical laser operates at several longitudinal modes (discrete wavelengths separated by $\Delta\lambda$) centered on the peak of the gain spectrum. This is shown schematically in Figure 2.2. For a semiconductor laser gain medium, the gain increases with increasing current (I). It can be approximated as [1–4, 11]

$$g = b\,(I/I_0 - 1), \tag{2.7}$$

where $I_0 \sim 1$ kA/cm^2 and $b \sim 100$ cm^{-1} [11]. As current is increased, the gain in a semiconductor laser increases until it reaches a certain current (called threshold current) when the laser begins to operate as a laser. Also,

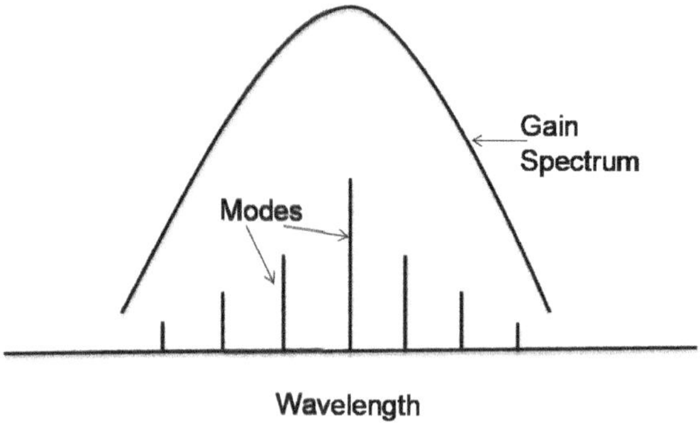

Figure 2.2. Schematic of the gain spectrum and a typical laser spectrum. The longitudinal mode near the maximum gain generally has higher intensity. The mode spacing is $\Delta\lambda$ (Equation 2.6). Many oscillating modes within the gain spectrum are needed for mode locking.

n_r changes with current. The z dependence of the electric field in a laser is given by (using Equation 2.1)

$$E(z) = E_0 \exp(-i\beta z) \quad \text{with} \quad \beta = n_r k_0 + i\,(\Gamma g - \alpha)/2. \tag{2.8}$$

For currents above threshold current (I_{th}), the total laser output (from two mirrors) at a current I is approximately given by

$$P_{out} = (h\nu/e)\,(I - I_{th})\,\eta_d \text{ with } \eta_d = \alpha_m/(\alpha + \alpha_m). \tag{2.9}$$

The quantity η_d is called the differential quantum efficiency. For a real laser, the measured differential quantum efficiency (η_d) needs to be multiplied by a quantity called internal efficiency (η_i). The internal efficiency represents the fraction of injected carriers that recombine radiatively. In some cases, $\eta_i \sim 1$. For a typical single-mode semiconductor laser, the dimensions are 0.1 μm is the thickness of the active region, 1 μm is the width of the active region, and 250 μm is length (L) of the active region. Thus, the area of gain region = 10^{-9} cm^2. For a laser with cleaved facets ($R_1 = R_2 = 35\%$), $\alpha_m = (1/2L)\ln(1/R_1R_2) = 21$ cm^{-1}. Typically, $\alpha = 20$ cm^{-1}, thus $\eta_d = 0.51$. For hν = 0.96 eV, i.e., for laser emitting near 1300 nm (~1 eV photon energy), the power output per unit current above the threshold is ~0.25 mW/mA/facet. The measured slope of power output vs. current above threshold is also often called differential efficiency. For reasons, such as leakage current and heating [4], the power output is not linearly proportional to current at high currents. The average intracavity power (P_{in}) is related to the intracavity photon density (N_{ph}) as

$$P_{in} = h\nu V N_{ph}/\tau_p, \tag{2.10}$$

where $h\nu$ is the photon energy, V is the volume of the active region, and τ_p is the photon lifetime in the cavity given by $\tau_p^{-1} = v_g\,(\alpha + \alpha_m)$ where v_g is the group velocity. Since the photons escape the laser cavity from the output facets at the rate of $v_g\alpha_m$, the power output from each facet for equal reflectivity is given by

$$P_{out} = (1/2)\,h\nu\, v_g\alpha_m\, VN_{ph}. \tag{2.11}$$

For many applications, the laser facets are coated with a dielectric material so that most of the power comes out of one facet. For facet reflectivity of R_1 and R_2, the ratio of the output power (P_1 and P_2) from the respective facets is given by [1]

$$\frac{P_1}{P_2} = \frac{(1-R_1)\sqrt{R_2}}{(1-R_2)\sqrt{R_1}}. \tag{2.12}$$

The sum of the power outputs $P_1 + P_2 = 2P_{out}$ is given by Equation (2.11).

2.3 Semiconductor Laser Materials

Extensive work on semiconductor laser fabrication was carried out in the 1970s using the GaAs/AlGaAs material system. These lasers had GaAs active region surrounded by p–$Al_{1-x}Ga_xAs$ and n–$Al_{1-x}Ga_xAs$ cladding layers. Much of the work on laser for communication has been carried out using the $In_{1-x}Ga_xAs_yP_{1-y}$ material system with the optical emission centered around 1300 or 1550 nm. These materials are grown over a GaAs or an InP substrate using a process known as epitaxy. In this process, the lattice of the grown layer matches with the lattice on which the layer is grown (often called the substrate). Liquid phase epitaxy growth of AlGaAs/GaAs and InGaAsP/InP heterostructures were studied by several researchers in the 1960s and 1970s [13–35]. Much of this work enabled the determination of material parameters [for a review see, 1–4, 11].

Various epitaxial growth techniques (LPE, MOCVD, and MBE) have been used to fabricate the semiconductor laser material. In liquid phase epitaxy (LPE), the epitaxial layer is grown by cooling a saturated solution of the composition of the layer to be grown while that solution is in contact with the substrate [36, 37]. In MBE, the epitaxial layer growth is achieved by the reaction of atomic or molecular beams of the constituent elements (of the layer to be grown) with a crystalline substrate held at high temperature in ultrahigh vacuum [38]. In VPE (vapor phase epitaxy), the epitaxial layer is grown by the reaction of gaseous elements or compounds at the surface of a heated substrate. The VPE technique has also been called chemical vapor deposition (CVD) depending on the constituents of the reactants. A variant of the technique is metal-organic chemical vapor deposition (MOCVD) [39], which has been very successful for lasers in which metal alkyls are used as the compound source. This technique is currently used by many manufacturers of lasers.

2.3.1 *Metal-organic chemical vapor deposition (MOCVD)*

Metal-organic chemical vapor deposition (MOCVD), also known as metal-organic vapor phase epitaxy (MOVPE), uses metal alkyls as sources from which epitaxial layers are formed [40–49]. The low pressure MOCVD technique where the gas pressure is ~0.1 atm has been used for the growth of AlGaAs and InGaAsP. MOCVD is a very versatile technique; it has been used to grow a wide variety of II–VI and III–V compound semiconductors.

Figure 2.3 shows the schematic of a low-pressure MOCVD system. The quartz reaction chamber contains the InP substrate placed on a

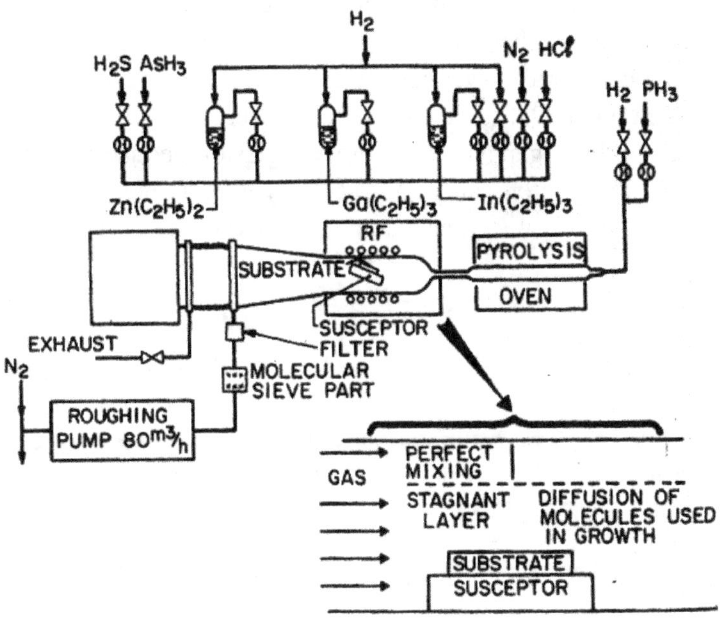

Figure 2.3. Schematic illustration of a metal-organic chemical vapor deposition (MOCVD) growth system [48].

radio frequency (RF)-heated (~600°C) carbon susceptor. Group III alkyls ($Ga(C_2H_5)_3$ and $In(C_2H_5)_3$) and group V hydrides (AsH_3 and PH_3) are introduced into the reaction chamber. The gas flow near the substrate is laminar with velocities in the range of 1–15 cm/s for a pressure of ~0.1 atm. At the hot substrate surface, the metal alkyls and hydrides decompose, producing elemental In, Ga, As, and P. The elemental species deposit on the surface forming an epitaxial layer. The gas flow rates are controlled to produce InGaAsP layers of different compositions. $Zn(C_2H_5)_3$ and H_2S are used as sources for *p*-type (Zn-doped) doping and *n*-type (S-doped) doping, respectively.

2.4 Optical Gain and Optical Waveguide

The radiative recombination of the electrons and holes is responsible for stimulated optical emission or gain. Sufficient numbers of electrons and holes must be present in the semiconductor for net optical gain. This is accomplished using a *p–n* double heterostructure [10].

The basis of light emission in semiconductors is the recombination of an electron in the conduction band with a hole from the valence band; the excess energy is emitted as a photon (light quantum). Consider the transition shown in Figure 2.4 for a photon whose energy $h\nu = E = E_g + E_c + E_v$, where E_g is the band gap and E_c, E_v are energies of the electron and hole, respectively. The photon can be absorbed creating an electron of energy E_c and a hole of energy E_v. The stimulated emission rate is given by

$$R_e = Bf_c f_v \rho(E), \tag{2.13}$$

where B is the transition probability, $\rho(E)$ is the density of states of photons of energy E, and $f_c(E_c)$ and $f_v(E_v)$ are the Fermi factors which are the probability that electron and hole states of energy E_c and E_v are occupied. The quantities $f_c(E_c)$ and $f_v(E_v)$ are given by

$$f_c(E_c) = \frac{1}{1 + \exp[(E_c - E_{fc})/kT]}, \tag{2.14}$$

$$f_v(E_v) = \frac{1}{1 + \exp[(E_v - E_{fv})/kT]}, \tag{2.15}$$

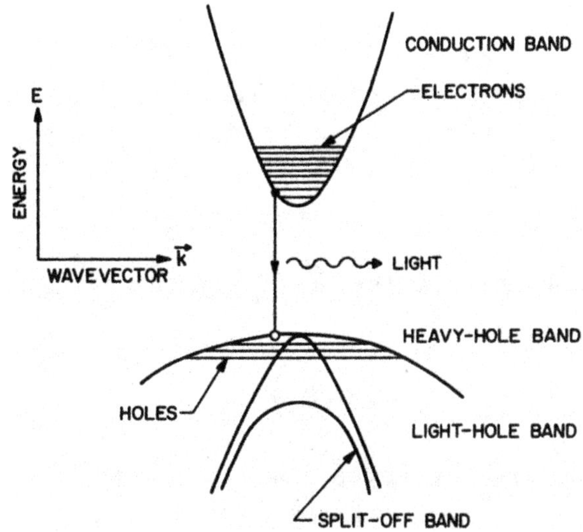

Figure 2.4. Energy versus wavevector of the four major energy bands for a zinc blende type of direct gap semiconductor.

where k is the Boltzmann constant, T is the temperature, and E_{fc}, E_{fv} are the quasi-Fermi levels of the electrons and holes, respectively. The stimulated emission process involves the recombination of an electron and hole pair, and the absorption process creates an electron and hole pair. The absorption (R_a) rate for photons of energy E is given by

$$R_a = B(1 - f_c)(1 - f_v)\ \rho(E). \tag{2.16}$$

The condition for net stimulated emission or gain is

$$R_e > R_a. \tag{2.17}$$

Using Equations (2.13) and (2.16), this becomes

$$f_c + f_v > 1. \tag{2.18}$$

Using Equations (2.14) and (2.15) for f_c and f_v, the condition for gain at photon energy E becomes

$$E_{fc} + E_{fv} > E - E_g. \tag{2.19}$$

This condition was first derived in Ref. [49]. The condition for zero net gain or zero absorption at photon energy is

$$E_{fc} + E_{fv} = E - E_g. \tag{2.20}$$

Since the minimum value of E is E_g, the condition for gain at any energy is

$$E_{fc} + E_{fv} = 0. \tag{2.21}$$

This is often known as the condition for transparency. With increasing injected carrier densities (electrons and holes) due to current injection, both E_{fc} and E_{fv} increase. Figure 2.5 shows the variation of the injected carrier density for transparency n_t as a function of temperature for undoped $\lambda = 1300$ nm ($E_g = 0.96$ eV) InGaAsP. The parameter values used in the calculation are $m_c = 0.061\ m_0$, $m_{hh} = 0.45\ m_0$, and $m_{lh} = 0.08\ m_0$, where m_0, m_{hh}, and m_{lh} are the free-electrons, heavy-hole, and the light-hole mass, respectively [4]. Figure 2.5 shows that n_t is considerably smaller at low temperatures. The threshold carrier density is ~20–30% higher than n_t.

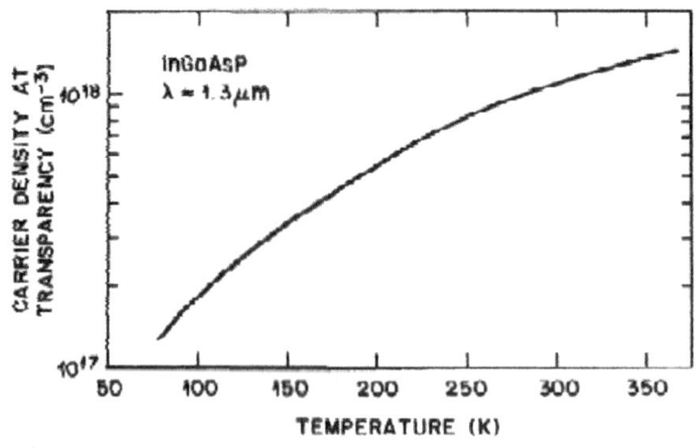

Figure 2.5. The calculated injected carrier density for transparency as a function of temperature for undoped $\lambda = 1300$ nm InGaAsP.

2.4.1 *Gain calculation*

The quantities associated with a radiative recombination are the absorption spectrum, emission spectrum, gain spectrum, and total radiative emission rate. The calculation of the above for transitions between parabolic bands (i.e., when both conduction and valence bands are assumed to be parabolic in E vs. $\mathbf{k}$) is described in Appendix 5 of [50]. For the general case of nonparabolic bands and exciton states, the optical absorption or gain for a transition between the valence band and the conduction band at an energy E is given by [4, 11]

$$\alpha(E) = \frac{e^2 h}{2\varepsilon_0 m_0^2 cnE} \int_{-\infty}^{+\infty} \rho_c(E')\rho_v(E'')|M(E',E'')|^2 \times [f(E''=E'-E) - f(E')]dE', \tag{2.22}$$

where m_0 is the free electron mass, h is Planck's constant, e is the electron charge, ε_0 is the permittivity of free space, E is the photon energy, n is the refractive index at energy E, and ρ_c and ρ_v are the densities of state per unit volume per unit energy in the conduction and valence band, respectively. $f(E')$ is the probability that a state of energy E' is occupied by an electron and M is the effective matrix clement between the conduction band state of energy E' and the valence band state of energy E''. Several models for the density of states and matrix element for transition between band-tail states exist [51–54]. The latest of such models that take into account the contributions of the parabolic bands and impurities is a Gaussian fit of the Kane form to the Halperin–Lax model of band tails [53]. The matrix element M may be expressed as a product of two terms: $M = M_b\, M_{\text{env}}$. The quantity M_b arises from the band-edge Bloch functions and M_{env} arises from the envelope wave functions. For III–V semiconductors using the Kane model [51],

$$|M_b|^2 = \frac{m_0^2 E_g}{12 m_c} \frac{E_g + \Delta}{E_g + \frac{2}{3}\Delta}, \tag{2.23}$$

where m_c is the conduction-band effective mass, m_0 is the free electron mass, E_g is the energy gap, and Δ is the spin–orbit coupling. The envelop matrix element for the band-tail states has been calculated by Stern [52].

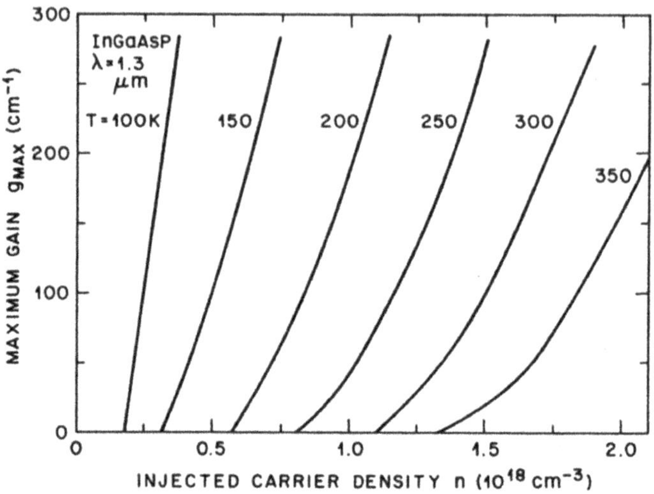

Figure 2.6. The maximum gain as a function of injected carrier density for undoped InGaAsP ($\lambda = 1300$ nm) at different temperatures [4].

The spectral dependence of absorption or gain at various injected carrier densities for InGaAsP ($\lambda \sim 1300$ nm) with acceptor and donor concentrations of 2×10^{17} cm^{-3} has been calculated [4] from which maximum gain can be obtained. Figure 2.6 shows the maximum gain as a function of injected carrier density at different temperatures. The material parameters used in the calculation are $m_c = 0.061\ m_0$, $m_{hh} = 0.45\ m_0$, $m_{lh} = 0.08\ m_0$, $\Delta = 0.26\ eV$, $E_g = 0.96\ eV$, and $\varepsilon = 11.5$. Note that considerably lower injected carrier density is needed at a lower temperature to achieve the same optical gain. This is the origin of the lower threshold current at low temperature. Similar results are obtained for different materials. It is often convenient to use a linear relationship between the gain g and the injected carrier density n of the form

$$g = a(n - n_0), \tag{2.24}$$

where a is the gain coefficient and n_0 is the injected carrier density at transparency.

2.4.2 *Spontaneous emission rate*

At unity quantum efficiency, the total spontaneous radiative recombination rate R equals the excitation rate. The latter is usually expressed in terms of the nominal current density J_n [4, 51–54]:

$$J_n[A/(\text{cm}^2\ \mu\text{m})] = eR \tag{2.25}$$

with $R = \int r_{\text{spon}}(E)dE$ where e is the charge of an electron, and the thickness of the active region is assumed to be $1\,\mu$m. $r_{\text{spon}}(E)$ is the spontaneous emission rate at a photon energy E. It is given by

$$r_{\text{spon}}(E) = \frac{4\pi n e^2 E}{m_o^2 \varepsilon_0 h^2 c^3} \int_{-\infty}^{+\infty} \rho_c(E')\rho_v(E'')\left|M(E',E'')\right|^2 f(E')[1-f(E'')]dE'. \tag{2.26}$$

The calculated maximum optical gain g as a function of the nominal current density at various temperatures is shown in Figure 2.7 for InGaAsP ($\lambda \sim 1300$ nm). The calculation is for an undoped lightly compensated material with 10^{17} cm^{-3} of acceptors and donors, respectively. Note that gain varies linearly with J_n above a certain gain. From Figure 2.7, the

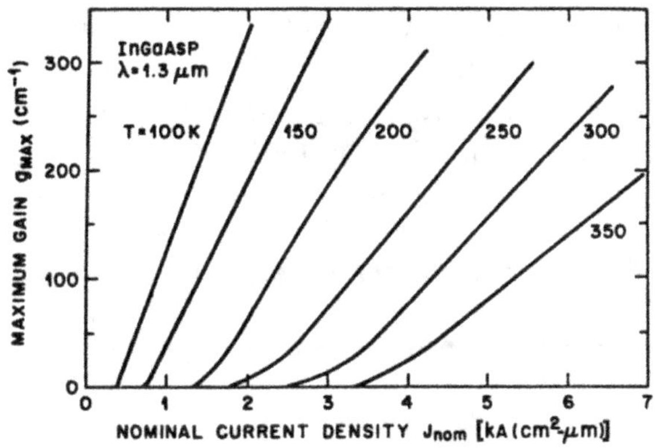

Figure 2.7. The maximum gain as a function of nominal current density for InGaAsP ($\lambda = 1300$ nm) at different temperatures [11].

relationship between gain and nominal current density for InGaAsP ($\lambda \sim$ 1300 nm) at 300 K is approximately given by

$$g = 0.08(J_n - 3100)\text{cm}^{-1}, \tag{2.27}$$

where J_n is expressed in amperes per square centimeter per micron. The total spontaneous radiative recombination rate R can be approximated by

$$R = Bnp, \tag{2.28}$$

where B is the radiative recombination coefficient and n, p are the electron and hole densities, respectively. For undoped semiconductors, Equation (2.27) becomes $R = Bn^2$. For GaAs, the measured $B = 1 \times 10^{-10}$ cm^3/s. Calculation of the radiative recombination rate shows that B decreases with increasing carrier density [4]. This has been confirmed by carrier lifetime measurements [4]. In double heterostructure materials, the injected carriers can recombine by both radiative and nonradiative recombination. The injected current density J is simply the sum of the radiative R and nonradiative R_{nr} recombination rates in the absence of carrier leakage:

$$J = e(R + R_{\text{nr}})\, d = J_r + J_{\text{nr}}\; (J_r = J_n d), \tag{2.29}$$

where e is the charge of an electron, d is the active layer thickness, and J_r, J_{nr} are the radiative and nonradiative components of the current density, respectively. For a GaAs double heterostructure material where the quantum efficiency is believed to be close to unity, that is, most of the injected carriers recombine radiatively, $J \sim J_r$. The relation between the optical gain and the nominal current density for the GaAs material system has been calculated by Stern [52]. It is given by

$$g = 0.045(J_n - 4200)\text{cm}^{-1}, \tag{2.30}$$

where J_n is expressed in amperes per square centimeter per micron.

It is useful to make the connection with the operation of a GaAs laser for which the nonradiative recombination is believed to be very small. For a laser, the optical gain provided by current injection must equal optical loss. The latter includes mirror losses. This is expressed as

$$\Gamma g_{\text{th}} = \alpha_a \Gamma + (1-\Gamma)\alpha_c + \frac{1}{L}\ln\frac{1}{R}, \tag{2.31}$$

where g_{th} is the threshold gain in the active region, α_a, α_c are the absorption losses in the active and cladding regions, respectively, L is the cavity length, and R is the mirror facet reflectivity. Typically, $L \sim 250\ \mu$m and $R \sim 0.3$. For a 0.2-μm-thick active layer, Γ (TE) is ~0.6, and using $\alpha_a = \alpha_c =$ 20 cm^{-1}, the calculated $g_{\text{th}} \sim 114$ cm^{-1}.

The threshold current of a 250-μm-long GaAs–A1GaAs double heterostructure laser with a 2 μm × 0.2 μm active region can now be calculated using Equations (2.29) and (2.30). The calculated $J_{\text{th}} = 1.35$ kA/cm^2 (for a gain $g_{\text{th}} \sim 114$ cm^{-1} i.e., in this case, gain ~84 cm^{-1}/(kA/cm^2)), and the threshold current is 6.7 mA. This compares well with the experimentally observed threshold currents in the range 5–10 mA for GaAs–A1GaAs double heterostructure lasers with good current confinement.

Figure 2.7 shows that the current density needed to achieve a certain optical gain increases with temperature. The same figure also illustrates that the threshold current density of a laser is expected to increase with increasing temperature. For InGaAsP lasers, a significant amount of nonradiative recombination is believed to be present at high carrier densities needed for gain. The nonradiative recombination can increase the current needed for gain and also its temperature dependence.

The temperature dependence of the threshold current (I_{th} (T)) of a laser is usually expressed by a relation:

$$I_{\text{th}}(T) = I_{\text{tho}} \exp(T/T_o), \tag{2.32}$$

where I_{tho} is a constant. T_o is a parameter known as characteristic temperature. T_o is low if the temperature dependence of the threshold current is high. T_o has a value of ~200 K for GaAs (850 nm) lasers and $T_o \sim 60$ K for InGaAsP 1300 nm and 1550 nm lasers. The high-temperature dependence of the threshold current (low T_o) of long wavelength InGaAsP lasers is generally attributed to nonradiative Auger recombination [55–60]. The band gap of a semiconductor decreases with increasing temperature, and the rate is about 0.35 nm/C.

An electron–hole pair can recombine nonradiatively, meaning that the recombination can occur through any process that does not emit a photon An example of such a process is Auger recombination [55–60]. The effect of nonradiative recombination on the performance of lasers is to increase the current needed for a certain optical gain. If τ_{nr} is the carrier lifetime associated with the nonradiative process, the increase in current density is given approximately by

$$J_{nr} = \frac{end}{\tau_{nr}}. \tag{2.33}$$

2.4.3 *Optical waveguide*

Conceptually, stimulated emission in a semiconductor laser arises from electron–hole radiative recombination in the active region, and the light generated is confined and guided by a dielectric waveguide (Figure 2.1). The threshold current of an injection laser was historically reduced using a double heterostructure for carrier confinement [10, 11]. The region of optical gain is determined by the thickness of the active region in double heterostructure lasers. In addition, the refractive index of the lower-gap active region is higher than that for the *n*-type and *p*-type higher-gap confining (also known as cladding) layers. The three layers form an optical waveguide. The energy distribution of the fundamental mode of the waveguide is also sketched in Figure 2.8. A fraction of the optical mode is confined in the active region. Two types of fundamental transverse modes can propagate in the waveguide: the transverse electric (TE) and the transverse magnetic (TM) modes.

The expression for the confinement factor (Γ), the fraction of mode in the active region, has been previously calculated [11, 12]. Figure 2.9 shows the calculated Γ as a function of active layer thickness for the TE and TM modes for the $\lambda = 1300$ nm InGaAsP double heterostructure with *p*-InP and *n*-InP cladding layers. For a three-layer slab waveguide with cladding layer index n_2 and active layer index n_1, the confinement factor of the fundamental mode can be approximated by [12]

$$\Gamma = D^2/(2 + D^2) \tag{2.34}$$

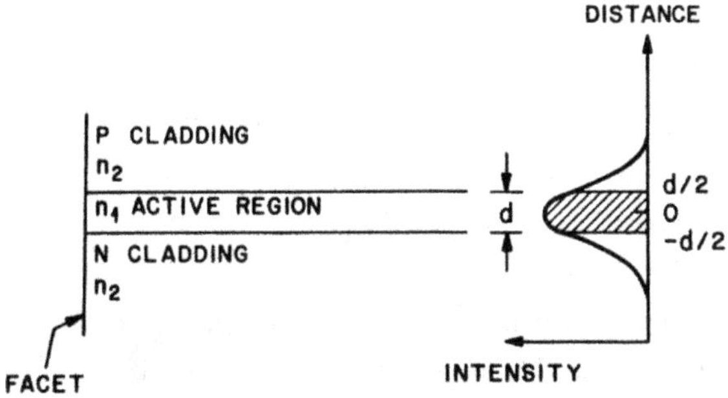

Figure 2.8. The dielectric waveguide of a semiconductor laser. n_2 is the refractive index of the cladding layers and n_1 is that of the active region. $n_1 > n_2$. The cladding layers are of a higher band gap material than the active region. The intensity distribution of the fundamental mode is shown. The cross-hatched region represents the fraction of the mode (Γ) within the active region.

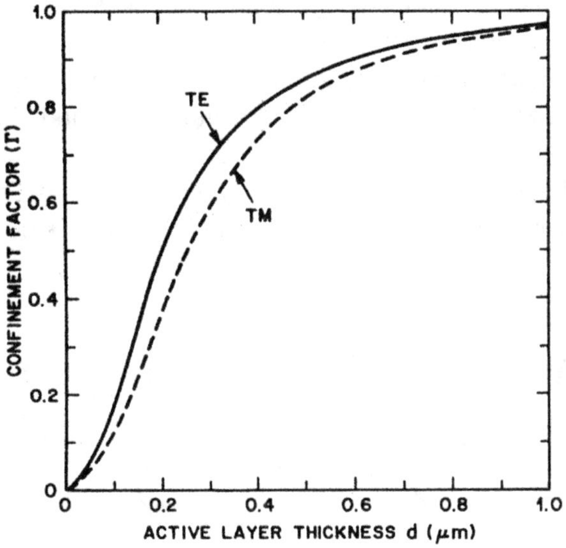

Figure 2.9. Confinement factor of the fundamental TE and TM modes for a waveguide with InGaAsP (λ = 1300 nm) active layer and InP cladding layers as a function of the thickness of the active region [11]. The refractive indexes of the active and cladding regions are 3.51 and 3.22, respectively.

with $D = k_0 (n_1^2 - n_2^2)^{1/2} d$, where d is the thickness of the active region and $k_0 = 2\pi/\lambda_0$ where λ_0 is the wavelength in free space.

The optical gain of a signal in the fundamental mode traveling through the amplifier is given by

$$g_m = \Gamma g. \tag{2.35}$$

The mode gain (g_m) is proportional to the mode confinement factor Γ. The semiconductor LED, laser, and optical amplifiers used in current applications utilize a double heterostructure for carrier injection and carrier confinement.

2.5 Laser Designs and Emission Characteristics

The double heterostructure material can be processed to produce lasers in several ways. Perhaps the simplest is the broad area laser (Figure 2.10) which involves putting contacts on the p- and n-side and then cleaving. Such lasers do not have transverse mode confinement or current confinement which leads to high threshold and nonlinearities in light vs. current characteristics.

Several laser designs have been developed to address these problems. Among them are the gain-guided laser, the weakly index-guided laser, and

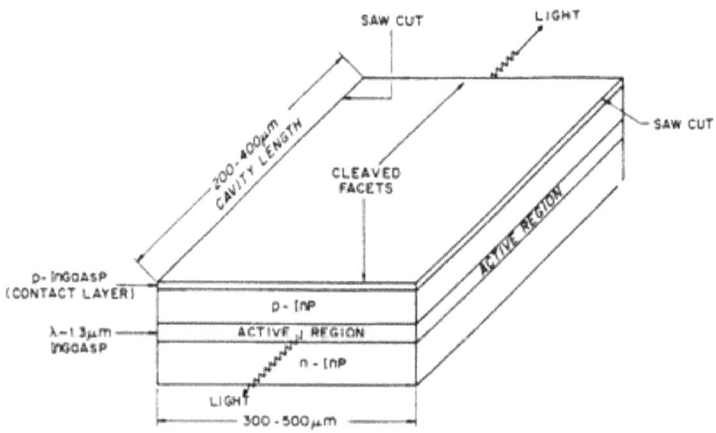

Figure 2.10. Schematic of a broad area laser.

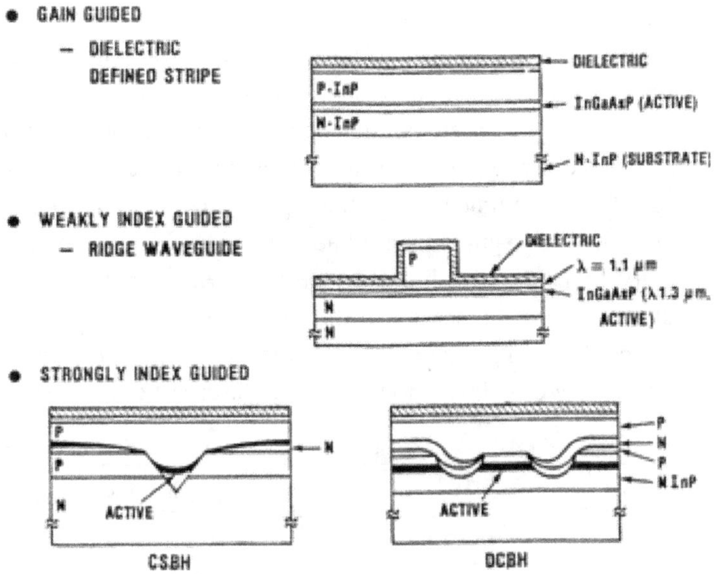

Figure 2.11. Several laser designs using InGaAsP/InP material system.

the buried heterostructure (strongly index-guided) laser. A typical version of these laser structures is shown in Figure 2.11. The gain-guided structure uses a dielectric layer for current confinement. The current is injected in the opening in the dielectric (typically 6–12 μm wide), which produces gain in that region and hence the lasing mode is confined to that region. The weakly index-guided structure has a ridge etched on the wafer; a dielectric layer surrounds the ridge. The current is injected in the region of the ridge, and the optical mode overlaps the dielectric (which has a low index) in the ridge. This results in weak index guiding.

The strongly index-guided, buried heterostructure, design shown in Figure 2.11 has the active region surrounded (buried) by lower index layers. The fabrication process of the double channel planar buried heterostructure (DCBH) laser involves growing a double heterostructure, etching a mesa using a dielectric mask, and then regrowing the layer surrounding the active region using a second epitaxial growth step. The second growth can be a single Fe-doped InP (Fe:InP) semi-insulating layer or a combination of *p*-InP, *n*-InP, and Fe:InP layers. Generally, MOCVD growth process is used for the growth of the regrown layer. Researchers have often given different names to the particular buried heterostructure laser design that they discovered [4, 11]. For the structure of Figure 2.12, the Fe-doped InP layer provides both optical confinement to the lasing mode and current confinement to the active region. Buried heterostructure lasers are used in communication system applications because a properly designed strongly index-guided buried heterostructure design has superior mode stability, higher bandwidth, and superior linearity in light vs current (*L* vs. *I*) characteristics compared to the gain-guided and weakly index-guided designs. Scanning electron micrograph of a capped mesa buried heterostructure (CMBH) laser along with the laser structure is shown in

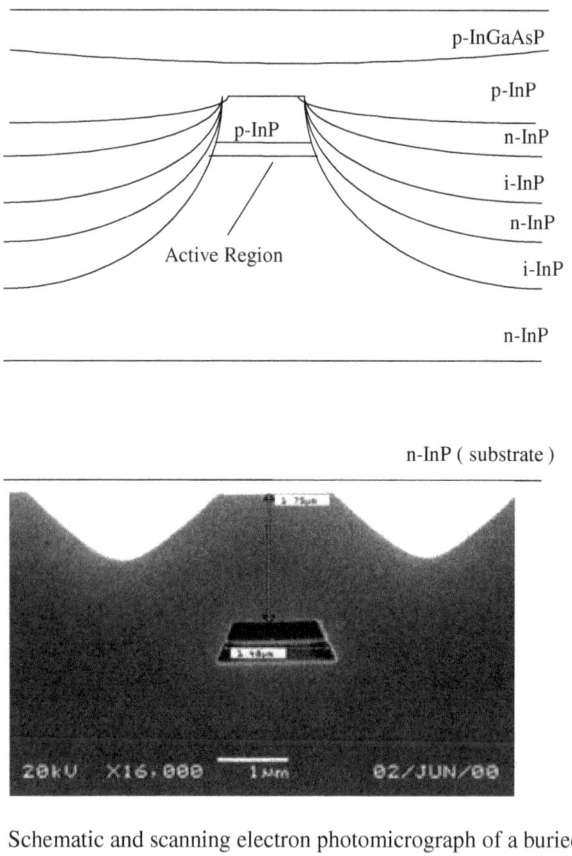

Figure 2.12. Schematic and scanning electron photomicrograph of a buried heterostructure (BH) laser.

Figure 2.12. The current blocking layers in this structure consist of *i*-InP (Fe-doped InP), *n*-InP, *i*-InP, and *n*-InP layers.

These sets of blocking layers have the lowest capacitance and are therefore needed for high-speed operation of the laser. An optimization of the thickness of these layers is needed for highest speed performance. The laser fabrication involves the following steps. First, the double heterostructure consisting of the active region and cladding layers is grown on an InP substrate. One-micron-wide mesas are etched on the wafer and the current blocking layers consisting of *i*-InP, *n*-InP, *i*-InP, and *n*-InP layers are grown on the wafer with an oxide layer on top of the mesa in place. The oxide layer is then removed and a third growth of *p*-InP cladding layer and *p*-InGaAs contact layer is carried out. The wafer is then processed using standard lithography, metallization, and cleaving techniques to produce the lasers.

The light vs. current characteristics at different temperatures of an InGaAsP BH laser emitting at 1.3 μm are shown in Figure 2.13. The typical threshold current of a BH laser at room temperature is in the 5–10 mA range. For gain-guided and weakly index-guided lasers, typical room temperature threshold currents are in the 25–50 mA and 50–100 mA ranges, respectively. The external differential quantum efficiency defined as the derivative of the L vs. I characteristics above threshold is ~0.25 mW/mA/facet for a cleaved uncoated laser emitting near 1300 nm. A semiconductor laser with cleaved facets generally emits in a few longitudinal modes of the cavity. A typical spectrum of a laser with cleaved facets is also shown in Figure 2.13. The discrete emission wavelengths are separated by the longitudinal cavity mode spacing which is ~10 A for a laser ($\lambda \sim 1300$ nm) with 250 μm cavity length. Lasers can be made to emit in a single frequency using frequency selective feedback, for example, using a grating internal to the laser cavity as described in later sections. For buried heterostructure (BH) devices, the confinement of the injected current to the active region is achieved through reverse-biased heterojunctions or forward-biased InP homojunctions (which turn on at a voltage higher than that needed for laser action). The difference between the total current I injected into the lasers and the current going through the active region (I_a) is the leakage current (I_l) i.e.,

$$I_l = I - I_a. \tag{2.36}$$

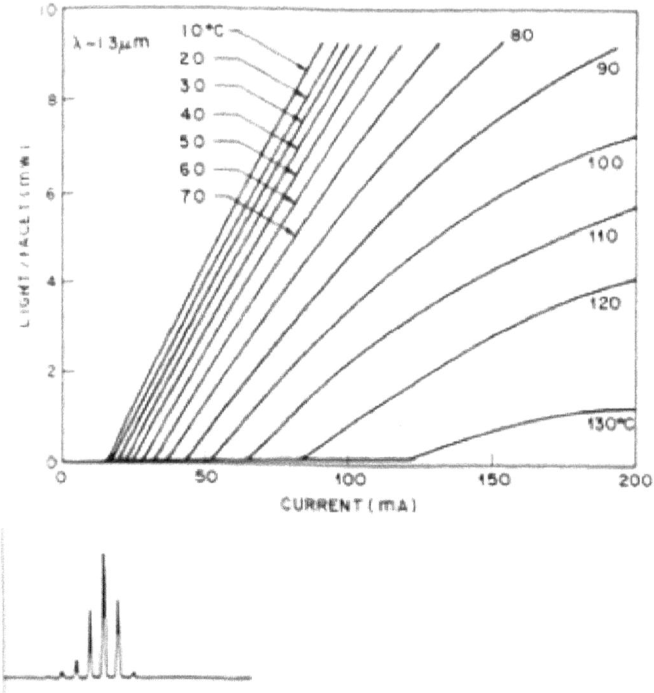

Figure 2.13. Light vs. current characteristics of a semiconductor laser fabricated using InGaAsP/InP material system emitting near 1.3 μm. The figure also shows the emission spectrum at 3 mW. The various peaks represent different longitudinal modes of the cavity separated by ~10 A. The center wavelength is 1300 nm.

A large leakage current would lead to low optical gain at high injected currents and is therefore not desirable. The current-blocking layers in the lasers are designed to minimize the leakage current [4].

Some lasers have gratings fabricated next to the active region. These gratings provide frequency selective feedback. As a result, these lasers (called distributed feedback or DFB lasers) emit at a single wavelength. They are not suitable for "mode locking" but they are important for many applications [4, 11]. Mode locking requires the presence of many longitudinal modes (Figure 2.2). Sources emitting in a single wavelength help reduce the effects of chromatic dispersion (for fiber communication systems) and are therefore used in most systems operating at high data rates (>1.5 Gb/s).

The single wavelength laser source used in most commercial transmission systems is the distributed feedback (DFB) laser [see, for example, Ref. 4] where a diffraction grating etched on the substrate close to the active region provides frequency selective feedback which makes the laser emit in a single wavelength. If Λ is the period of the grating and n is the effective index of the optical mode, the grating reflects light at the free space wavelength (λ) of

$$\lambda = 2n\Lambda. \tag{2.37}$$

For $\lambda = 1550$ nm and $n \sim 3.3$, this corresponds to $\Lambda \sim 240$ nm. The schematic of a DFB laser structure emitting near 1550 nm is shown in Figure 2.14. The fabrication of the device involves the creation of a grating with a periodicity of 240 nm on a (100)-oriented n-InP substrate using optical holography and wet chemical etching. Rest of the fabrication follows the steps described in previous sections. The final laser chips have antireflection coating (<1%) at one facet and high reflection coating (~65%) at the back facet. The asymmetric facet coatings help single-mode operation by removing the degeneracy between the two modes in the stop band [4].

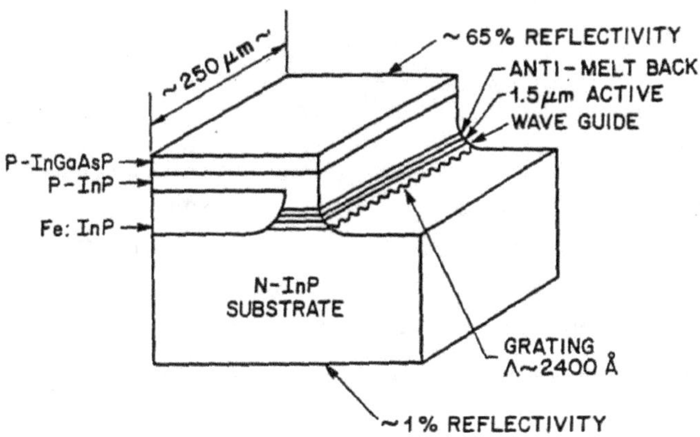

Figure 2.14. Schematic of a distributed feedback laser.

2.6 Multiquantum Well Lasers

Beginning with the 1980s, lasers with very thin active regions, quantum well lasers, were being developed in many research laboratories [61–70]. Quantum well (QW) lasers have active regions of ~100 A thick which restrict the motion of the carriers (electrons and holes) in a direction normal to the well. This results in a set of discrete energy levels and the density of states is modified to a "two-dimensional-like" density of states. This modification of the density of states results in several improvements in laser characteristics, such as lower threshold current, higher efficiency, higher modulation bandwidth, and lower CW and dynamic spectral width. Many of these improvements were first predicted theoretically and then demonstrated experimentally [4, 11].

A regular double heterostructure laser consists of an active layer of thickness ~0.1–0.2 μm sandwiched between higher-gap cladding layers. When the thickness L_z of a narrow-gap semiconductor layer confined between two wide-gap semiconductors becomes comparable to the de Broglie wavelength ($\lambda \sim h/p \sim L_z$), quantum-mechanical effects are expected to occur. The energy levels of the carriers confined in the narrow-gap semiconductor can be determined by separating the Hamiltonian into a component normal to the layer (z component) and into the usual (unconfined) Bloch function components (x, y) in the plane of the layer. The resulting energy eigenvalues are

$$E(n,k_x,k_y) = E_n + \frac{\hbar^2}{2m_n^*}(k_x^2 + k_y^2), \tag{2.38}$$

where E_n is the nth confined-particle energy level for carrier motion normal to the well, m_n^* is the effective mass of the nth level, $\hbar$ is Planck's constant (h) divided by 2π, and k_x, k_y are the usual Bloch function wavevectors in the x and y directions. Figure 2.15 shows schematically the energy levels E_n of the electrons and holes confined in a quantum well. The confined particle energy levels E_n are denoted by E_{1c}, E_{2c}, E_{3c} for electrons, E_{1hh}, E_{2hh} for heavy holes, and E_{1lh}, E_{2lh} for light holes. The calculation of these quantities is a standard problem in quantum mechanics for a

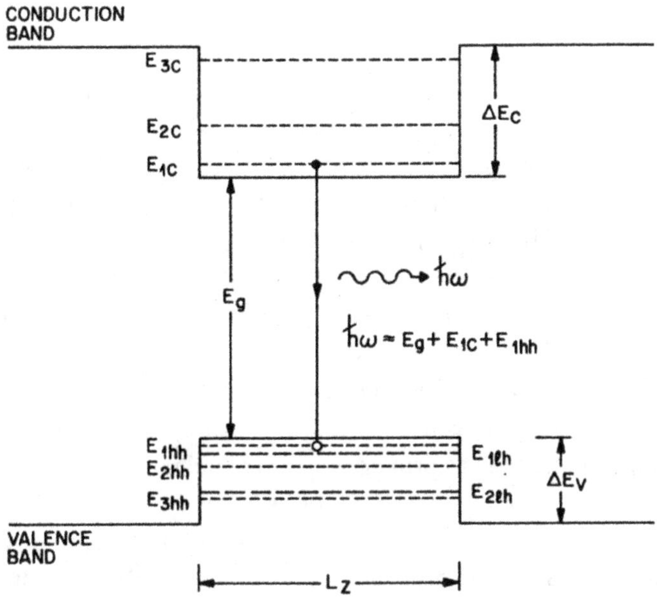

Figure 2.15. Energy levels in a quantum well structure [61, 62].

given potential barrier (ΔE_c, ΔE_v). For an infinite potential well, the following simple result is obtained:

$$E_n = \frac{h^2 n^2}{8L_z^2 m_n^*}. \tag{2.39}$$

The separation between the lowest conduction band level and the highest valence band level is given by

$$E_q = E_g + E_{1c} + E_{1hh} \approx E_g + \frac{h^2}{8L_z^2}\left(\frac{1}{m_c} + \frac{1}{m_h}\right). \tag{2.40}$$

It follows that in a quantum well structure, the energy of the emitted photons can be varied simply by varying the well width L_z. This has been observed experimentally for an InGaAs active region sandwiched between InP layers [4]. Lasers are often fabricated with multiple quantum wells next to each other. These are called multiquantum well (MQW) lasers.

The discrete energy levels result in a modification of the density of states in a quantum well to a "two-dimensional-like" density of states. This modification of the density of states results in several improvements in laser characteristics, such as lower threshold current, higher efficiency, higher modulation bandwidth, and lower CW and dynamic spectral width. Various types of quantum well laser and also amplifier active region designs have been reported. This is summarized in Refs. [4, 11]. Detailed discussion of semiconductor lasers can also be found in these references.

Semiconductor lasers continue to be the laser of choice for applications in optical communication systems, sensor systems, and many consumer applications, such as printers, displays, and barcode scanners. They are far more reliable than other types of lasers, such as gas lasers and dye lasers [71–74]. Semiconductor lasers are also used as high-power pump sources for solid-state lasers (slab-based or fiber-based) which are important for many applications. These are discussed in Chapters 4 and 7. With the development of materials and processing technologies, semiconductor lasers at a new range of wavelengths and with new designs will continue to emerge as important light sources for new applications.

References

1. H. Kressel and J. K. Butler, *Semiconductor Lasers and Heterojunction LEDs* (Academic Press, New York, 1977).
2. H. C. Casey, Jr. and M. B. Panish, *Heterostructure Lasers* (Academic Press, New York, 1978).
3. G. H. B. Thompson, *Physics of Semiconductor Lasers* (John Wiley and Sons, New York, 1980).
4. G. P. Agrawal and N. K. Dutta, Long Wavelength Semiconductor Lasers, (van Nostrand Reinhold Co. 1986; second ed. 1993).
5. J. J. Hsieh, *Appl. Phys. Lett.*, **28**, (1976) 283.
6. N. Holonyuk, Jr. and S. F. Bevacqua, *Appl. Phys. Lett.*, **1**, (1962) 82.
7. M. I. Nathan, W. P. Dumke, G. Burns, F. H. Dill, Jr., and G. Lasher, *Appl. Phys. Lett.*, **1**, (1962) 63.
8. T. M. Quist, R. H. Retiker, R. J. Keyes, W. E. Krag, B. Lax, A. L. McWhorter, and H. J. Ziegler, *Appl. Phys. Lett.*, **1**, (1962) 91; R. N. Hall, G. E. Fenner, J. D. Kingsley, T. J. Soltys, and R. O. Carlson, *Phys. Rev. Lett.*, **9**, (1962).

9. I. Hayashi, M. B. Panish, P. W. Foy, and S. Sumsky, *Appl. Phys. Lett.*, **47**, (1970) 109.
10. H. Kroemer, *Proc. IEEE*, **51**, (1963) 1782.
11. G. P. Agrawal and N. K. Dutta, *Semiconductor Lasers* (Van Nostrand Reinhold, New York, 1992. Sec. 2.5).
12. D. Botez, *IEEE J. Quantum Electron.*, **QE-17**, (1981) 178.
13. J. M. Woodall, H. Rupprecht, and W. Reuter, *J. Electrochem. Soc.*, **116**, (1969) 899.
14. H. Nelson, U.S. Patent 3,565,702 (1971).
15. I. Hayashi, M. B. Panish, P. W. Foy, and S. Sumski, *Appl. Phys. Lett.*, **17**, (1970) 109.
16. M. Feng, L. W. Cook, M. M. Tashima, and G. E. Stillman, *J. Electron. Mater.*, **9**, (1980) 241.
17. I. Ladany and F. Z. Hawrylo, *J. Cryst. Growth*, **54**, (1981) 69.
18. J. J. Hsieh, *Appl. Phys. Lett.*, **28**, (1976) 283.
19. R. J. Nelson, *Appl. Phys. Lett.*, **35**, (1979) 654.
20. J. J. Hsieh, *J. Cryst. Growth*, **27**, (1974) 49.
21. D. L. Rode, *J. Cryst. Growth*, **20**, (1973) 13.
22. W. A. Tiller, *J. Cryst. Growth*, **2**, (1968) 69.
23. M. B. Small and I. Crossley, *J. Cryst. Growth*, **27**, (1974) 35.
24. R. L. Moon, *J. Cryst. Growth*, **27**, (1974) 62.
25. P. Besomi, J. Degami, N. K. Dutta, W. R. Wagner, and R. J. Nelson, *J. Appl. Phys.*, **56**, (1984) 2879.
26. A. Doi, N. Chinone, A. Aiki, and R. Ito, *Appl. Phys. Lett.*, **34**, (1979) 393.
27. M. A. DiGuiseppe, H. Temkin, and W. A. Bonner, *J. Cryst. Growth*, **58**, (1982) 279.
28. A. R. Clawson, W. Y. Lum, and G. W. McWilliams. *J. Cryst. Growth*, **46**, (1979) 300.
29. G. A. Antypas, *Appl. Phys. Lett.*, **37**, (1980) 64.
30. P. Besomi, R. B. Wilson, W. R. Wagner, and R. J. Nelson, *J. Appl. Phys.*, **54**, (1983) 535.
31. G. A. Antypas and R. L. Moon, *J. Electrochem. Soc.*, **120**, (1973) 1574.
32. G. A. Antypas and J. Edgecumbe, *J. Cryst. Growth*, **34**, (1976) 132.
33. K. Nakajima, T. Tanahashi, K. Akita, and T. Yamaoka, *J. Appl. Phys.*, **50**, (1979) 4975.
34. K. Nakajima, T. Kusunoki, K. Akita, and T. Kotani, *J. Electrochem. Soc.*, **125**, (1978) 123.
35. R. E. Nahory, M. A. Pollack, W. D. Johnston Jr., and R. L. Barns, *Appl. Phys. Lett.*, **33**, (1978) 659.
36. H. Nelson, *RCA Rev.*, **24**, (1963) 603.

37. W. F. Finch and E. W. Mehal, *J. Electrochem. Soc.*, **111**, (1964) 814.
38. A. Y. Cho, *J. Vac. Sci. Technol.*, **8**, (1971) 531; J. R. Arthur, *J. Appl. Phys*, **39**, 4032 (1968).
39. R. D. Dupuis, *J. Crystal Growth*, **55**, (1981) 213.
40. H. M. Manasevit and W. I. Simpson, *J. Electrochem. Sac.*, **116**, (1969) 1725.
41. Y. Seki, K. Tanno, K. Iida, and E. Ichiki, *J. Electrachem. Soc.*, **122**, (1975) 1108.
42. R. D. Dupuis, P. D. Dapkus, R. D. Yingling, and L. A. Moudy, *Appl. Phys. Lett.*, **31**, (1978) 201.
43. P. D. Dapkus, *Ann. Rev. Mater. Sci.*, **12**, (1982) 243.
44. J. P. Hirtz, M. Razeghi, M. Bonnet, and J. P. Duchemin, Chap. 3 in T. P. Pearsall (ed.), *GalnAsP Alloy Semiconductors* (John Wiley & Sons, New York 1982); S. Hersee and J. P. Duchemin, *Annu. Rev. Mater. Sci.*, **12**, (1982) 65.
45. J. P. Hirtz, J. P. Larivain, J. P. Duchemin, T. P. Pearsall, and M. Bonnet, *Electron. Lett.*, **16**, (1980) 415.
46. J. P. Hirtz, J. P. Larivain, D. Leguen, M. Razeghi, and J. P. Duchemin, in *Gallium Arsenide and Related Compounds 1980*, Conf. series Vol. 56, pp. 29–35 (Institute of Physics, Bristol, 1981).
47. J. P. Hirtz, M. Razeghi, J. P. Larivain, S. Hersee, and J. P. Duchemin, *Electron. Lett.*, **17**, (1981) 113.
48. M. Razeghi, S. Hersee, J. P. Hirtz, R. Blondeau, B. de Cremoux, and 1. P. Duchemin, *Electron. Lett.*, **19**, (1983) 336.
49. M. G. A. Bernard and G. Duraffourg, *Phys. Status Solidi*, **1**, (1961) 699.
50. N. K. Dutta, *Optoelectronics Devices* (World Scientific, Singapore, 2018).
51. E. O. Kane, *J. Phys. Chem. Solids*, **1**, (1957) 249.
52. F. Stern, *J. Appl. Phys.*, **47**, 5382 (1976); *IEEE J. Quantum Electron*, **QE-9**, (1973) 290.
53. C. J. Hwang, *Phys. Rev. B*, **2**, (1970) 4117.
54. N. K. Dutta, *J. Appl. Phys.*, **51**, (1980) 6095; **52**, (1981) 55.
55. A. R. Beattie and P. T. Landsberg, *Proc. R. Soc. Lond*, **16**, (1959) 249; *Ser. A*, **258**, (1960) 486.
56. Y. Horikoshi and Y. Furukawa Jap, *J. Appl. Phys.*, **18**, (1979) 809.
57. G. H. B. Thomson and G. D. Henshall, *Electron. Lett.*, **16**, (1980) 42.
58. N. K. Dutta and R. J. Nelson, (a) *Appl. Phys. Lett.*, **38**, (1981) 407; (b) *J. Appl. Phys.*, **53**, (1982) 74 and references therein.
59. A. Sugimura, *IEEE J. Quantum Electron.*, **QE-17**, (1981) 627.
60. A. Haug, *Appl. Phys. Lett.*, **42**, (1983) 512.

61. N. Holonyak, Jr., R. M. Kolbas, R. D. Dupuis, and P. D. Dapkus, *IEEE J. Quantum Electron.*, **QE-16**, (1980) 170.
62. N. Holonyak, Jr., R. M. Kolbas, W. D. Laidig, B. A. Vojak, K. Hess, R. D. Dupuis, and P. D. Dapkus, *J. Appl. Phys.*, **5**, (1980) 1328.
63. W. T. Tsang, *Appl. Phys. Lett.*, **39**, (1981) 786.
64. W. T. Tsang, *IEEE J. Quantum Electron.*, **QE-20**, (1986) 1119.
65. S. D. Hersee, B. DeCremoux and J. P. Duchemin, *Appl. Phys. Lett.*, **44**, (1984).
66. F. Capasso and G. Margaritondo (eds.), *Heterojunction Band Discontinuation: Physics and Applications* (North Holland, Amsterdam, 1987).
67. N. K. Dutta, S. G. Napholtz, R. Yen, R. L. Brown, T. M. Shen, N. A. Olsson, and D. C. Craft, *Appl. Phys. Lett.*, **46**, (1985) 19.
68. N. K. Dutta, S. G. Napholtz, R. Yen, T. Wessel, and N. A. Olsson, *Appl. Phys. Lett.*, **46**, (1985) 1036.
69. P. J. A. Thijs, L. F. Tiemeijer, P. I. Kuindersma, J. J. M. Binsma, and T. van Dongen, *IEEE J. Quantum Electron.*, **27**, (1991) 1426.
70. P. J. A. Thijs, L. F. Tiemeijer, J. J. M. Binsma, and T. van Dongen, *IEEE JQE*, **QE-30**, (1994) 477.
71. The reliability requirements of a submarine lightwave transmission system are discussion in a special issue of *AT&T Tech. J.*, **64**, (1985) 3.
72. O. Ueda, S. Yamakoshi, S. Komiya, K. Akita, and T. Yamaoka, *Appl. Phys. Lett.*, **36**, (1980) 300.
73. E. I. Gordon, F. R. Nash, and R. L. Hartman, *IEEE Electron Device Lett.*, **ELD-4**, (1983) 465.
74. R. L. Hartman and R. W. Dixon, *Appl. Phys. Lett.*, **26**, (1975) 239.

Chapter 3

Pulse Generation and Mode-Locked Semiconductor Lasers

3.1 Introduction

There are several methods to generate short optical pulses from semiconductor diode lasers. This includes mode locking [1], Q switching [2], and gain switching [3–7]. Gain switching relies upon the switching of the optical gain through the modulation of the drive current. Gain switching does not need external cavity. By modulating any kind of diode laser, distributed feedback (DFB), distributed Bragg reflector (DBR), and Fabry–Perot (FP) lasers, for instance, one can obtain short pulses if the modulation power is high enough. Another advantage of gain switching over mode locking is that one can arbitrarily vary the repetition rate by changing the frequency of the modulation current. However, pulses obtained using mode locking are shorter than those obtained using gain switching.

3.2 Gain Switching

Optical pulse generated by gain switching is considerably shorter than the electrical pulse. The gain-switching phenomena have been extensively studied [3–7]. In order to understand the dynamics of gain-switched pulses in diode lasers, we need to study a set of nonlinear rate equations of a diode laser shown in the following:

$$\begin{aligned}\frac{dn}{dt} &= \frac{I(t)}{eV} - \frac{\Gamma g_d(n-n_0)}{1+\varepsilon S}S - \frac{n}{\tau_e},\\ \frac{dS}{dt} &= \frac{\Gamma g_d(n-n_0)}{1+\varepsilon S}S - \frac{S}{\tau_{ph}} + \frac{\beta\Gamma n}{\tau_e},\end{aligned} \tag{3.1}$$

where n and S are carrier density and photon density, respectively, e is the charge of an electron, β is the spontaneous emission coupling factor, Γ is the optical confinement factor, V is the volume of the laser's active cavity, g_d is the differential gain coefficient, ε is the nonlinear gain coefficient which plays an important role in gain-switching dynamics [8], and τ_e and τ_{ph} are the carrier lifetime and the photon lifetime, respectively. These simple rate equations assume the distributions of carriers and photons inside the laser cavity are uniform. Also, it assumes that the laser operates in a single spatial mode. However, the equations can also be written for the multimode case [7]. In Equations (3.1), bias current $I(t)$ is taken as a train of electrical pulses. The laser is modulated using a sinusoidal microwave signal. So, $I(t)$ is written as

$$I(t) = I_0 + I_m \sin(2\pi f_m t), \tag{3.2}$$

where I_0 is the DC bias current. The carrier lifetime τ_e is dependent on carrier density n:

$$\tau_e = (A + Bn + Cn^2)^{-1}, \tag{3.3}$$

where the recombination rate A is due to mechanisms such as trap or surface recombination, B is the radiative recombination coefficient, and C is related to Auger recombination. The photon lifetime is expressed as

$$\tau_{\text{ph}} = 1/(v_g\,\alpha), \tag{3.4}$$

where v_g is the group velocity and α is the total optical loss inside the active layer. If we ignore the nonlinear gain effect ε, loss, and spontaneous emission, the second Equation of (3.1) can be written as [4]

$$dS/dt = S/\tau_r = S\Gamma g_d\,(n - n_0). \tag{3.5}$$

The pulse is then approximately described as a combination of two exponential curves with time constant τ_r on the rising edge and $\tau_f \sim (2\text{~}3)\ \tau_r$ on the trailing edge. When the peak carrier density is higher, the rise time is smaller, and the pulse width becomes shorter.

Analytical solutions to (3.1)–(3.3) are not possible without any further approximation. Numerically solving the rate Equations (3.1) together with (3.3)–(3.5), various behaviors of the gain-switched pulses under different conditions are obtained. In the following section, the numerical calculations are compared with experimental results under different conditions.

The Runge–Kutta method has been used to get numerical solutions. Values of the parameters used in the simulation are listed in Table 3.1.

Figure 3.1 illustrates the simulation of a typical optical response for the gain-switched laser. The dashed line is the carrier density versus time, and the solid line shows the photon density as a function of time. The repetition rate is f_m = 5GHz, DC bias current is 24 mA, and modulation

Table 3.1. Parameter values used in the calculation. These are typical values for a semiconductor laser.

Parameters	**Value**
Wavelength	$\lambda = 1.54917\ \mu m$
Cavity length	$L = 250\ \mu m$
Cavity width	$w = 1.2\ \mu m$
Cavity thickness	$d = 0.1\ \mu m$
Optical confinement factor	$\Gamma = 0.35$
Differential gain coefficient	$g_d = 2.5 \times 10^{-6}\ cm^3\ s^{-1}$
Transparency carrier density	$n_0 = 1.023 \times 10^{18}\ cm^{-3}$
Carrier recombination parameters	$A = 2 \times 10^8\ s^{-1}$ $B = 10^{-10}\ cm^3\ s^{-1}$ $C = 3 \times 10^{-29}\ cm^6\ s^{-1}$
Spontaneous coupling factor	$\beta = 10^{-4}$
Total loss	$\alpha = 65\ cm^{-1}$
Nonlinear gain coefficient	$\varepsilon = 1.55 \times 10^{-17}\ cm^{-3}$
Calculated threshold current[13]	$I_{th} = 3.5\ mA$
Linewidth enhancement factor	$\beta_c = 2$

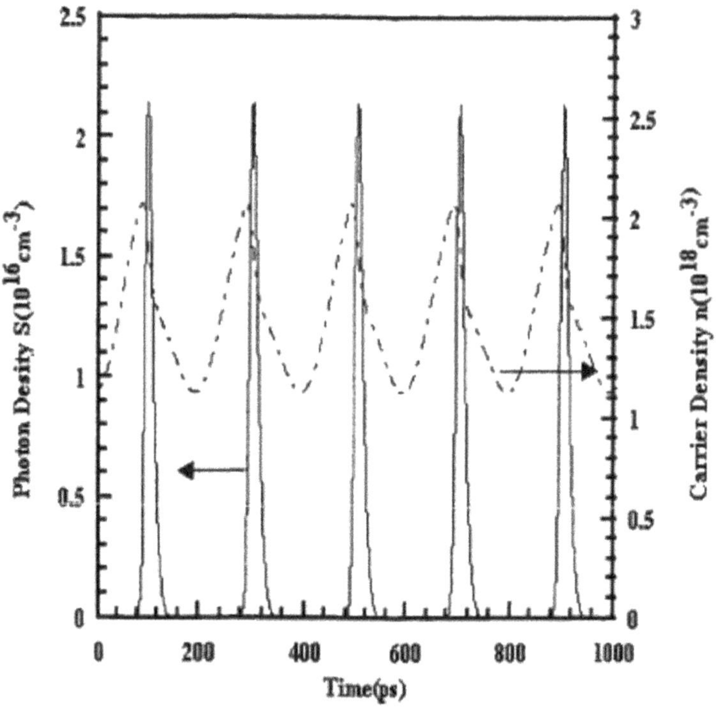

Figure 3.1. Calculated photon density S (solid line) and carrier density n (dashed line) versus time. The repetition rate is 5GHz, DC bias I_o = 24 mA, and modulation power is 20 dBm or 100 mW (odulation current I_m = ~63 mA).

power is 20 dBm or 100 mW corresponding to I_m = 63 mA. Figure 3.1 also illustrates the typical temporal dependence of the photon density and the carrier density. The carrier density increases without emission even when it exceeds the threshold. After it reaches a critical value, the carrier density drops quickly, which produces short pulses. The calculated pulse shape is asymmetrical as observed in the experiments i.e., the rise time is shorter than the fall time.

3.2.1 *Gain switching results*

The experimental results on gain switching using a distributed feedback (DFB) diode laser are now described. The CW laser wavelength is 1549 nm. The threshold of this DFB laser was 10 mA. The coupled power

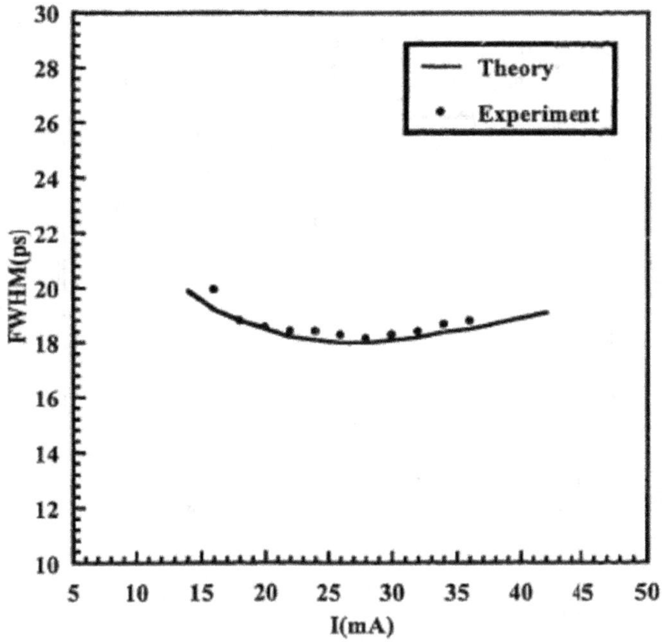

Figure 3.2. Pulse width as a function of DC bias current. Repetition rate is 5 GHz and modulation power is 20 dBm or 100 mW (I_m = 63 mA).

into the fiber is more than 1 mW at the bias current of ~20 mA. The modulation signal is provided by a synthesizer. The modulation power is about 20 dBm. The autocorrelator was used to measure the width of the output pulses.

The gain switching is strongly dependent on the DC bias current. One can observe that if the bias is too low, the gain-switched pulses are not stable, and, if the bias current is too high, the pulses are broad. One has to optimize the DC bias current in order to get the shortest pulses. Figure 3.2 illustrates that the pulse width (FWHM) changes with DC bias current. The dots are experimental results. The solid line is from numerical computation. The theoretical calculation fits the experiments.

It is of interest to mention that the optimum bias current changes if the modulation frequency (repetition rate) changes (Figure 3.3). When the modulation frequency increases, the carrier dynamics in the laser needs to increase to respond to the modulation signal. The modulation speed of the

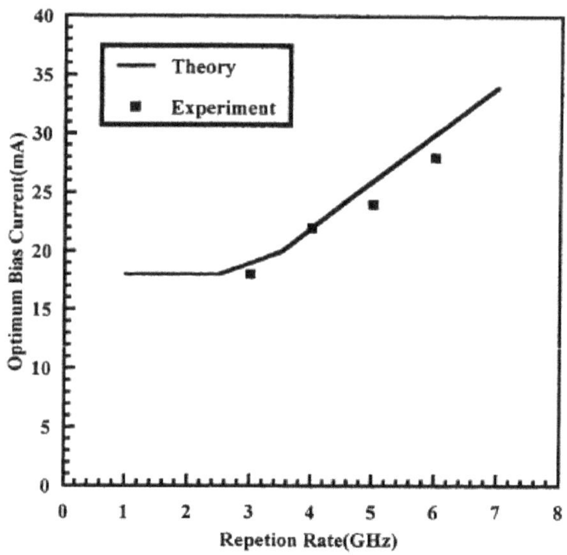

Figure 3.3. Optimum bias current increases with repetition rate. P_m = 20 dBm or 100 mW.

diode laser is higher when the bias is higher (Section 3.4). That is the reason why the bias current needs to be increased in order to get the shortest pulses when the repetition rate of the modulation signal increases. According to the calculation, at low repetition rates, the optimum DC bias does not increase with repetition rate because carrier dynamics in the laser is fast enough. At high repetition rates, the optimum DC bias current increases linearly with the modulation frequency.

When the modulation power is low, the pulses are broad. It decreases fast when the modulation power increases. According to Equation (3.5), the pulse rise time is roughly proportional to $I/(\Gamma g_d(n_p-n_0)$, where n_p is the peak value of carrier density shown in Figure 3.1. When modulation power is higher, the peak carrier density is also higher, then the pulse width is smaller. The peak power of gain-switched pulses increases when the modulation power increases and it eventually saturates (Figure 3.4).

3.2.2 *Gain-switched laser spectrum*

The spectrum of a gain-switched single-mode DFB laser is analyzed here using Equations (3.1) and a phase equation. It is well known that the

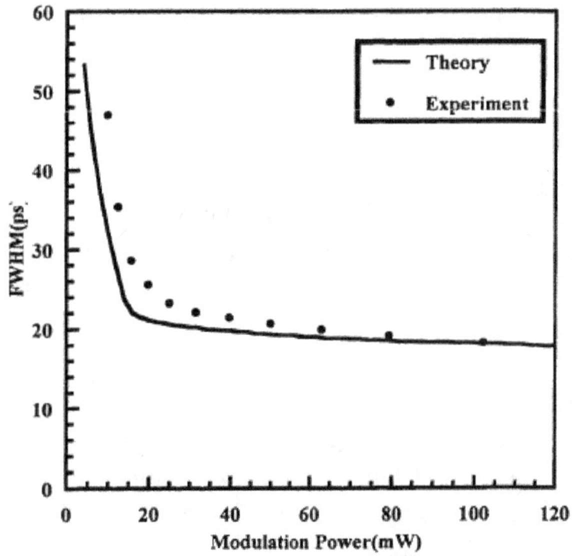

Figure 3.4. Relationship between modulation power and gain-switched pulse width.

refractive index in the laser depends on carrier density [9]. It was shown in Section 3.2.1 that the carrier concentration experiences a large variation during the emission of a gain-switched pulse (see Figure 3.1). Hence, the refractive index should vary as well. The spectral width under gain switching is much larger than that under CW operation. The time dependence of wavelength ($\lambda(t)$) is related to the time dependence of carrier density, which is expressed as

$$\lambda(t)/\mu(t) = 2\Lambda/m = \lambda_0/\mu_0, \tag{3.6}$$

where $\mu(t)$ is the refractive index of the laser cavity at time t, Λ is the grating period, and m is the order of the grating ($m = 1$ for first-order grating and $m = 2$ for second-order grating). Λ and m are chosen appropriately to make the laser emit near 1550 nm. λ_0 and μ_0 are the wavelength and index under CW operation. The time-dependent wavelength is related to the time-dependent carrier density i.e.,

$$\Delta\mu\,(t) = \Delta n(t)\; d\mu/dn, \tag{3.7}$$

where $d\mu/dn$ (rate of change of index with carrier density) is assumed to be a constant. For InGaAsP laser, $d\mu/dn$ is ~$-(2\text{–}3) \times 10^{-20}/\text{cm}^3$. From

Equations (3.6) and (3.7), the wavelength variation (chirp) $\Delta\lambda$ during the pulse is given by

$$\Delta\lambda \approx \frac{\lambda_0}{\mu_0}\frac{d\mu}{dn}\Delta n, \tag{3.8}$$

where Δn is the carrier density change during the emission of a gain-switched pulse. According to Figure 3.1, at the leading edge, Δn is positive and $\Delta\lambda$ is negative. That is the "blue shift" at the leading edge. Similarly, at the falling edge, there is a "red shift". So, the wavelength of the leading edge is shorter than the wavelength of the falling edge. This is called "negative chirp". The wavelength chirp has been observed in time-resolved spectra and time-averaged spectra of gain-switched lasers [10, 11]. Osinski and Adams [10] have presented a numerical analysis of gain-switched laser spectra. Using Equations (3.1–3.8), the gain-switched laser spectrum is simulated. In addition to the rate equations for carrier density and photon density (Equation 3.1), there are phase equations. The slow optical envelope is given by

$$E = S^{1/2} \exp(i\phi), \tag{3.9}$$

where ϕ is the phase. The phase rate equation is [Ref. 7, see also Section 3.4]

$$\frac{d\phi}{dt} = -(\omega_0 - \omega_{th}) + \frac{1}{2}\beta_c[\Gamma g_d(n - n_0) - v_g\alpha], \tag{3.10}$$

where ω_o is the frequency of the laser, ωth is the longitudinal mode frequency corresponding to the threshold value of mode index, and β_c is the linewidth enhancement factor of the laser, which is expressed as

$$\beta_c = -(4\pi/\lambda_0)\,(d\mu/dn)/(dg/dn). \tag{3.11}$$

Equation (3.1) can be solved together with Equation (3.10). Then, a Fourier transform is taken in order to get the spectral density of the optical field:

$$\left|\int_{-\infty}^{+\infty} Ee^{-t(\omega-\omega_0)t}dt\right|^2 = \left|\int_{-\infty}^{\infty} \sqrt{S}e^{-i\phi}e^{-1(\omega-\omega_0)t}dt\right|^2. \tag{3.12}$$

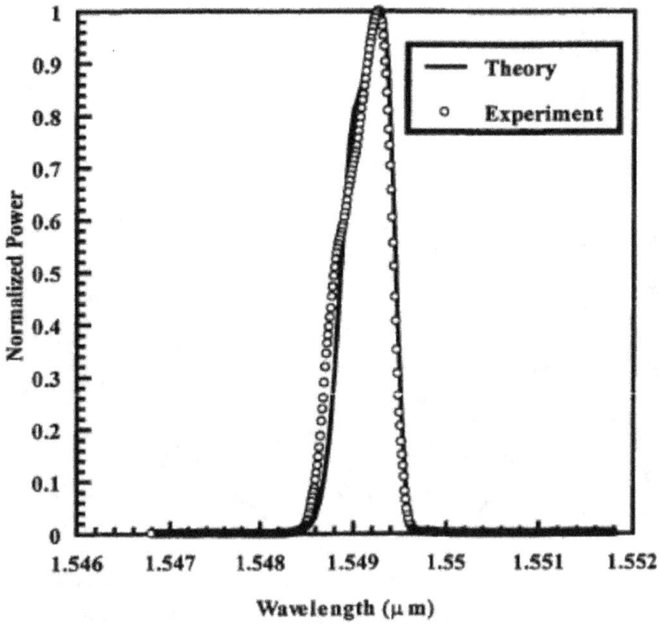

Figure 3.5. Optical spectrum of gain-switched DFB laser. I_o = 24 mA, P_m = 20 dBm or 100 mW (I_m = 63 mA), and f_m = 5GHz. Circles are experimental data. Solid curve is theoretical result.

Figure 3.5 shows time-averaged spectra obtained from the experimental measurement and the theoretical calculation (solid line). The simulation fits the experimental data quite well.

3.2.3 *Compression of gain-switched pulses*

Pulse compression using optical fiber is described in detail in Chapter 6. Here, the result of modeling and experiment for a gain-switched laser pulse is presented. The principle is as follows: since the front of the pulse is blue-shifted and the back of the pulse is red-shifted, by injecting the pulse into a fiber where the blue-shifted light travels a little faster than the red-shifted light, the pulse will shorten. This is the case for a dispersion compensating fiber (also known as DCF). Under some approximation, the

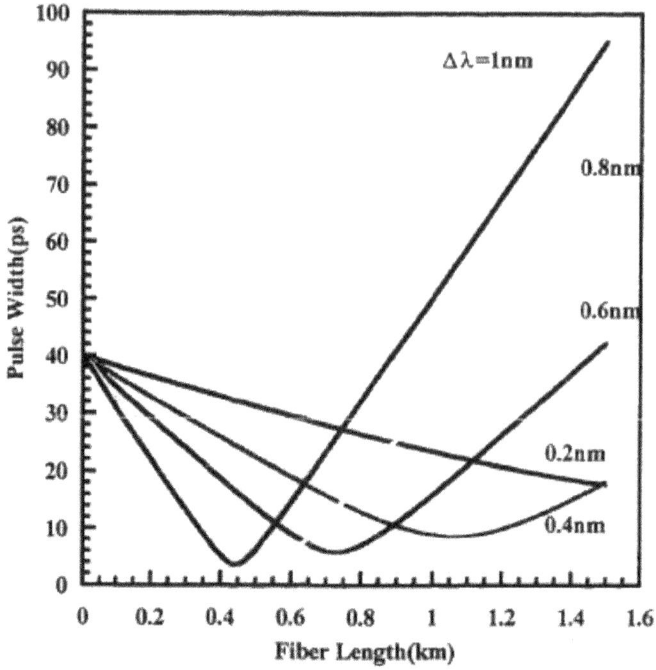

Figure 3.6. Pulse width as a function of fiber length (propagation distance in fiber).

pulse width at the output ($\Delta\tau_1$) is related to the pulse width at the input ($\Delta\tau_0$) by the following formula (see Chapter 6):

$$\frac{\Delta\tau_1}{\Delta\tau_0} = \left[\left(1+\frac{D\Delta\lambda z}{\Delta\tau_0}\right)^2 + \left(2\ln 2\frac{D\lambda^2 z}{\pi c\Delta\tau_0^2}\right)^2\right]^{1/2}, \tag{3.13}$$

where D is the dispersion of the fiber, $\Delta\lambda$ is the chirp of the incident light, z is the length of the fiber, λ is the wavelength, and c is the speed of light. Figure 3.6 shows the plot of Equation (3.13) for different spectral widths ($\Delta\lambda$) of the initial pulse. The quantity $D = -90$ ps/km-nm and the initial pulse width $\Delta\tau_0 = 40$ ps in the figure. As shown in the figure, as the pulse travels, it becomes narrower first and then eventually becomes broader. There is an optimum condition for a narrow pulse. This happens when

$$DL = -\frac{\Delta\tau_0}{\Delta\lambda}, \tag{3.14}$$

and the narrowest pulse is given by

$$\Delta t_{\min} = 2\ln 2 \frac{\lambda^2}{m\Delta\lambda}. \tag{3.15}$$

Measured data for a gain-switched laser is shown in Figure 3.7.

The modeling and experiments show that gain-switched pulses at various repetition rates can be easily generated using a semiconductor laser.

3.3 Mode Locking

Mode locking is a technique used to generate short pulses from a laser [12–17]. In this technique, the phases of the longitudinal modes are locked when the gain or loss of the laser is modulated at the inverse of the round-trip time of the optical cavity. The principle of mode locking is described in Chapter 1.

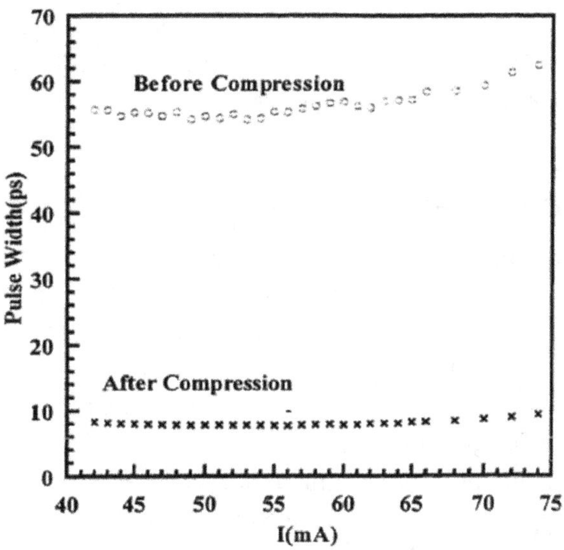

Figure 3.7. Pulse compression using a dispersion compensating fiber, $L = 1$ km, $D = -90$ ps/km-nm. Modulation power of DFB laser is 20 dBm (63 mA).

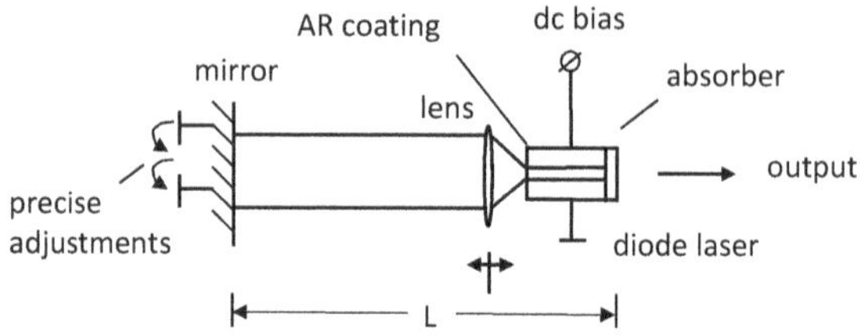

Figure 3.8. Schematic of an external cavity mode-locked laser diode.

3.3.1 *External cavity mode-locked laser*

Many demonstrations of mode-locked semiconductor lasers have been carried out in an external cavity configuration shown in Figure 3.8. An AlGaAs (or any other semiconductor e.g., InGaAsP) laser chip is used as the gain element. It has an antireflection coating on the side with the optical cavity.

A saturable absorber is created on one facet by oxygen ion bombardment. The implantation depth is about 10 μm. Round-trip time in the cavity is 1.2 ns. The laser has a DC bias and the current is modulated at a frequency corresponding to the inverse round-trip time. Precise adjustment of the laser cavity is needed for stable mode-locked operation. Using proton bombardment as an absorber, a pulse with 560 fs pulse width at ~625 MHz repetition rate is obtained [12].

3.3.2 *Compact mode-locked lasers*

Mode-locked semiconductor lasers at transmission data rates (~10 Gb/s and higher) are important for a wide range of fiber transmission systems. These include soliton generation and clock recovery in very high-speed transmission systems. Compact, single-chip, mode-locked lasers are desirable for practical systems since they are likely to be more stable than the external cavity designs [14–16].

The schematic of a compact mode-locked laser on a chip is shown in Figure 3.9 [14]. The laser is fabricated using the InGaAsP/InP material system. A saturable absorber is placed in the middle of the optical cavity.

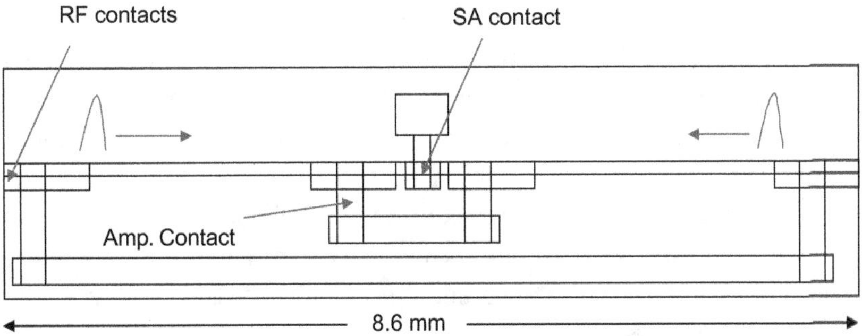

Figure 3.9. Schematic of a colliding pulse mode-locked laser. SA: Saturable absorber. The laser has multiquantum well-active region. The waveguide region is InGaAsP (λ ~ 1300 nm) [14].

The mode-locked pulses from opposite directions arrive at the saturable absorber at the same time, which enhances the saturation and hence a shorter pulse could be obtained. This principle is called colliding pulse mode locking (CPM). The round-trip time in CPM configuration is $t = nL/c$, where L is the length of the linear cavity (8.6 mm in Figure 3.9), n is the index, and c is the speed of light in vacuum.

The colliding-pulse mode-locking (CPM) scheme has been widely used to generate ultrashort transform-limited pulses in dye lasers [see Chapter 1]. Instead of having a single pulse passing through the saturable absorber at one time, CPM lasers utilize the coherent interaction of two counter-propagating pulses colliding at the saturable absorber to produce shorter pulses. The laser operates with two counter-propagating pulses colliding simultaneously at the saturable absorber because this operation has the lowest loss at the saturable absorber. This enhanced effective saturation cross-section of the absorber helps stabilize and shorten the pulses.

The laser has four gain sections. Two sections are in the middle and two sections are at each end of the optical cavity. These sections are joined by a waveguide layer through which the light propagates and are pumped with enough current to produce gain. The saturable absorber is in the middle. All sections have both DC and RF contacts. The modulation is applied to the saturable absorber section. The waveguide below the active region exists over the entire length of the chip. The light generated in the active region travels through this waveguide and is reflected back. The optical cavity is formed by the cleaved facets. The four active gain sections are ~350 μm long each, and the saturable absorber section is ~150 μm long. The total length of the chip is ~8.6 mm. The estimated round-trip time is 0.1 ns which corresponds to a repetition rate of about 10 GHz for the mode-locked pulses. The laser emits near 1560 nm. The laser has a threshold current of 65 mA and a differential efficiency of 0.04 mW/mA when coupled into a single-mode fiber.

The fabrication of the device involves the following steps. First, the waveguide region (~1.3 μm, *n*-InGaAsP), a thin InP region (which acts as a stop etch), the multiquantum well active region with four quantum wells, a *p*-InP cladding region, and a thin *p*-InGaAs region are grown on a (100)-oriented InP substrate. The multiquantum well active region consists of 100-A-thick active region wells (InGaAs) and 80-A-thick barrier regions (~1.3 μm InGaAsP). The quantum wells are also present in the saturable absorber region. The thin *p*-InGaAs layer is etched away, the wafer is patterned with dielectric, and the wafer is etched down passed the active region in some parts. A second set of layers consisting of *p*-InP cladding layer and *p*-InGaAs contact layer are then grown on the wafer. The wafer is processed using standard photolithography, metallization, and cleaving procedures to fabricate the laser chip shown in Figure 3.9. The laser is bonded to a submount, which is further mounted on a thermoelectric-controlled heat sink for testing.

The mode-locking behavior of the laser is investigated by modulating the laser at close to 10 GHz. The pulse width is measured using an autocorrelator. The measured mode-locked pulse width plotted as a function of modulation frequency is shown in Figure 3.10. The total bias current on center amplifiers is 110 mA (55 mA on each), 50-mA bias current on each facet amplifier, 0.2-V bias on saturable absorber, and modulation power of 1 W on the saturable absorber. The minimum pulse width is 1 ps at a frequency of 9.61 GHz.

The measured pulse width as a function of bias is shown in Figure 3.11. The pulse width is low (~1 ps) for a range of bias currents and it increases beyond that. Since the refractive index varies with the injected current, this may be explained by the variation of round-trip time (and hence resonant

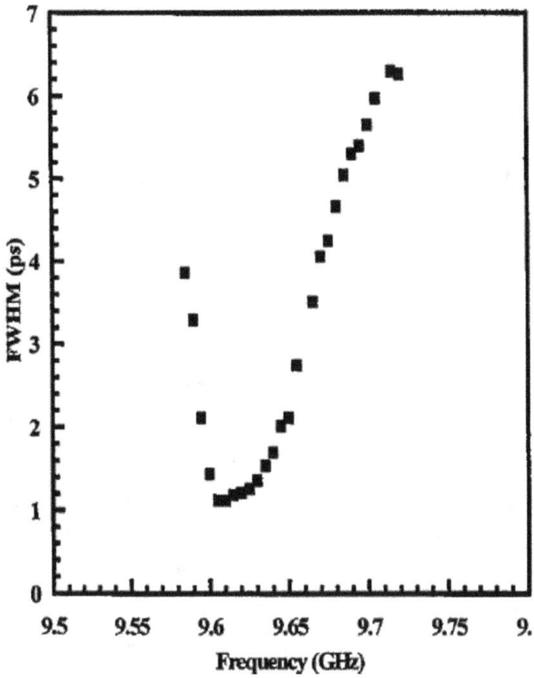

Figure 3.10. The measured mode-locked pulse width is plotted as a function of modulation frequency [14].

frequency) with injection current. The pulse width was not found to vary with RF modulation power to the saturable absorber.

Mode-locked lasers that operate at a higher speed have been reported using a colliding pulse scheme and a short cavity. The laser schematic is shown in Figure 3.12 [15]. It has two gain sections around a saturable absorber region along the optical cavity. The laser is fabricated using the InGaAsP/InP material system and it emits near 1520 nm. The Fe:InP semi-insulating layer is needed to reduce the capacitance. Both laser contacts are on the top surface. Lasers with various cavity lengths were fabricated. The short cavity lasers operated at a higher speed. The shortest lasers were 534 μm in linear length; this corresponds to a round-trip time of 2.87 ps. Transform-limited pulses with 640 fs pulse width at 350 GHz were obtained. The results of pulse width measurement and spectral measurement are shown in Figure 3.13. The measured optical spectra of the

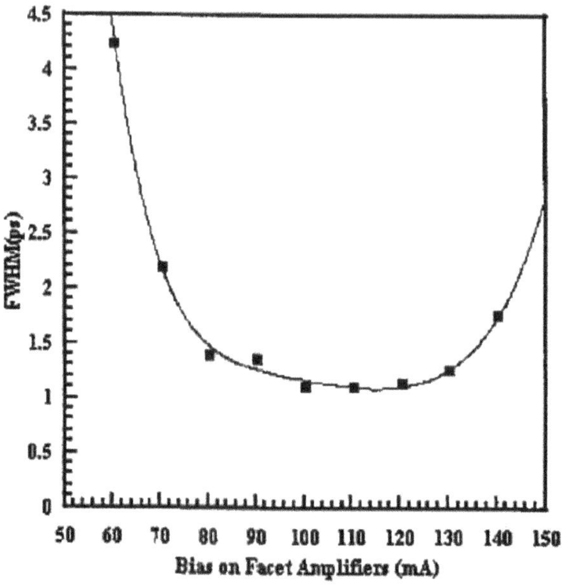

Figure 3.11. Pulse width as a function of the bias on facet amplifiers. Bias on center amplifiers is 110 mA. V sa = 0:2 V. Modulation power is 1 W. Modulation frequency is 9.61 GHz [14].

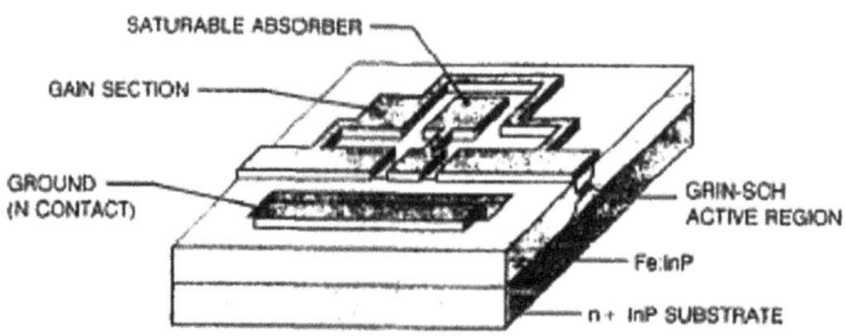

Figure 3.12. Schematic of a colliding pulse mode-locked laser fabricated using InGaAsP material system. The active region has a graded index separate confinement heterostructure (GRIN SCH) [15].

350 GHz pulses show an FWHM spectral width of 4 nm. This corresponds to a nearly transform-limited time–bandwidth product of 0.32.

For effective pulse shaping in passive mode-locked lasers, it is necessary for the absorber to saturate faster than the gain [Ref. 18, 19, see also Chapter 6]. This sharpens the leading edge of the pulses, while the

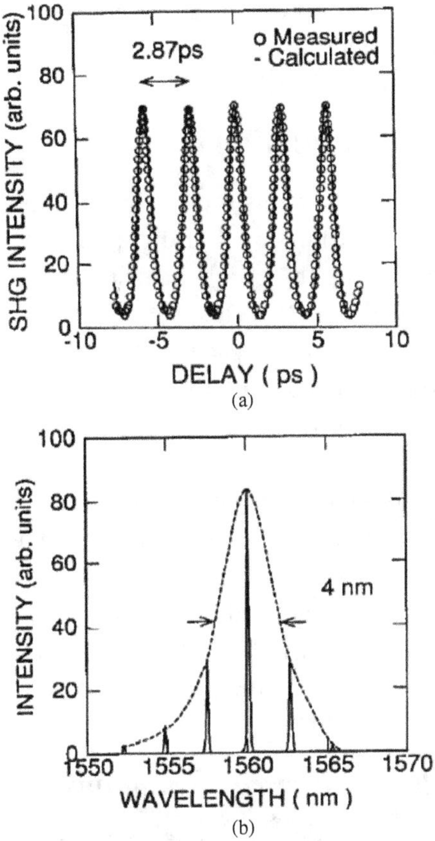

Figure 3.13. (a) Measured (dots) and simulated (solid curves) SHG traces of a passive CPM laser indicate optical pulses with an FWHM pulse width of 640 fs at 350 GHz. (b) The measured optical spectra of the 350 GHz pulses show an FWHM spectral width of 4 nm. This corresponds to a nearly transform-limited time–bandwidth product of 0.32 [15].

saturation of gain reduces the trailing edge. The pulse shortening mechanism continues as the pulse makes multiple round trips inside the laser cavity. Eventually, dispersive effects prevent the pulse from becoming narrower. To sharpen optical pulses, the physical length of the saturable absorber should be shorter than the optical pulses. It takes almost 1.5 ps to traverse the absorber and the gaps (~70 μm) for the laser in Figure 3.12. Hence, the minimum pulse width in this case is limited by the physical size of the absorber and the dispersion from the active region. Optimization of the length and saturation property of the absorber and low dispersion may reduce the pulse width.

3.4 Pulse Code Modulation

Most optical transmission systems use two-level amplitude (or intensity) modulated signals for transmission. For digital transmission systems, the "1s" and "0s" are determined by a high intensity (S_1) and low intensity (S_0) of the transmitted signal [16–21]. This is called "pulse code modulation" (PCM) i.e., a digital current pulse pattern (1100 10...) is converted to an optical signal pattern by modulating the laser with the current. For high transmission systems, this requires that the laser be capable of high-frequency modulation i.e., its modulation characteristics (defined as the ratio of the modulated light intensity as a function of current modulation frequency) be sufficiently high [20]. A 3dB bandwidth is often defined as the frequency at which the modulated light power is half of its value at low frequencies. This 3dB bandwidth must be sufficiently high e.g., for 20 Gb/s PCM, the 3dB bandwidth needs to be higher than 15 GHz. The following describes a theory of 3dB bandwidth of lasers.

3.4.1 *Semiconductor laser rate equations*

The rate equations involving the phase of the laser are examined in order to obtain information on chirp, linewidth, phase noise, intensity noise, etc. The wave equation is

$$\nabla^2\vec{E} = \frac{1}{c^2}\frac{\partial^2}{\partial t^2}(\epsilon\vec{E}). \tag{3.16}$$

For the light propagating along $\hat{z}$ (laser cavity direction), polarized along $\hat{x}$ (along the junction),

$$\bar{E}(x,y,z,t) = \frac{1}{2}\hat{x}\varphi(x)\phi(y)\sum_j \sin(k_j z)E_j(t)e^{-i\omega t} + c.c., \tag{3.17}$$

where $\varphi(x)$ and $\phi(y)$ are lateral and transverse mode profiles and c.c. is complex conjugate.

$$k_j = \frac{2\pi}{\lambda_j} = \frac{\pi m_j}{\mu L}, \tag{3.18}$$

where λ_j is the jth longitudinal mode that satisfies $2\,\mu L = m_j\lambda_j$, $m_j = 0, 1, 2, \ldots, k_j$ are discrete values of wavenumber, and μ is the refractive index. For simplicity, we consider one longitudinal mode and drop the subscript j,

$$E(x,y,z,t) = \frac{1}{2}\hat{x}\varphi(x)\phi(y)\sin(k\,z)E(t)e^{-i\omega t} + c.c. \tag{3.19}$$

Substituting Equation (3.19) in Equation (3.16) and integrating over x, y, we get

$$\frac{2i\omega}{c^2}\left(\langle\epsilon\rangle + \frac{\omega}{2}\frac{\partial\langle\epsilon\rangle}{\partial\omega}\right)\frac{dE}{dt} + \left(\frac{\omega^2}{c^2}\langle\epsilon\rangle - k^2\right)E = 0, \tag{3.20}$$

where $\langle\epsilon\rangle = \int_{-\infty}^{+\infty}\int_{-\infty}^{+\infty}\epsilon(x,y)\varphi^2(x)\phi^2(y)dxdy$ is the average dielectric constant:

$$\begin{aligned}\langle\epsilon\rangle &= (\bar{\mu} + \Gamma\Delta\mu)^2 \\ &= \bar{\mu}^2 + 2\Gamma\bar{\mu}\Delta\mu,\end{aligned} \tag{3.21}$$

$$\Delta\mu = \Delta\mu_r + i\Delta\mu_i, \tag{3.22}$$

$\Delta\mu$ is the change in index (both real and imaginary) due to carrier injection and $\bar{\mu}$ is the mode index:

$$\Gamma\Delta\mu_i = \bar{\alpha}/2k_0, \tag{3.23}$$

where $\bar{\alpha}$ is the mode absorption coefficient

$$\bar{\alpha} = -\Gamma g + \alpha_{\text{int}} + \alpha_m \tag{3.24}$$

and α_{int} and α_m are scattering and mirror losses. Thus,

$$\langle\epsilon\rangle = \bar{\mu}^2 + 2\Gamma\bar{\mu}\Delta\mu_r + i\frac{\bar{\mu}\bar{\alpha}}{k_0}. \tag{3.25}$$

In Equation (3.20), now we use $\bar{k} = \bar{\mu}\Omega / c$, Ω is specific longitudinal mode frequency, $(\omega^2 - \Omega^2) \approx 2\omega(\omega - \Omega)$, and $\langle\epsilon\rangle \simeq \bar{\mu}^2$, we get

$$\frac{dE}{dt} = \frac{i\bar{\mu}}{\mu_g}(\omega - \Omega)E + \frac{i\omega}{\mu_g}\left(\Gamma\Delta\mu_r + \frac{i\bar{\alpha}}{2k_0}\right), \tag{3.26}$$

$$\langle\epsilon\rangle + \frac{\omega}{2}\frac{\partial\langle\epsilon\rangle}{\partial\omega} = \bar{\mu}\mu_g, \tag{3.27}$$

where μ_g is the group index corresponding to mode index $\bar{\mu}$. The electric field E can be represented by

$$E = A\exp(-i\phi). \tag{3.28}$$

A is amplitude and ϕ is phase. Separating into real and imaginary parts,

$$\frac{dA}{dt} = \frac{1}{2}v_g[\Gamma g - (\alpha_{\text{int}} + \alpha_m)]A, \tag{3.29}$$

$$\frac{d\phi}{dt} = -\frac{\bar{\mu}}{\mu_g}(\omega - \Omega) - \frac{\omega}{\mu_g}\Gamma\Delta\mu_r, \tag{3.30}$$

where $v_g = c/\mu_g$ is the group velocity. The amplitude equation above can be written in terms of photon number P defined as

$$P = \frac{\epsilon_0\bar{\mu}\mu_g}{2\hbar\omega}\int |E|^2 dV. \tag{3.31}$$

Thus, $P \propto A^2$, we get

$$\begin{aligned} \frac{dP}{dt} &= (G - \gamma)P + \beta R_{\text{sp}}, \\ G &= \Gamma v_g g, \\ \gamma &= v_g(\alpha_m + \alpha_{\text{int}}) = 1/\tau_p, \end{aligned} \tag{3.32}$$

where τ_p is defined as the photon lifetime. βR_{sp} is the spontaneous emission in the stimulated photon mode. β is the spontaneous emission factor. The α factor is defined as

$$\alpha = \frac{dn_R / dN}{dn_I / dN}$$
$$= -2k_0\left(\frac{\partial n_R / \partial N}{\partial g / \partial N}\right), \tag{3.33}$$

$n = n_R + in_I$, n_R and n_I are the real and imaginary parts of the index, and g is the gain. Thus,

$$\Delta\mu_r = \Delta n_R = -\left(\frac{\alpha}{2k_0}\right)\Delta g. \tag{3.34}$$

If $\Delta G = \Gamma v_g \Delta g$ is replaced by $G - \gamma$, the following is obtained:

$$\Delta\mu_r = -\left(\frac{\alpha}{2k_0}\right)\frac{(G-\gamma)}{\Gamma v_g}, \tag{3.35}$$

$$\frac{d\phi}{dt} = -\frac{\bar{\mu}}{\mu_g}(\omega - \Omega) + \frac{1}{2}\alpha(G-\gamma). \tag{3.36}$$

The steady-state solution of the above is

$$\omega \simeq \Omega + \frac{\mu_g}{\bar{\mu}}\frac{1}{2}\alpha(G-\gamma). \tag{3.37}$$

Assuming the laser is oscillating in a single mode, the rate equations for power output (P) and the total carrier density (N) are given by Equations (3.29 and 3.1), rewritten here as (with $\varepsilon = 0$)

$$dP/dt = (G - \gamma)P + \beta R_{sp}, \tag{3.38}$$
$$dN/dt = I/e + \gamma_e N - GP, \tag{3.39}$$

where G is the optical gain which depends on N, γ is the total optical loss, βR_{sp} is the spontaneous emission rate, I is the injected current, and γ_e is the inverse of the lifetime (τ_e^{-1}) of the carriers. γ_e is a function of n, where $n = N/V$ is the carrier density per unit volume. Equation (3.39) may be written as

$$\frac{dn}{dt} = \frac{I}{eV} - \gamma_e(n)n - GP/V. \tag{3.40}$$

with $\gamma_e(n) = A_{nr} + Bn + Cn^2$, where A_{nr} is the nonradiative recombination (constant lifetime) rate and the other two terms are due to radiative and Auger recombinations. Often a constant lifetime recombination model is used; in such cases, $\gamma_e = \tau^{-1}$, where τ is the average lifetime which is independent of n. Below the threshold ($P = 0$), the following equation for carrier density (n) is obtained:

$$\frac{dn}{dt} = \frac{I}{eV} - n/\tau. \tag{3.41}$$

When the current is turned on, it takes a certain amount of time (τ_d called delay time) for the carrier density to reach threshold (nth) and hence for laser action to begin. Using $\gamma_e = \tau^{-1}$, the following simple expression for τ_d is obtained by integrating Equation (3.41) from $n = 0$ to $n = n$th:

$$\tau_d = \tau \ln\left(\frac{I}{I - I_{\text{th}}}\right), \tag{3.42}$$

where Ith = eV nth/τ is the threshold current. If the laser has a CW bias current of I_0 (= eV n_0/τ), the above equation is modified to

$$\tau_d = \tau \ln\left(\frac{I - I_0}{I - I_{\text{th}}}\right). \tag{3.43}$$

A semiconductor laser exhibits damped periodic oscillation before settling to a steady state value. The steady-state solutions (P_s, N_s) of (3.38) is (3.39) are

$$Ps = bRsp/(\gamma - G),\ N_s = (GP_s - I/e)/\gamma_e. \tag{3.44}$$

The rate Equations (3.38) and (3.39) are now linearized using $P = P_s + p_1$, and $N = N_s + n_1$ and using $= G(N_s) + G_n (N - N_s)$, where $G_n = dG/dN$

$$dp_1/dt = G_N P_s n_1 - \Gamma_P p_1, \tag{3.45}$$

$$dn_1/dt = \Gamma_N n_1 - G_s p_1,$$

where

$$\Gamma_N = \gamma_e + G_N P_s,$$
$$\Gamma_P = R_{\text{sp}} / P_s, G_s = G(N_s).$$

Using $p_1(t) = p_0 \exp(-ht)$ and $n_1(t) = n_0 \exp(-ht)$ and using $h = \Gamma_R + i\Omega_R$, the following values are obtained for Γ_R and Ω_R:

$$\Gamma_R = \frac{1}{2}(\Gamma_N + \Gamma_P),$$
$$\Omega_R = (GG_N P)^{1/2}. \tag{3.46}$$

Since the power output P is proportional to $(I–I_{\text{th}})$, the relaxation oscillation frequency Ω_R varies as $(I–I_{\text{th}})^{1/2}$. In terms of the previously described gain coefficient a (i.e., $g = a\,(n–n_o)$ in Section 2.4), the quantity G_n is given by $G_n = \Gamma v_g a/V$, where Γ is the confinement factor of the mode, v_g is the group velocity, and V is the volume of the active region.

3.4.2 *Small signal modulation and bandwidth*

The small signal modulation response of a laser is obtained using Equations (3.36)–(3.39). Assume a small modulation ($I_m(t)$) over a CW bias current (I) applied to a laser i.e., $I(t) = I + I_m\,(t)$. Assume $I > I$th, then (3.38) and (3.39) for $I_m << I$ becomes

$$\delta \dot{P} = -\Gamma_P \delta P + (G_N P)\delta N,$$
$$\delta \dot{N} = -\Gamma_N \delta N - G\delta P + \frac{I_m(t)}{e},$$
$$\delta \dot{\phi} = \frac{1}{2}\beta_c G_N \delta N. \tag{3.47}$$

Now consider $I_m(t) = I_p \exp(-i\omega_m t)$ i.e., $I_m(t)$ is sinusoidal with frequency ω_m. Then, the above equations can be solved and the following expression for the real-valued modulated power is obtained:

$$\delta P(t) = \delta P_0 \sin(\omega_m t + \phi). \tag{3.48}$$

with

$$\delta P_0 = \frac{G_N P I_P / e}{\left[(\omega_m^2 - \Omega_R^2 - \Gamma_R^2)^2 + 4\omega_m^2 \Gamma_R^2\right]^{1/2}},$$

$$\frac{\delta P_0(\omega_m = \Omega_R)}{\delta P_0(\omega_m = 0)} \simeq \frac{\Omega_R}{2\Gamma_R}.$$

Thus, the modulation response is relatively flat for $\omega_m << \Omega_R$, peaks near $\omega_m \sim \Omega_R$, and drops off sharply for $\omega_m >> \Omega_R$, as $\delta P_0 \sim \omega_m^{-2}$. The small-signal modulation response of an InGaAsP/InP-buried heterostructure laser with semi-insulating (Fe:InP) current blocking layers is shown in Figure 3.14.

The time-dependent phase change ($\delta\phi(t)$) results in a variation in frequency $\delta\nu(t)$ around the frequency of laser oscillation (ν_0):

$$\delta v(t) = \frac{\delta\dot{\phi}(t)}{2\pi}. \tag{3.49}$$

The time-dependent frequency shift is called frequency chirping. An expression for frequency chirp can be obtained using the Fourier relation

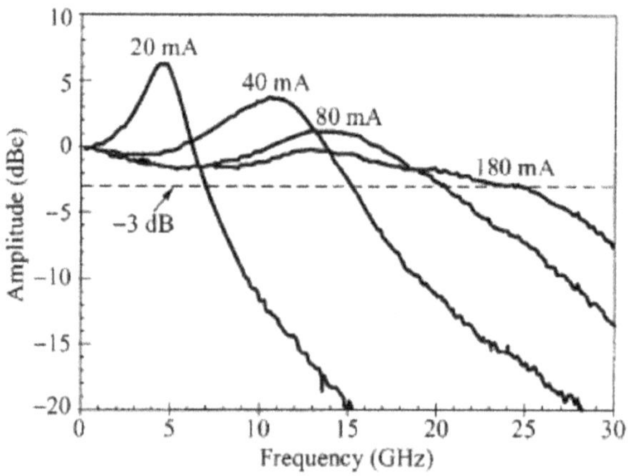

Figure 3.14. The measured modulation response of a buried heterostructure laser emitting near 1.55 μm. Note that there is a peak (relaxation oscillation) in response in agreement with the model described earlier [21].

$$\delta\dot{\phi}(t) = \frac{1}{2\pi}\int_{-\infty}^{\infty} i\omega\phi(\omega)\exp(i\omega t)d\omega. \tag{3.50}$$

For the sinusoidal modulation, discussed earlier, an expression for $\phi(\omega)$ can be obtained by using Equation (3.47). The variation of frequency $\delta\nu(t)$ is given by

$$\delta v(t) = \delta v_0 \sin(\omega_m t + \phi).$$

The expression for $\delta\nu_0$ is

$$\delta v_0 = \frac{\beta_c I_P G_N}{4\pi e}\left(\frac{\omega_m^2 + \Gamma_P^2}{(\omega_m^2 - \omega_R^2 - \Gamma_R^2)^2 + 4\omega_m^2\Gamma_R^2}\right). \tag{3.51}$$

Using $\omega_m << \Omega_R$ and $\Gamma_R << \Omega_R$ and $\Omega_R^2 = GG_nP$, $\delta\nu_0$ has the following form:

$$\delta v_0 = \frac{\beta_c I_p}{4\pi eGP}(\omega_m^2 + \Gamma_P^2)^{1/2}. \tag{3.52}$$

Thus, $\delta\nu_0$ (amplitude of the frequency chirp) is proportional to the amplitude of the modulation current. The transmission system results using pulse code modulation (PCM) are described in a later chapter.

Advances in materials technologies have led to novel semiconductor lasers, such as nanostructure lasers and compact mode-locked lasers. These lasers demonstrate not only advances in semiconductor laser research but also new directions in applications. These advances are expected to continue.

References

1. J. P. van der Ziel and R. M. Mikulyak, *J. Appl. Phys.*, **51**, (1980) 3033.
2. D. Z. Tsang and J. N. Walpole, *IEEE J. Quantum Electron.*, **QE-19**, (1983) 145.
3. J. A. Yeung, *Appl. Phys. Lett.*, **38**, (1981) 308–310.

4. P. M. Downey, J. E. Bowers, R. S. Tucker, and E. Agyekum, *IEEE J. Quantum Electron.*, **QE-23**, (1987) 1039–1046.
5. H. Liu, M. Fukazawa, Y. Kawai, and T. Kamiya, *IEEE J. Quantum Electron.*, **25**, (1989) 1417.
6. P. Vasil'ev, *Ultrfast Diode Lasers: Fundamentals and Applications* (Artech House, Boston, 1995).
7. G. P. Agrawal and N. K. Dutta, *Semiconductor Lasers* (Van Nostrand Reinhold, New York, 1993).
8. J. E. Bowers, B. R. Hemenway, A. H. Gnauck, and D. P. Wilt, *IEEE J. Quantum Electron.*, **22,** (1986) 833.
9. B. R. Bennett, R. A. Soref, and J. A. Del Alamo, *IEEE J. Quantum Electron.*, **26**, (1990) 113.
10. M. Osinski and M. J. Adams, *IEEE J. Quantum Electron.*, **QE-21**, (1985) 1929.
11. C. T. Lin, T. P. Lee, and C. A. Burms, *Appl. Phys. Lett.*, **42**, (1983) 141.
12. J. P. van der Ziel, R. A. Logan, and R. M. Mikulyak, *Appl. Phys. Lett.*, **39**, (1981) 867.
13. P. B. Hansen, G. Raybon, U. Koren, B. I. Miller, M. G. Young, M. Newkirk, M. D. Chien, B. Tell, and C. A. Burrus, *Appl. Phys. Lett.*, **62**, (1993) 1445.
14. H. Fan, C. Wu, M. El-Aasser, N. K. Dutta, U. Koren, and A. B. Piccirilli, *IEEE Photon. Tech. Lett.*, **12**, (2000) 972.
15. Y. K. Chen, M. C. Wu, T. Tanbun-Ek, R. A. Logan, and M. A. Chin, *Appl. Phys. Lett.*, **58**, (1991) 1253, *IEEE Photon. Tech. Lett.*, **3**, (1991) 824.
16. P. P. Vasilev, V. N. Morozov, Y. M. Popov, and A. B. Sergeev, *IEEE J. Quantum Electron.*, **QE-22**, (1986) 149.
17. P. A. Morton, V. Mizrahi, and T. Tanbun-Ek, *Electron. Lett.*, **28**, (1992) 2156.
18. R. Kaiser and B. Huttl, *IEEE J. Sel. Top. Quantum Electron.*, **13**, (2007) 127.
19. H. A. Haus and Y. Silberberg, *J. Opt. Soc. Am.*, **B 2**, (1985) 1237.
20. G. P. Agrawal and N. K. Dutta, *Semiconductor Laser* (van Nostrand Reinhold, NY, 1993).
21. P. Morton, R. A. Logan, and T. Tanbun-Ek, *Electron. Lett.*, **28**, 2156 (1992).

Chapter 4

Fiber Laser

4.1 Introduction

The optical fiber used for transmission has a core about 10 μm in diameter which is surrounded by a cladding layer about 100 μm in diameter. The core region has a slightly higher index (~0.3%) which allows the light to be guided in the core. The cladding is further surrounded by a polymer jacket which provides both protection from the environment and strength. The fiber laser has a doped fiber core. The doping material (Er, Yb, Nd, Ho, Tm, etc.) provides the gain needed for laser operation when pumped by an external pump laser. The concentration of rare-earth dopants in rare-earth-doped fiber (used for optical amplification) can vary considerably. It can be as much as ~1000 ppm (parts per million) in muticomponent glasses, to as small as <100 ppm in erbium-doped fibers. In erbium-doped fibers used for distributed amplification (over the entire transmission path), it can be a few ppm.

Optical fibers can be fabricated by several methods [1–8]. The principles among them are modified chemical vapor deposition (MCVD), plasma-assisted chemical vapor deposition (PCVD), axial vapor deposition (VAD), and outside vapor deposition (OVD). Rare-earth dopants for silica fiber are often introduced during the growth of the fiber itself. Thus, the fabrication of rare-earth-doped fiber is very much linked to the fabrication of regular low-loss transmission fiber. A typical rare-earth dopant level is ~100 ppm. For a detailed study, please see Ref. [9].

4.1.1 *Fiber laser fundamentals*

In Chapter 2, the semiconductor laser fundamentals and the condition for gain in a semiconductor laser medium are discussed. In this section, the fundamentals of gain and laser operation in rare-earth-doped glass fiber are described. The considerations, such as population inversion, for rare-earth-doped glasses are similar to that for lasers based on an atomic or a molecular transition in a gaseous medium. Examples of the latter class of lasers are Ar^+ lasers, He–Ne lasers, CO_2 lasers, etc. For simplicity, a two-level system (shown in Figure 4.1) is considered.

Figure 4.1 shows the two relevant energy levels involved in the laser transition i.e., the atom makes a transition from the higher energy level (2) to lower energy level (1) with the emission of a photon. Let N_2 and N_1 be the population density of levels 2 and 1 respectively and σ is the cross section of stimulated emission for transition from 2 to 1. The cross-section for absorption for transition from 1 to 2 is also σ. The typical σ values are in the 3×10^{-21} to 3×10^{-20} cm^2 range. Then the following equations are obtained for an incident light at intensity I (expressed in number of photons/cm^2/s):

$$\begin{aligned} \text{Stimulated emission rate} &= N_2\sigma I, \\ \text{Absorption rate} &= N_1\sigma I, \\ \text{Net stimulated emission rate} &= (N_2 - N_1)\sigma I. \end{aligned} \tag{4.1}$$

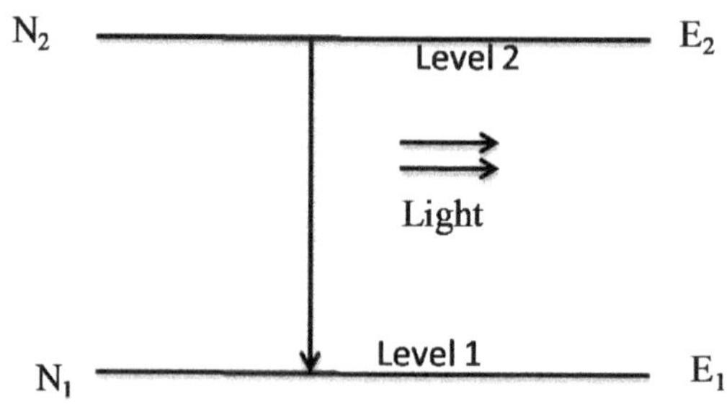

Figure 4.1. The two relevant energy levels (noted as E_2 and E_1) of a two-level laser are shown. N_2 and N_1 are the population densities of levels 2 and 1, respectively.

Thus, for net stimulated emission, N_2 must be larger than N_1 i.e., the population density of the high-energy state must be higher than that for the lower-energy state. This condition is known as "population inversion" since under normal thermal equilibrium the population of the higher-energy state is lower than that of the lower-energy state. The population inversion is created by a pump source which produces atoms in the higher-energy state. For rare-earth-doped atoms (e.g., in fiber lasers), the pump source is another laser. For He–Ne laser, the pump source is an energy transfer mechanism by the collision of the excited He atom to the Ne atom, which produces Ne atoms in a high-energy state. The quantity

$$g = (N_2 - N_1)\sigma \tag{4.2}$$

is the optical gain of the laser medium. In addition to stimulated emission, spontaneous emission (and other transfers) also causes a decrease in the population of high-energy levels. Assuming τ is the lifetime of level 2 due to these processes, the rate equation for N_2 is given by

$$\frac{dN_2}{dt} = P - (N_2 - N_1)\sigma I - \frac{N_2}{\tau}, \tag{4.3}$$

where P is the pump rate and $N_2 + N_1 = N$ where N is the total number density of atoms. Typical τ values are in the 0.5–3 ms range. In steady state, the following equations for N_2 and gain (g) are obtained:

$$N_2 = \frac{P\tau + NI/2I_S}{1 + I/I_S},$$

$$g = \frac{g_0}{1 + I/I_S}, \tag{4.4}$$

with $g_0 = (2P\tau - N)\sigma$ and $I_S = 1/2\tau\sigma$, where g_0 is known as the unsaturated gain and I_s is known as the saturation intensity. The rate equation for intensity (I) is given by

$$\frac{dI}{dt} = gI - \alpha_t I, \tag{4.5}$$

where α_t is the total optical loss. In the context of Section 2.1, $\alpha_t = \alpha + \alpha_m$ i.e., it is the sum of scattering and mirror losses (see Chapter 2 also). In steady state, gain = loss i.e., $g = \alpha_t$, this leads to the following equation for I:

$$I = I_S(g_0/\alpha_t - 1)$$
$$= \frac{1}{\alpha_t}(P - P_0), \tag{4.6}$$

where $P_0 = N/2\tau + \alpha_t/2\tau\sigma$.

The equation shows stimulated intensity, $I > 0$ for $P > P_0$. The quantity P_0 is called the threshold pump rate. At pump rates higher than the threshold pump rate, the laser intensity increases linearly with pump power. This is similar to that of a semiconductor laser output, which increases linearly with current for currents higher than the threshold current.

4.1.2 *Energy levels of rare-earth dopants*

The commonly used rare-earth dopants in glass fibers are Er, Yb, Nd, Tm, Ho, etc. Each of these materials has its characteristic energy level, the optical transitions between these levels are responsible for the absorption and emission of light, and, subsequently, laser action if the population of atoms in the higher energy level is higher than that for the lower energy level as described in Section 4.1.1.

The energy levels of these Er, Yb, Nd, and Tm are shown in the next set of figures (Figures 4.2–4.5). The principle optical gain transitions are noted in the figure captions in each case [9].

Er-doped glass amplifiers have been extensively studied for application in optical fiber communication systems [2, 8, 9]. The first Er-doped glass laser using the ${}^4I_{13/2}$ to ${}^4I_{15/2}$ transition was demonstrated in 1965 (Figure 4.2). The laser operated near 1500 nm. For optical communication applications, the region of interest for fiber amplifiers is 1550 nm. A common mechanism for exciting Er ions to the excited state (${}^4I_{13/2}$) is the absorption of 980 nm or 1480 nm light from a pump laser.

Yb-doped glass amplifiers have been extensively studied by various authors for high-power laser sources and amplifiers in the 1100 nm

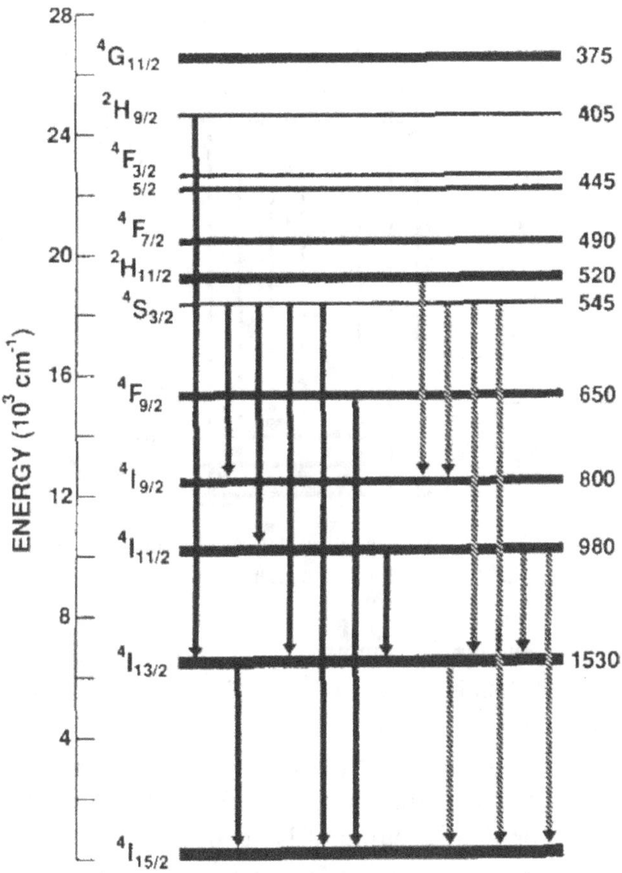

Figure 4.2. Energy levels of Er^{3+} in silicate glass. The solid lines indicate observed lasing transitions. The numbers on the right are the absorption wavelength peaks for absorption from the ground state [10].

spectral region [9]. Yb ion in silica has two energy-level manifolds. They are the ground-state manifold ($^2F_{7/2}$) and the excited-state ($^2F_{5/2}$) manifold (Figure 4.3). The numbers on the right at each excited state are the energy in cm^{-1} from the ground state. The solid lines are absorption transition which terminates at that level. The dashed arrows represent the observed fluorescence/lasing transitions. The transitions between the sublevels are not fully resolved for Yb ions in a glass at room temperature because of

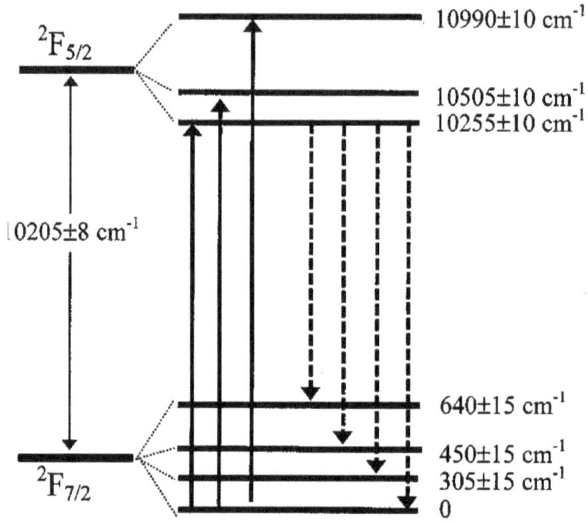

Figure 4.3. Energy levels of Yb^{3+} in silicate glass. The solid lines indicate observed absorption transitions and the dashed lines are laser/fluorescence transitions [11].

the strong homogeneous and inhomogeneous broadening. The Yb-doped silicate glass laser using the $^2F_{5/2}$ to $^4F_{7/2}$ transition was demonstrated in 1988. The laser operated near 1100 nm.

Nd-doped glass laser was reported in 1961 [14]. Nd has a strong absorption spectrum along the entire wavelength range from UV to near-infrared. The absorption spectrum of Nd-doped in ZBLAN (Zr, Ba, La, Al, and Na) flourozirconate glass is shown in Figure 4.4. The sharp spectral lines are indicative of low influence of crystal field interaction [9].

The main interest in Tm arises from emission bands in the 1400–2700 nm range, particularly for energies where gaps exist between Er and Nd (Figure 4.5). Tm is generally doped in the usual glasses, such as fluorides, tellurites, phosphates, and silica. The emission strengths of different transitions depend on the glass host.

4.2 Glass Fiber for Optical Transmission

The optical fiber used for transmission has a core (Ge-doped SiO_2) about 10 μm in diameter which is surrounded by a cladding (SiO_2) about 100 μm in diameter. The addition of Ge in SiO_2 increases the index. A discussion

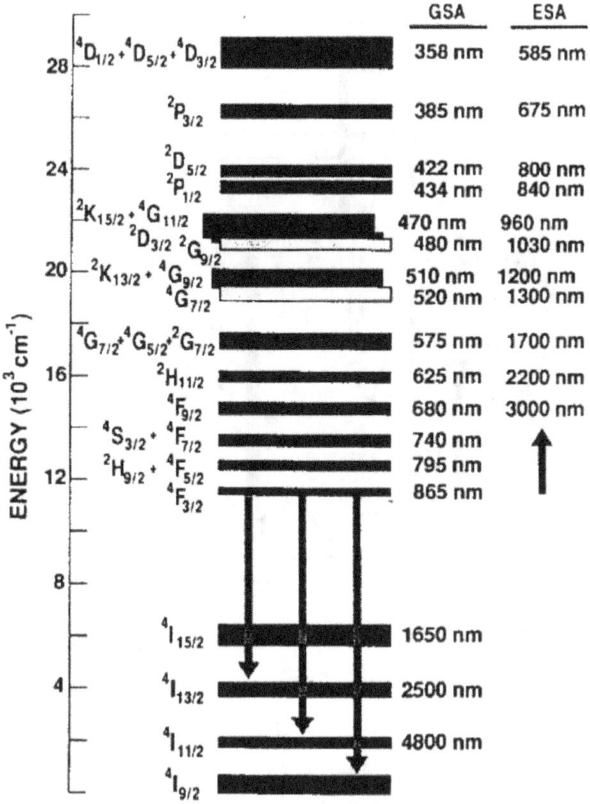

Figure 4.4. Energy levels of Nd^{3+} in flourozirconate glass. The solid lines indicate observed lasing transitions. The numbers on the right are the absorption wavelength peaks for absorption from the ground state [12, 13].

of fiber fabrication is important because several pulse compression techniques use nonlinear optical fibers and dispersion-tailored fibers which are not doped with rare-earth atoms.

4.2.1 *Optical fiber fabrication*

The optical fiber used in optical transmission systems is based on silica (SiO_2) material. Low-loss silica fibers were first proposed in 1966 [1], and by the early ~1970s, low-loss fibers had an attenuation of 20 dB/km. Since then, the manufacturing processes have been refined and developed so that

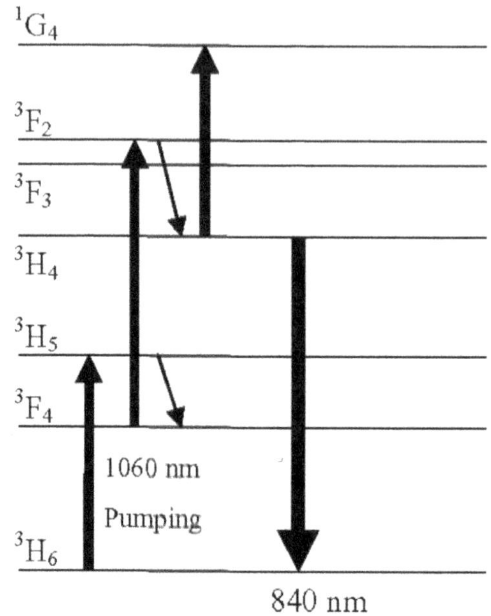

Figure 4.5. Schematic of energy levels of Tm-doped in silica [15].

transmission fibers have typical losses of <0.15 dB/km in the 1550 nm region of the optical spectrum [1–8].

The principal methods of optical fiber fabrication are (i) MCVD, modified chemical vapor deposition, (ii) PCVD, plasma chemical vapor deposition, (iii) VAD, axial vapor deposition, and (iv) OVD, outside vapor deposition [1–8, 16–24]. The MCVD method is described in the following section. The methods for the creation of SiO_2 are hydrolysis (reaction with H_2O) or oxidation (reaction with O_2) of $SiCl_4$. The hydrolysis and oxygen reactions are

$$SiCl_4 + 2H_2O = SiO_2 + 4HCl, \tag{4.7}$$

$$SiCl_4 + O_2 = SiO_2 + 2Cl_2. \tag{4.8}$$

The hydrolysis method is accomplished by flowing $SiCl_4$ vapor into a hydrogen flame and the resulting silica particles get deposited on a rotating target. The oxidation method involves a reaction of oxygen and chloride inside a substrate tube heated to 1200 C. Other materials such as GeO_2 can be incorporated into the silica by using a reaction of $GeCl_4$ with the hydrogen flame or oxygen.

4.2.2 *MCVD Process*

The MCVD process was invented around 1980 [3]. It involves vapor deposition of high-purity material on the inner surface of a tube. The tube may provide part of the cladding, and additional materials for the cladding and core get deposited on the inner surface of the tube. In the MCVD process, first a tube of suitable dimension and purity is chosen. The tube is mounted on a lathe, which rotates the tube. It has a chemical (chlorides) delivery system at one end and the other end is the exhaust. A heat torch traverses along the tube which causes the entrants to react forming SiO_2 and depositing it on the walls of the tube.

The deposition process is the high-temperature gas phase oxidation of chlorides which creates SiO_2 (oxides) and the deposition of these oxides as submicron particles on the tube [4, 7, 8]. These particles are converted to a clear glass film by a process called fusion. The entrants into the tube are controlled amounts of chemical reagents which are generated by passing (O_2 or He) through liquid sources, such as $SiCl_4$, $GeCl_4$, or $POCl_3$. Sometimes, gaseous sources such as SiF_4 are also mixed. In MCVD, first a high-quality cladding is deposited and then the core is deposited. The total number of deposited layers and their composition depend on the fiber design. The schematic of an MCVD process is shown in Figure 4.6.

4.2.3 *Doping and fiber drawing*

It is necessary to add dopants to the primary glass component SiO_2, to change its index. This allows the fabrication of different waveguide designs. Index-raising dopants are germanium, phosphorus, and aluminum and index-lowering dopants are boron and fluorine. They are introduced as halides or gases in the reaction chamber. The halides are carried by oxygen and inert gas. The oxidation process is controlled by the thermodynamic equilibrium established during the doping reaction,

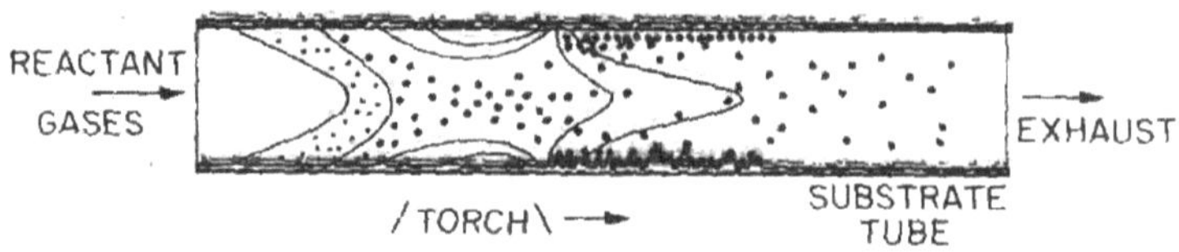

Figure 4.6. Schematic of an MCVD reactor [4].

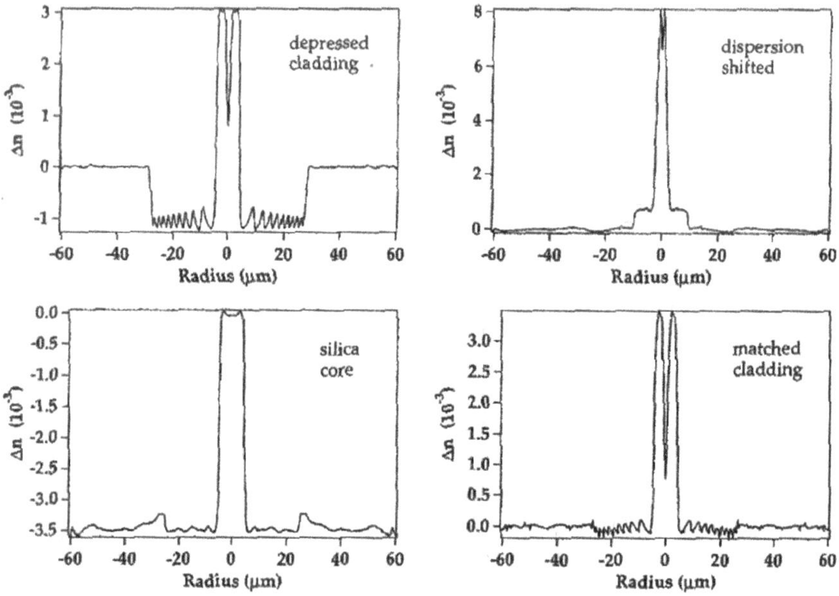

Figure 4.7. Schematic of various index profiles that have been fabricated. Plotted is change in index from SiO_2 as a function of distance from the center.

deposition, and sintering. Typical fiber index profiles that have been produced are shown in Figure 4.7. Different index profiles are used for the fabrication of fibers with different chromatic dispersion characteristics.

A major step in the fabrication of optical fiber is fiber drawing and the associated fiber coating with polymer [4, 25–27]. This step is important in determining the resultant optical, dimensional, and mechanical properties of the final optical fiber. Figure 4.8 shows the essential features of a fiber drawing facility for drawing silica-based fiber from preforms. The preform is essentially the collapsed tube in the MCVD process after the deposition of the core and cladding materials. A tall structure is used

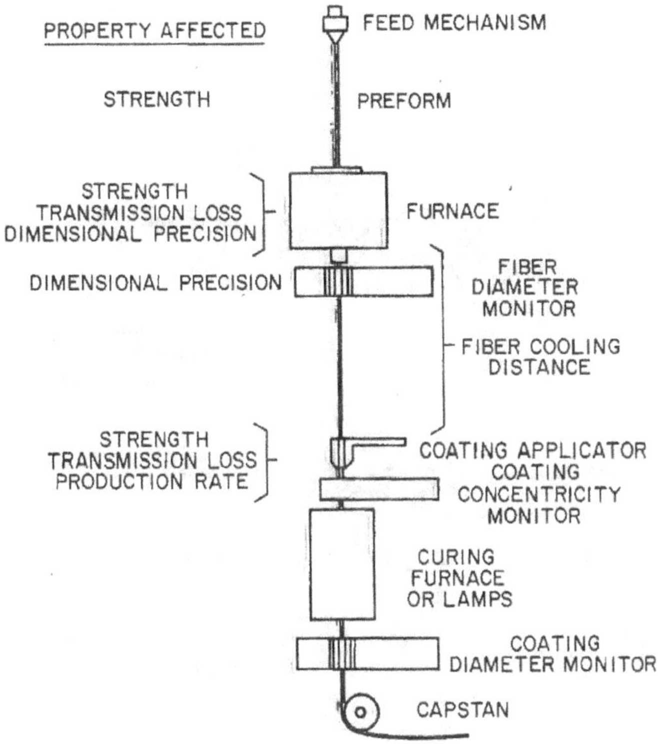

Figure 4.8. Schematic of a fiber drawing and coating apparatus [4].

for supporting the control equipment. The glass preform is fed into a high-temperature furnace by a preform feed mechanism where it is aligned and centered with respect to the furnace. For silica fiber drawing, the temperature of the furnace is set in the 1950–2300°C range. This allows proper viscosity for drawing. The tip of the preform softens as it is fed to the high-temperature region and both gravity and an applied tensile force cause it to shrink to a small diameter. A controlled tensile force is provided by a fiber pull mechanism (capstan). The preform feed rate and the capstan rotation rate determine the ratio of the drawn-down radius of fiber to preform. Typically, preforms with diameter in the 10–70 mm range are drawn down to a fiber 100–225 μm in diameter. 125 μm is the diameter of a typical transmission fiber. In-line coating is a key step in the fiber drawing process. Both glass fiber diameter and coated fiber diameter are measured and controlled by feedback loops.

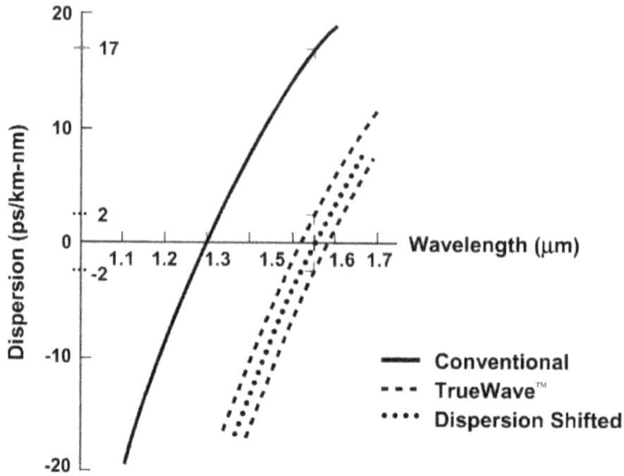

Figure 4.9. Chromatic dispersion as a function of wavelength for three fiber types [28].

The modern optical fiber used in transmission has zero dispersion near 1300 nm and lowest loss (~0.15 dB/km) at 1550 nm. However, using suitable cladding geometries such as that shown in Figure 4.7, the dispersion of a fiber can be tailored for a specific wavelength range. The dispersion is the sum of material dispersion and waveguide dispersion. The latter is varied by varying the cladding layer geometry. The chromatic dispersion as a function of the wavelength of three types of fibers is shown in Figure 4.9. The conventional fiber has zero dispersion at 1300 nm. The dispersion-shifted fiber has zero dispersion at 1550 nm. The true wave fiber has a dispersion of ~1 ps/km–nm near 1550 nm. The latter is important for reducing four wave-mixing-based nonlinear effects in wavelength division multiplexed (WDM) transmission. Measured optical loss of a single-mode fiber is shown in Figure 4.10.

4.2.4 *Rare-earth doping for fiber lasers and fiber amplifiers*

The fiber used in fiber amplifiers is similar to regular optical fibers except that the core is doped with rare-earth dopants. These are typically Er, Yb, Nd, Er–Yb, etc. A fiber laser is produced by adding feedback to a fiber amplifier. As an example, an emission cross-section of Er-doped material for different glass host types is shown in Figure 4.11.

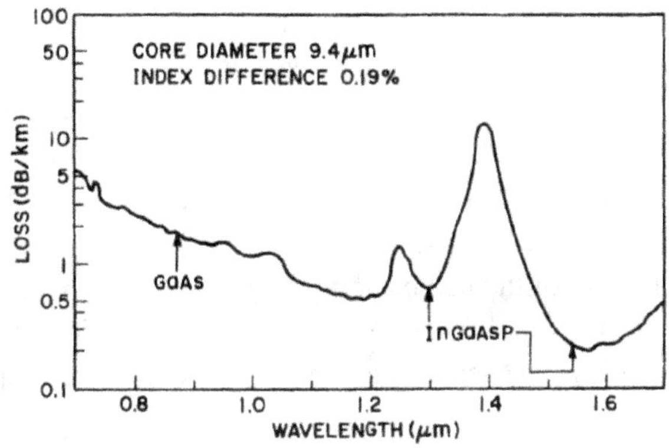

Figure 4.10. Measured optical loss of a single-mode fiber as a function of wavelength. Typical wavelength for optical communication along with the semiconductor material for lasers is indicated. The wavelengths are 850, 1300, and 1550 nm, respectively.

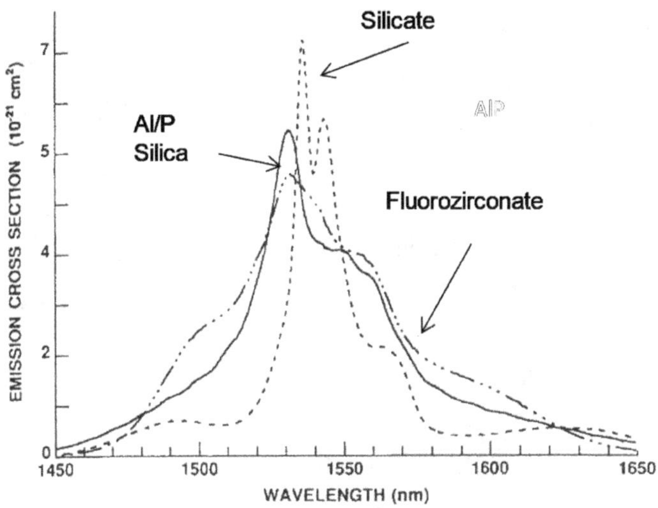

Figure 4.11. Emission cross-section spectrum of Er near 1500 nm for different glass hosts [12]. Fluorozirconate is also known as ZBLAN.

Methods of incorporating rare-earth dopants in the preform have been developed for all the perform fabrication techniques stated earlier. The rare-earth dopants are sent to the reaction region along with the index-controlling dopants. Rare earths generally have a low vapor pressure. This is usually compensated by placing the vapor source close to the reaction zone or by delivering the material as an aerosol or high vapor pressure organic compound (Figure 4.12). The heated frit source is made by soaking previously deposited porous soot with a rare-earth chloride solution (Figure 4.12). From the heated source, the rare-earth chloride is directly injected (after dehydration) close to the reaction chamber (B). An attractive feature of the heated source injector (C) is that the rare-earth source (in the small ampule) is isolated from potential unwanted reactions with other chlorides. The aerosol delivery method does not require heated

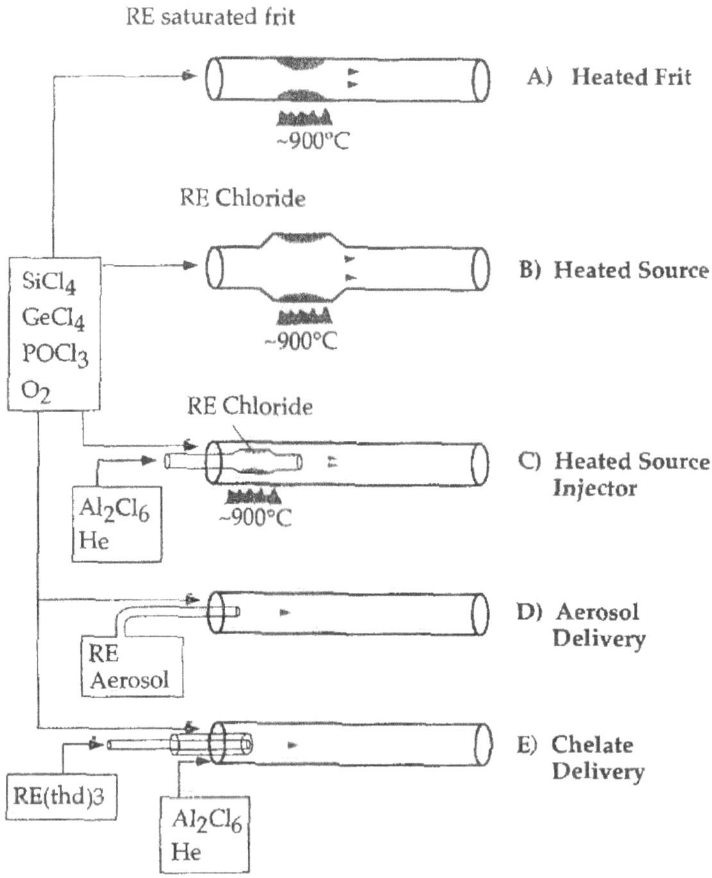

Figure 4.12. Low vapor pressure dopant such (as rare earth) delivery methods for MCVD [27, 29]. RE: rare-earth element.

source compounds because it generates vapor at the reaction site (D). The organic compound of rare earths (E) provides higher vapor pressure for rare earths than the chlorides, so this is another potential delivery method.

So far, silica-based fiber fabrication has been described. There are several reasons (for example, lower loss than silica at long wavelengths) to consider other types of glass hosts. These glass hosts are selenides, tellurites, germinates, fluorides, chlorides, iodides, and sulfides. The glass host influences the transition rates and hence the gain [9, 12, 27].

4.3 Pump Laser

A high-power laser is needed to excite the rare-earth ions to the upper laser level. In the context of fiber laser, this is generally a semiconductor laser and is often called a pump laser. Typical semiconductor lasers used in communication or consumer system applications emit a few mW of power. High-power (>100 mW) semiconductor lasers are important as pump lasers for rare-earth-doped fiber laser systems which are also used for printing and industrial (metal cutting) applications. This section describes high-power lasers which have been used as pump lasers for rare-earth-doped fibers to fabricate fiber lasers and fiber amplifiers. The typical laser wavelengths are 1480 nm (for a pump source for Er-doped fiber), 980 nm (for a pump source for Er-doped fiber), 960 nm (for a pump source for Yb-doped fiber), and 808 nm (for a pump source for Nd-doped fiber).

4.3.1 *1480 nm laser*

A high differential quantum efficiency, low threshold current, and low leakage current are important for high power operation [30–32]. These quantities are determined by the active region design. In addition, the laser must have low thermal impedance (low resistance contacts) and be bonded to a good heat sink so that its temperature does not increase significantly with the application of high current.

The design used for a high-power 1480 nm pump has a multiquantum well (MQW) active region with a graded index region (GRIN) for optical confinement. The scanning transmission electron micrograph (TEM) of an MQW active region is shown in Figure 4.13. The parameters that need to be optimized are the doping of layers, number of wells, graded index (GRIN) design, and optical cavity length. There are many trade-offs. For example, long cavity length reduces the thermal impedance but increases the threshold current and decreases quantum efficiency due to higher cavity optical loss.

Similarly, higher doping reduces resistance but increases optical loss. Increasing the number of active layer wells increases the threshold current beyond a certain number of wells. An experimental investigation has been carried out using the active region band diagram shown in Figure 4.14.

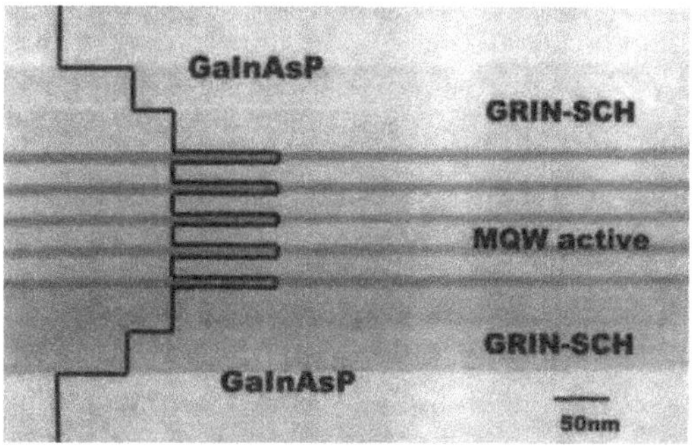

Figure 4.13. Scanning transmission electron micrograph (TEM) of the active region of a 1480 nm pump laser [30].

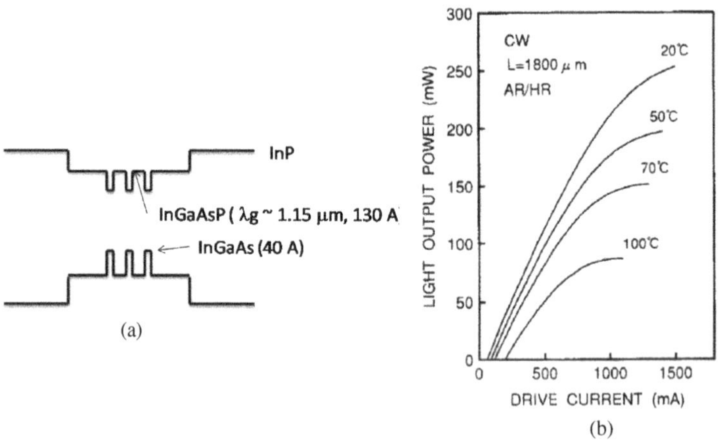

Figure 4.14. (a) Band diagram of the MQW active region. (b) Light vs. current characteristics of a 1480 nm pump laser at various temperatures [31].

The active layer is made up of 40 A thick InGaAs quantum wells with 130 A thick InGaAsP ($\lambda \sim 1.15\ \mu$m, $E_g = 1.08$ eV) barriers. The width of the entire active region is 2000 A. The wafers are grown by MOCVD and strongly index guided lasers are fabricated. Lasers with 2, 3, 5, and 7 quantum wells are fabricated. Lasers with 5 quantum wells were found to have the best high-power performance. The light vs. current characteristics of an 1800 μm long laser with five quantum wells at various temperatures are shown in Figure 4.14. The laser has an antireflection coating (AR ~ 4%) at one facet and high reflectivity (~90%) at the other facet. The laser emits more than 250 mW near room temperature. ~500 mW of single-mode output power has been reported from similar designs [30]. Typical far-field divergence of such a laser is 25° and 20° normal and parallel to the junction plane, respectively [30].

4.3.2 *980 nm laser*

For ~980 nm laser operation, thin layers of $In_{0.2}Ga_{0.8}As$ are grown on GaAs using either MOCVD or MBE growth technique [33–37]. The lattice mismatch introduces a strain between the grown layer ($In_{0.2}Ga_{0.8}As$) and the substrate (GaAs). As a result, such lasers are often called strained quantum well lasers. A typical design of a single mode 980 nm laser consists of one or two quantum wells of $In_{0.2}Ga_{0.8}As$ surrounded by GaAs barrier layers. The exact well thickness needed for a 980 nm laser is often determined experimentally. It is generally in the 7–10 nm range. For good transverse mode confinement, a graded index (GRIN) design is often used. The wafer is generally fabricated into a ridge waveguide-type laser with a ridge width of ~4 μm for single transverse mode operation.

Specific designs of active region for high-power kink-free operation or circular mode (i.e., nearly equal beam divergences along and normal to junction plane) have been reported.

The weak index-guided ridge waveguide design allows a large area of the emitted light at the facet, thus, for a given output power, the laser power density is lower at the facet compared to that of a strongly index-guided buried heterostructure design. Lower optical power density at the facet reduces the probability of catastrophic mirror damage (COMD) at the facet. Various schemes have been developed to reduce or eliminate the COMD. The schematic of a ridge waveguide laser is shown in Figure 4.15.

This active region design provides high kink-free output power. The output has an antireflection (AR ~ 4%) coating and the rear facet has a high reflection (HR ~ 93%) coating of dielectric materials. The light vs. current characteristics of a 980 nm laser at different temperatures are shown in Figure 4.16. Also shown is the voltage–current (V–I) and slope (*dP/dI–I*) vs. current characteristics. The slope efficiency near room temperature is ~0.9W/A. The far-field patterns are shown in Figure 4.17.

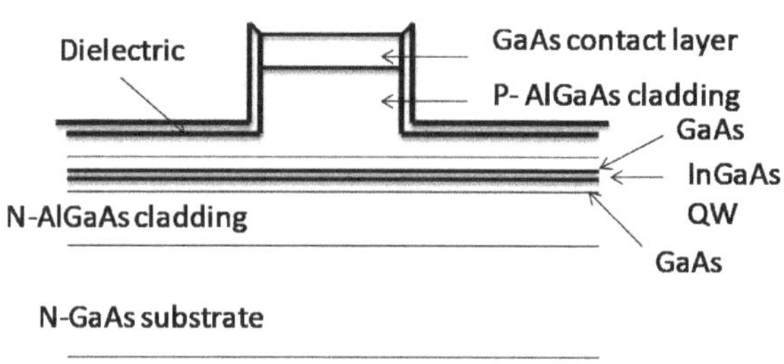

Figure 4.15. A simple schematic of a ridge waveguide laser emitting near 980 nm. The device of Ref. [33] has graded index layers surrounding the active InGaAs quantum well (QW) region.

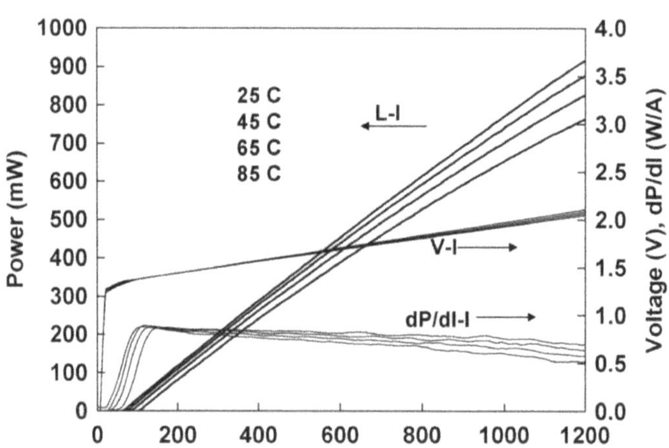

Figure 4.16. Light vs. characteristics of a 980 nm laser [34].

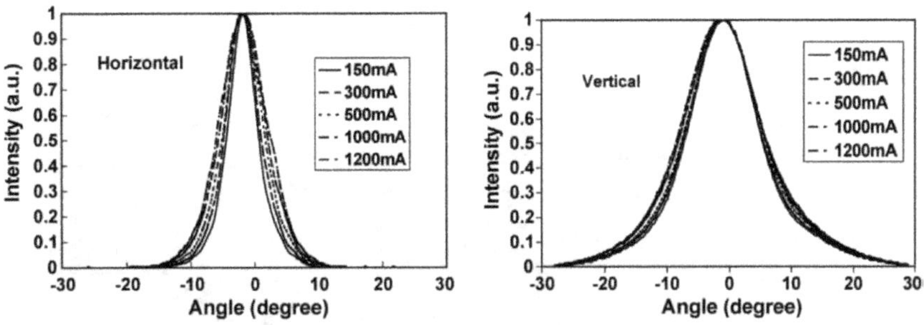

(a) (b)

Figure 4.17. The far-field patterns along (horizontal) and normal (vertical) to the junction plane are shown. Single-lobed pattern illustrates single transverse mode operation [34]. The full width at half maximum of beam divergences along the horizontal and vertical planes are ~12° and 18°, respectively.

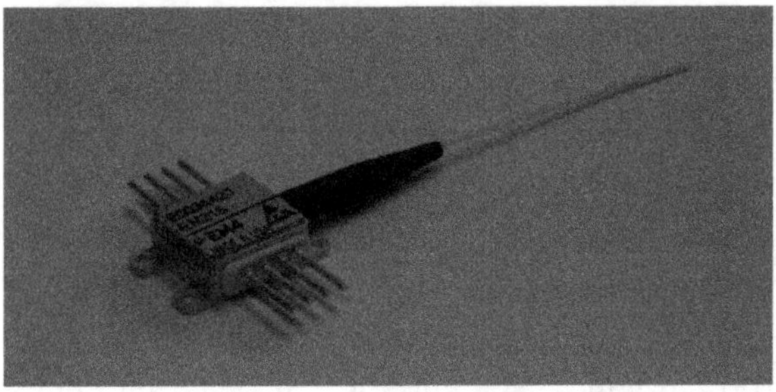

Figure 4.18. Single-mode pump laser package. The output is coupled to a single-mode fiber. Regular low-power lasers used for communication systems are also packaged in a similar fashion.

Reliability of the 980 nm pump laser has been extensively studied. The lasers are found to be reliable for communication system applications. Specialized facet coating in vacuum or impurity-induced disordering has been used to reduce catastrophic mirror damage at high powers.

Single-mode lasers are typically packaged with a fiber-coupled output (Figure 4.18). The package has a thermoelectric cooler which keeps the laser at a certain temperature. The temperature control is important for fixed wavelength operation and for power stability.

Lasers emitting near 960–970 nm are important as pump lasers for multimode large area Yb-doped fiber amplifiers and Er–Yb-doped double-clad fibers. These lasers are generally fabricated as 100 μm wide stripe

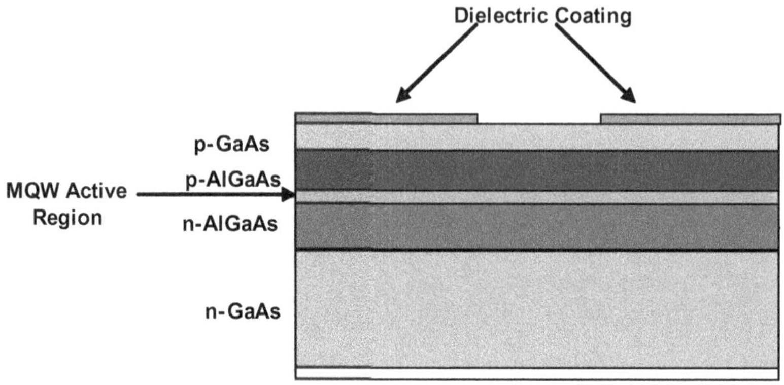

Figure 4.19. Schematic of a 100 μm stripe laser. The current is injected through an opening in the dielectric.

devices. A schematic of the device cross-section is shown in Figure 4.19. The active region is InGaAs with In composition (~0.2) chosen to produce emission in the 960–970 nm wavelength range. The output facet is AR (antireflection) coated and the rear facet is HR (high reflection) coated.

The laser cavity length is 2 mm. The long cavity length is important for reducing thermal impedance. The lasers typically emit ~9–10W at 10 A. The lasers are similar to broad-area lasers with an opening in the SiO_2 dielectric (~100 μm wide). The current is injected into the active region through this opening. This makes the mode area larger and hence lowers the power density at the facet which prevents high-power induced facet degradation. For higher-power operation, multiple openings (typically 50–100 μm wide) are present in the dielectric stripe. These openings are typically 50–100 μm apart. Lasers with 10 openings (10 stripes) emit in 10 spots along the active region. The combined output power can be >100 W. These devices are also called laser arrays. These laser arrays have been stacked in two dimensions and are known to produce nearly 1 kW in quasi-CW operation. These types of laser arrays have been used as a pump for Yb-doped fiber lasers.

4.3.3 *808 nm laser*

Semiconductor laser emitting near 808 nm is important as a pump source for Nd-doped rare-earth materials. Both single-mode and multimode broad stripe 808 nm lasers are fabricated using the AlGaAs/GaAs material system [35–37]. The Al composition in the active region $Al_xGa_{1-x}As$ is chosen so that the emission is near 808 nm. The band gap of $Al_xGa_{1-x}As$ varies with the Al composition (x). The band gap E_g is given by $E_g(x) = 1.424 + 1.247x$ (eV). The layers are grown by MBE or MOCVD growth technique. The layers are N-AlGaAs cladding layer, $Al_xGa_{1-x}As$ active layer, P-AlGaAs cladding layer, and P-GaAs contact layer. The grown wafer is fabricated into ridge waveguide lasers. The reliability of these types of lasers is generally limited by catastrophic facet degradation which occurs at high powers. A special facet coating process has been reported to lessen this problem. It involves the cleaving of the lasers in vacuum (to prevent facet oxidation) and applying a dielectric coating after cleaving. It is the same cleaving and coating process which lessens the degradation of lasers emitting near 980 nm.

For the single-mode device described in the following, the active region is a single quantum well and is grown by MBE growth technique. The light vs. current characteristics at various temperatures are shown in Figure 4.20. The laser has an antireflection coating at the emitting facet and a high reflectivity coating at the rear facet. The device emits >800 mW at 25°C @ ~900 mA. The slope efficiency is ~1.1 W/A up to ~600 mA

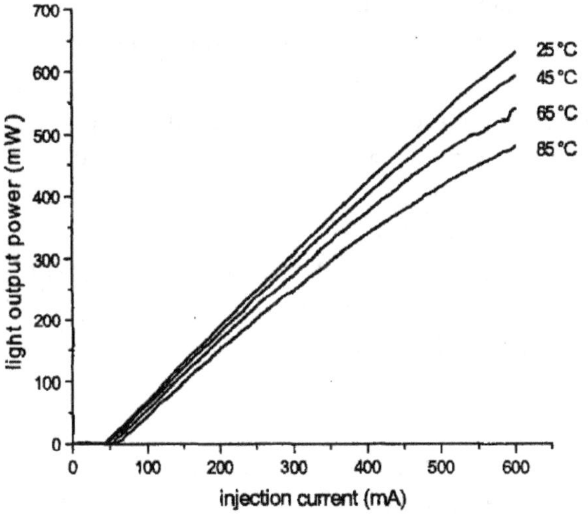

Figure 4.20. Light vs. current characteristics of a laser emitting near 808 nm [35].

at 25°C. At higher currents or higher temperatures, the slope efficiency decreases due to increase in junction temperature.

The single transverse mode nature of the emission is determined through a measurement of the far-field characteristics. The far-field pattern shows a full width at half maximum normal to the junction plane and along the junction plane of 27° and 7°, respectively.

Multimode devices with 100 μm stripe width have been reported. They produce several W (~12 W at 15 A [37]) of output power and generally fail after some time due to catastrophic mirror damage (COMD). A laser device with proper facet coating (e.g., with ZnSe in this case) operates to peak power without sudden degradation and rolls over due to heating [37]. Semiconductor lasers operating at high power have been fabricated at many wavelengths. They offer reliable turn-key pump sources for rare-earth-doped amplifiers and lasers. Advances in cooling schemes have resulted in >100 W of CW output power from rare-earth-doped lasers using semiconductor laser pump sources.

4.4 Fiber Gain Characteristics

Optical fiber amplifiers (OFAs) are typically ~10–30 m in length. They provide optical gain (Figure 4.21). Rare-earth-doped fiber lengths are similar for mode-locked fiber lasers. The small signal gain for two signal wavelengths as a function of pump power for a single-mode fiber amplifier is shown in Figure 4.22. The gain increases rapidly at pump powers near threshold and increases slowly at high pump powers where almost all the erbium ions along the length of the fiber are inverted.

Single-mode amplifiers are often tailored to specific performance characteristics by using different pump configurations and amplifier lengths. They are generally limited to ~500 mW output power.

An approach to a high-power fiber amplifier and fiber laser design is to use a single-mode guide for the signal surrounded by a multimode pump guide (Figure 4.23) [39]. The core is doped with rare-earth ions. Pump light is launched at one end into the undoped cladding, where it propagates with a fraction of the pump confined in the doped core, as it travels along the fiber. The fraction confined in the doped core is absorbed by the ions leading to amplification of the signal.

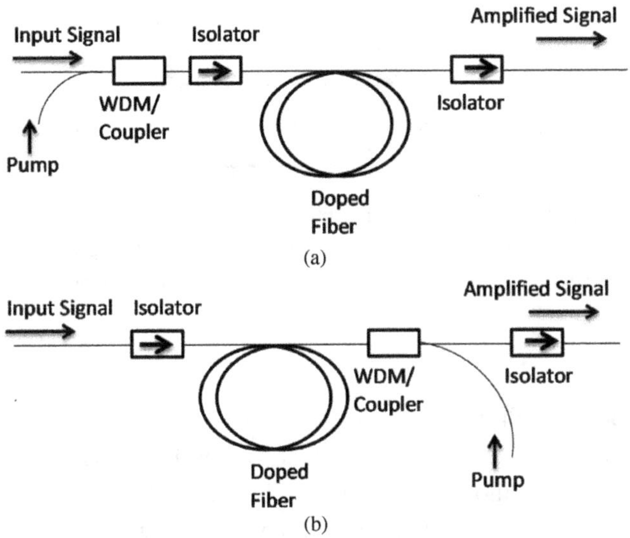

Figure 4.21. Schematic of an optical fiber amplifier. The isolators are needed for reducing the effect of amplified spontaneous emission noise. (a) Forward pumping or co-propagating pump, (b) backward pumping or counter propagating pump. Bidirectional pumping is also used.

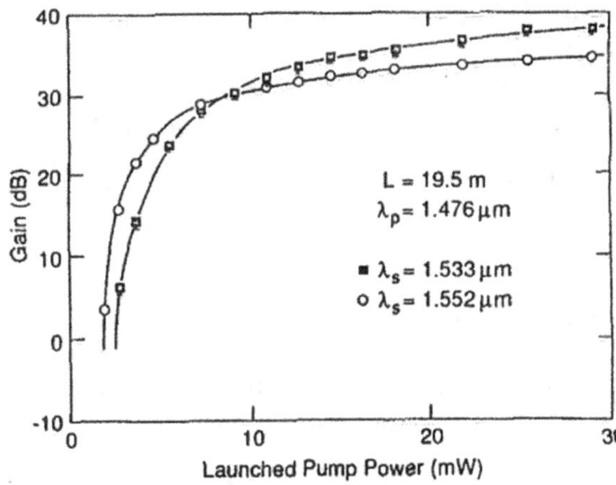

Figure 4.22. Gain as a function of input pump power [9, 38].

Results on a large core area Yb-doped clad fiber laser are described in the following. This double-clad fiber has been designed to produce high power levels (>1 kW). It is drawn from a preform that is fabricated by the MCVD (modified chemical-vapor deposition) process and

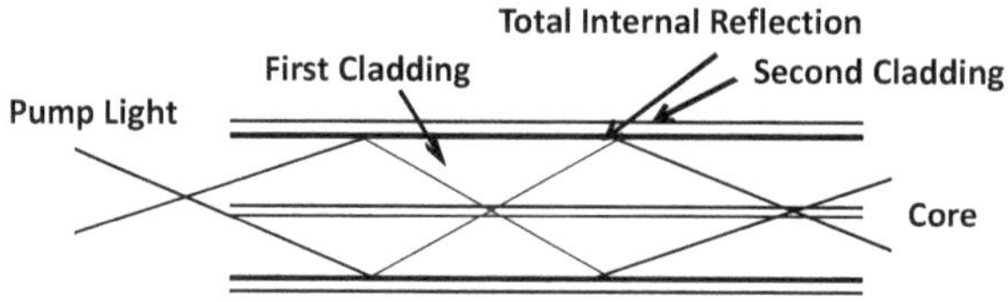

Figure 4.23. Cladding-pumped fiber amplifier design [9, 39]. The pump light is launched into an inner (first) cladding which is around the doped core. This inner cladding and core form a waveguide. An outer (second) cladding and the inner cladding form a waveguide for the pump.

solution-doping technique. The fiber has a 40-μm diameter Yb^{3+}-doped alumino-silicate-based glass core with an NA of 0.05. The mode field area for the fundamental mode is estimated to be ~0.9×10^3 μm^2. The D-shaped inner cladding had a 650/600-μm diameter for the longer/shorter axis. This diameter is chosen to enable efficient coupling of the large beams from high-power diode pump sources. The fiber is coated with a low-refractive-index polymer outer cladding which provides a nominal inner-cladding NA of 0.48. The fiber length used in the laser experiments is 12 m. The experimental setup is shown in Figure 4.24 [40]. The fiber was pumped at both ends using diode laser stacks which emit near 975 nm. The absorption coefficient in the inner cladding is ~1.5 dB/m at the pump wavelength of 975 nm. This corresponds to an Yb^{3+} concentration of ~6000 ppm by weight. ~1.3 kW of power was obtained using ~1.8 kW of pump power.

4.4.1 *Fiber Bragg gratings*

Fiber Bragg gratings (FBGs) are an important component of fiber lasers, particularly fiber Raman lasers. They are used for providing feedback to the gain region and as both input couplers and output couplers. For a multilayered planar dielectric medium with a small refractive index difference between the layers, reflections take place at every interface for a normally incident light. For a significant reflection from the dielectric stack, the reflections at each interface must add up in phase. A similar phenomenon must happen for an index grating close to a waveguide. For the schematic shown in Figure 4.25 (an index grating

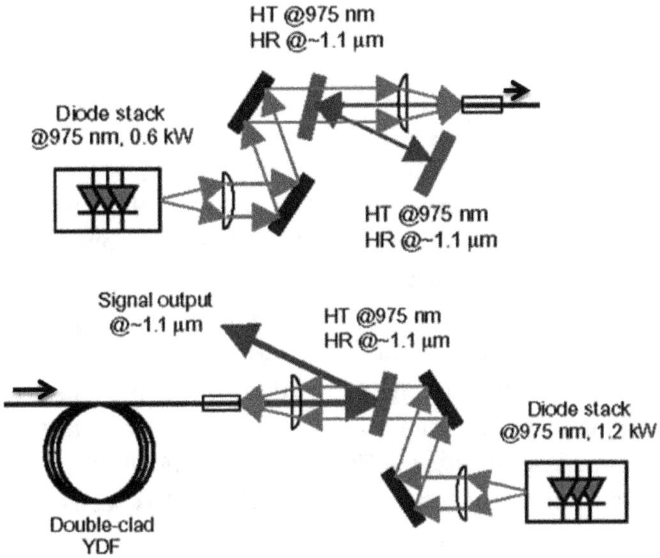

Figure 4.24. YDCF experimental setup. Note to diode laser stacks for end pumping. HR — high reflectivity, HT — high transmission [40]. The components in the top figure are input to the components in the lower figure.

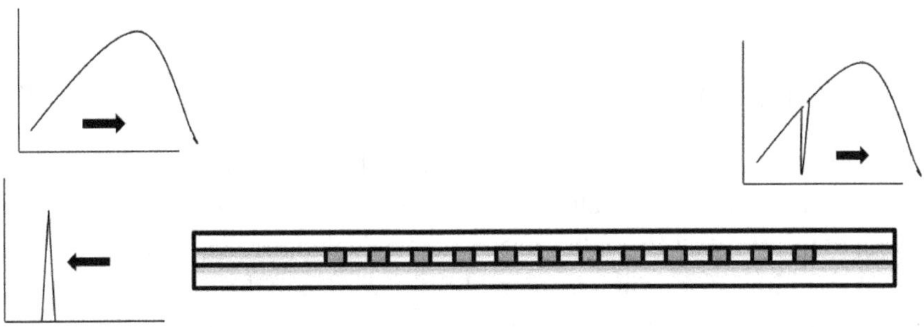

Figure 4.25. Schematic of an index grating and its operation. The grating reflects a specific wavelength. The grating period (in silica glass) is ~500 nm for reflection near 1550 nm. The reflected light spectrum and transmitted light spectrum are sketched for an incident light signal with the spectrum shown.

near a dielectric layer is a periodic region of alternating high and low indexes), the small reflections of the optical field outside the waveguide from the "crests" of the grating must add up in phase to produce a significant reflectivity.

This constructive interference occurs if the period of the grating (Λ) obeys the following equation:

$$2n\Lambda = m\lambda, \tag{4.9}$$

where n is the index of the material ($n \sim 1.5$ for silica glass), λ is the wavelength in free space, and m is an integer. Equation (4.9) is called the Bragg condition. For $m = 1$, the grating is called a first-order grating. For a first-order grating, the Bragg wavelength λ_B for a grating with a period Λ is $\lambda_B = 2n\Lambda$. Thus, for reflection at $\lambda \sim 1500$ nm, the grating period is ~500 nm for first-order gratings in silica ($n = 1.5$).

Typically, first-order gratings are used in FBG applications. In a laser cavity, the light travels both in the forward and backward directions. Thus, for an FBG in a laser cavity, the FBG reflection couples the forward and backward traveling waves. The reflectivity of FBG has been derived using coupled mode equations. The result for reflectivity, R, is [41–44]

$$R = \frac{K^2 \sinh^2(|S|L)}{\delta\beta^2 \sinh^2(|S|L) + |S|^2 \cos h^2(|S|L)} \text{ for } K^2 > \delta\beta^2$$

$$= -\frac{K^2 \sinh^2(|S|L)}{\delta\beta^2 - K^2 \cos h^2(|S|L)} \text{ for } K^2 < \delta\beta^2$$

with $S = \sqrt{[\delta\beta^2 - K^2]}$. (4.10)

In these equations, $\delta\beta = 2\pi n_e/\lambda - m\pi/\Lambda$, where n_e is the effective index of the waveguide. The quantity $\delta\beta$ is a measure of the mismatch between Bragg condition and the actual free space wavelength (λ). The quantity K is the coupling constant between the forward and backward traveling waves and L is the length of the grating region. For uniform sinusoidal modulation of the grating index

$$K = (\pi\delta n/\lambda_B)\eta, \tag{4.11}$$

where η is the overlap of the electric field in the waveguide core with the grating and δn is the difference between the maximum and minimum index in the grating region. If the grating is far from the core, K decreases. The maximum reflectivity occurs for $\delta\beta = 0$ i.e., when the Bragg condition is satisfied. The maximum reflectivity, R_{max} is given by

$$R_{max} = \tanh^2(KL). \tag{4.12}$$

For $KL > 3$, the reflectivity approaches 100%. For a significant reflectivity, few mm of grating length is needed. Another measure of an FBG is its bandwidth. A measure of this quantity is the wavelength difference ($\Delta\lambda$) between the two minimum on either side of the maximum reflection peak. It is given by (for $\lambda \sim \lambda_B$)

$$\Delta\lambda = \frac{\lambda_B^2}{\pi n_e L}\left[(kL)^2 + \pi^2\right]^{1/2}. \tag{4.13}$$

Thus, the reflection peak position, the width of the reflection spectrum, and the maximum reflectivity can be varied by varying the grating pitch and the grating length respectively.

There are several techniques for writing gratings. The principles among them are the dual beam holographic technique and the phase mask technique. These techniques are also used for fabricating grating on a semiconductor substrate for distributed feedback laser fabrication. In the holographic technique, the light from a UV laser is split up into two beams using a beam splitter. The beams are focused back using lenses on a fiber with the cladding removed. Cylindrical lenses are used to elongate the beams along the fiber length. The interfering light on the fiber produces an intensity pattern. Figure 4.26 shows the schematic of the holographic method.

The interference pattern writes a grating on the fiber. This is due to changes in absorption in fiber when exposed to UV light. An ideal silica

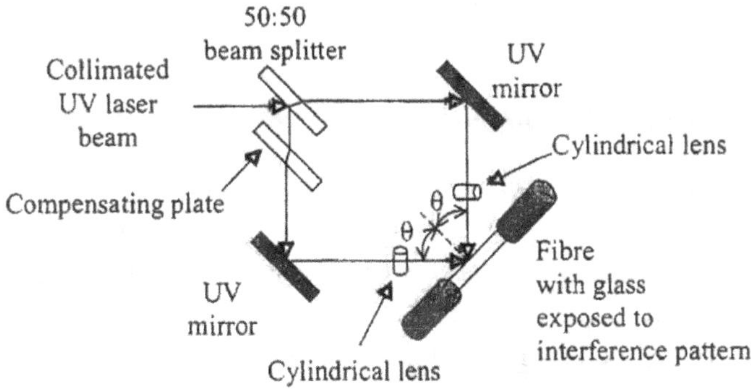

Figure 4.26. Schematic of grating writing using a dual beam holographic method [41, 42].

fiber would have SiO_2 bonds. When dopants, such as Ge, are introduced to change the index to form the fiber waveguide, defects are introduced which can take the form of Si–Si, Si–Ge, and Ge–Ge bonds. In the presence of UV light (~244 nm), these bonds break increasing the absorption. An index change is then caused through the Kramers–Kronig relationship. Thus, a periodic index change is produced by the periodic interference pattern of UV light. The grating period is given by

$$\Lambda = \frac{n_{\text{eff}}\lambda_{\text{laser}}}{n_{\text{laser}}\sin\theta}, \tag{4.14}$$

where λ_{laser} is the wavelength of laser, n_{laser} is the index at laser wavelength in the fiber, n_{eff} is the index of fiber near Bragg wavelength, and θ is the angle shown in Figure 4.26. The quality of the grating can be improved by increasing the photosensitivity of the fiber using different dopants. The dual beam holographic technique requires a highly coherent laser source, a stable grating writing system over a large area (few centimeters), and low alignment tolerances. These disadvantages are overcome using a phase mask technique [42]. The performance of fiber gratings and their use are discussed in later sections.

4.5 CW Fiber Lasers

Fiber laser is a device very similar to a fiber amplifier [41]. Fiber amplifiers amplify light using a rare-earth-doped gain medium. Fiber lasers also use a rare-earth-doped fiber gain medium and, in addition, has optical feedback.

4.5.1 *Single-mode fiber laser*

A typical Er-doped fiber laser is shown in Figure 4.27 [45]. In this design, the gain length is reduced by high doping, ~600 ppm vs. a typical value of ~10–30 ppm for a typical erbium-doped amplifier. The Ge-doped core is co-doped with Al to reduce concentration quenching effects.

The core diameter of the fiber is ~2.6 μm. The Er-doped fiber is spliced to two fiber Bragg gratings (FBGs). FBGs are refractive index gratings generated in the fiber by optical interference of two light (UV)

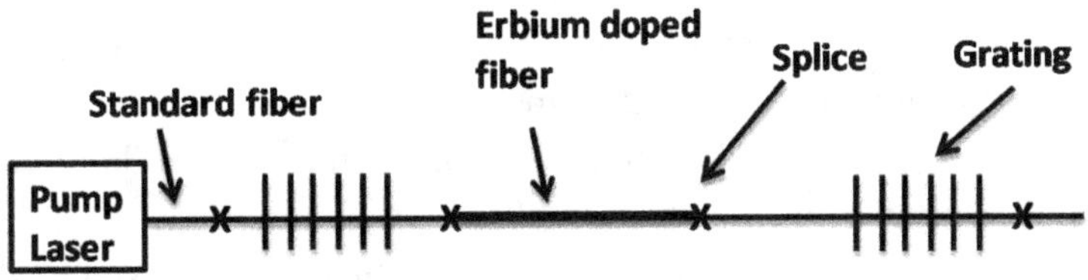

Figure 4.27. Schematic of an Er-doped fiber laser with fiber Bragg gratings (FBGs).

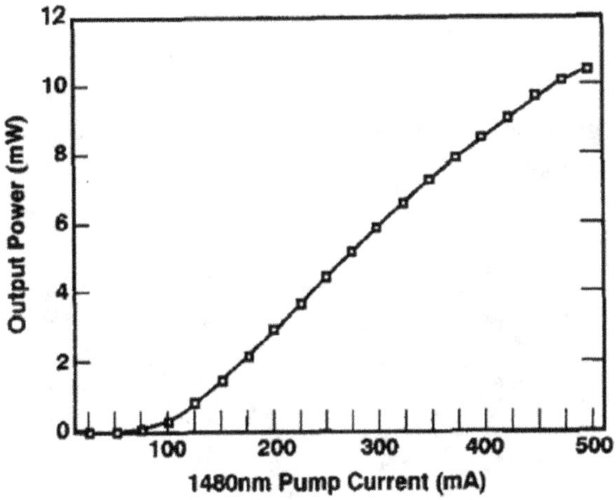

Figure 4.28. Laser output as a function of 1480 nm pump laser current [45].

beams (Section 4.4). They can have very high reflectivity at a specific wavelength. The output coupler grating and the rear reflector have reflectivities of 98% and >99%, respectively. Er-doped fiber lasers of the type shown in Figure 4.27 have been pumped using both 1480 nm and 980 nm pump lasers. The laser output as a function of pump power for the laser design of Figure 4.27 is shown in Figure 4.28. The laser emits near 1539 nm. The amplifier fiber length is 2 cm. These types of fiber lasers have been used as sources in optical transmission experiments.

4.5.2 *High-power fiber laser*

Yb-doped fiber lasers, which emit near 1100 nm, have been studied for a high-power laser source, which has many applications. As a result, many

Yb-doped fiber lasers are made in the cladding-pumped geometry [46–51]. The fiber used is a double-clad fiber with an inner cladding diameter of 240 μm and an outer cladding diameter of 355 μm.

The Yb-doped core is 10 μm in diameter with an NA of 0.06. The Yb doping is $1.4 \times 10^{21} cm^{-3}$ (about 12 wt.% of Yb_2O_3).

The Yb concentration is about 10 times higher than the typical values used. The active core, offset from the inner cladding and the outer cladding, is made of phosphate glass. An air hole of 10 μm diameter is added 30 μm from the core to increase pump absorption (Figure 4.29 which shows a microscopic image of the fiber end). The schematic of the experimental setup is shown in Figure 4.30. The pump laser emits at 940 nm.

The fiber gain medium is 71 cm long and the laser has a high reflector (bulk) at one end and the cleaved fiber end (reflectivity ~3–4%) serves as

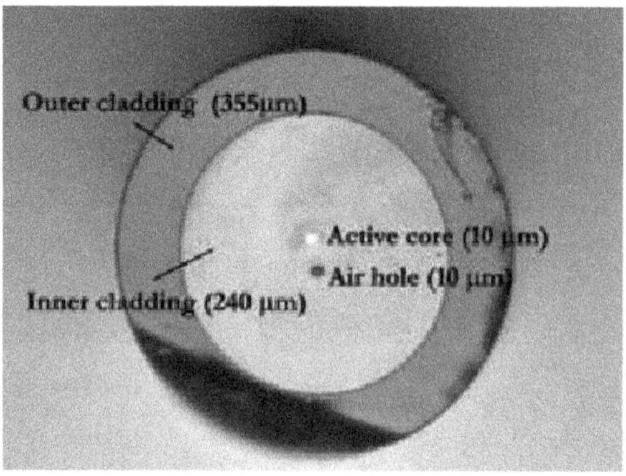

Figure 4.29. Image of the Yb-doped fiber end. The core has a diameter of 10 μm, inner cladding has a diameter of 240 μm, and the outer cladding has a diameter of 355 μm [51].

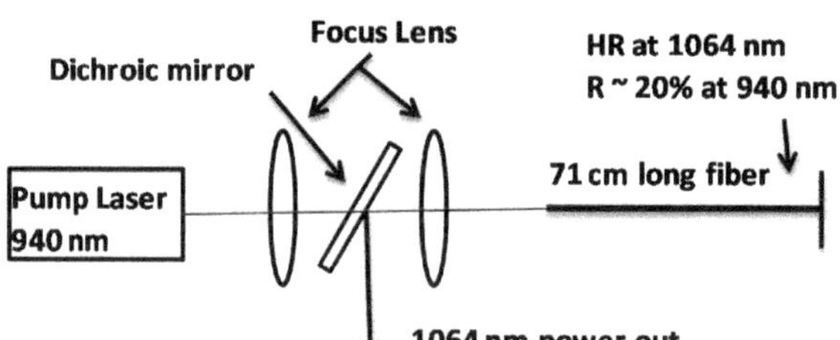

Figure 4.30. Schematic of the experimental setup for Yb-doped fiber [51].

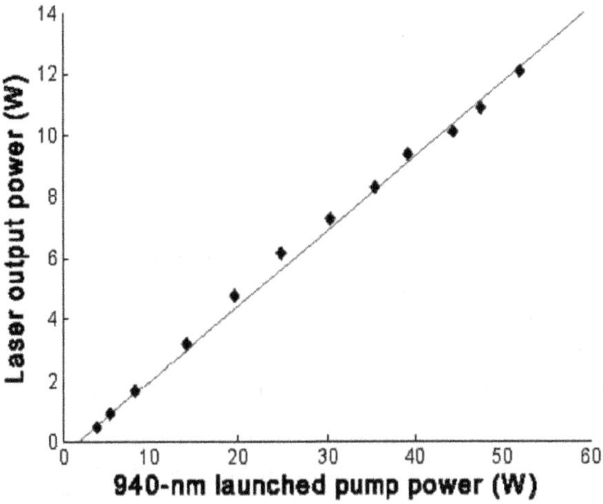

Figure 4.31. Yb laser output as a function of pump power [51].

the second reflector. The 940 nm pump light is coupled at the cleaved end using a dichroic mirror. The Yb laser output power as a function of pump power is shown in Figure 4.31. The laser emits near 1062 nm. The measured laser slope efficiency against launched power and absorbed power are 24% and 37%, respectively. Tunable fiber lasers have been demonstrated using ring geometry [52].

Mode-locked pulses have been reported using rare-earth-doped fiber lasers using both active and passive mode-locking schemes. The mode locking is often done with the laser in a ring configuration.

4.6 Fiber Raman Lasers and Amplifiers

Fiber Raman lasers are based on the nonlinear process of stimulated Raman scattering [53–57]. When a sufficiently intense light beam is incident on a material, light at a new of wavelength is generated by a process called Raman scattering. Typically, the energy of the photons ($h\nu_p$) at the incident wavelength is larger than the energy of Raman process generated photon ($h\nu_s$) by an amount equal to the separation between two energy levels (δE) in the material i.e., $h\nu_p - h\nu_s = \delta E$. The photons at the lower energy are generated by both spontaneous and stimulated emissions (Figure 4.32).

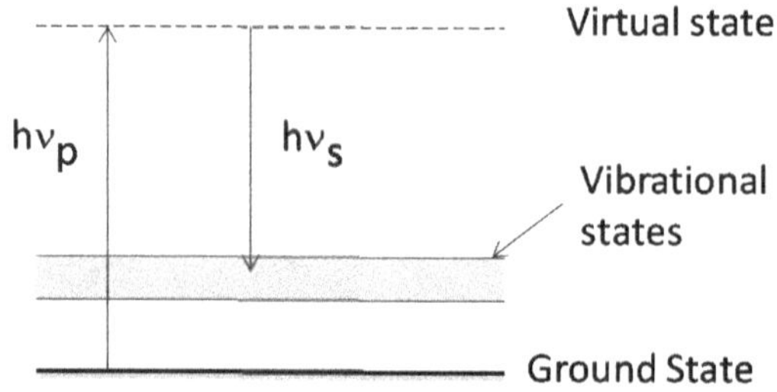

Figure 4.32. Schematic of quantum mechanical transition in a Raman process. The figure shows that a molecule is excited to a virtual state when a pump photon is incident and it quickly decays to a different energy level emitting a photon at a lower energy.

The stimulated Raman scattering produces Raman gain and hence they are called fiber Raman lasers. The energy levels in the material include electronic, vibration, and rotational energy levels. For typical fiber Raman lasers, the excitations are vibration energy in the SiO_2 glass matrix i.e., vibration energy of the Si–O bond. The Raman gain is proportional to the input pump power. The vibration energy is ~467 cm^{-1} and the Raman gain is 0.8 cm^{-1}/GW. Since Raman fiber lasers are based on stimulated Raman scattering (which produces gain at the energy of the Raman scattered light), the incident light (known as the pump) needs to be sufficiently intense. The Raman gain for the generated Raman signal is proportional to $P_p\sigma N$, where P_p is the intensity of the incident pump, σ is the stimulated Raman cross-section, and N is the number of participating molecules per unit volume. The quantity $\sigma N = g_R$ is often called the Raman gain coefficient. The difference in energy between the pump and signal photons (Raman scattering produced or injected) is determined by the molecular vibration of the host material. This frequency difference is called the Stokes shift. The amorphous nature of silica makes the Raman gain curve quite broad in optical fiber. The simplest Raman fiber laser consists of a single-mode fiber into which pump light is injected.

The single-mode fiber has fiber Bragg grating (FBG) attached at both ends (Figure 4.33(a)). The FBG has the unique property that it can be fabricated so that it reflects at a very narrow band of wavelengths (by proper choice of the periodicity of the grating) and can be easily spliced to a single mode fiber with low loss. The grating reflectivity can also be

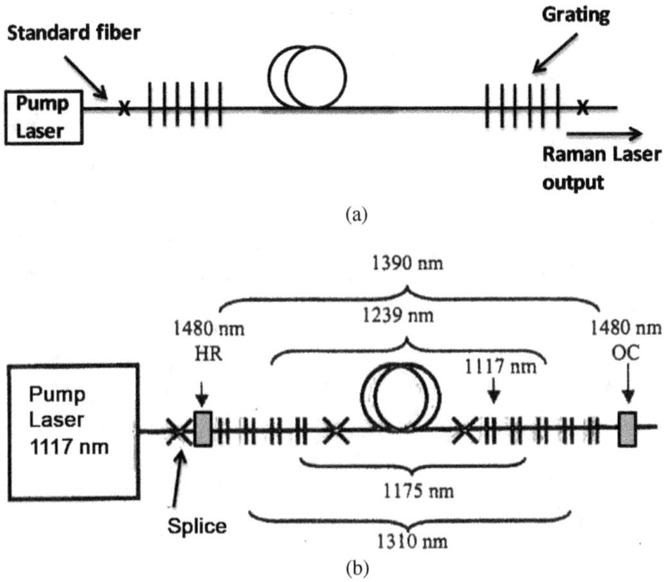

Figure 4.33. (a) Typical schematic of a fiber Raman laser with fiber Bragg gratings (FBGs). (b) A cascaded Raman laser design [58]. The laser emits at 1480 nm using a 1117 nm pump laser. HR — high reflection, OC — output coupler.

varied by varying the length of the FBG. At the input pump end, the FBG reflects the Raman signal light but lets the pump light through. At the output coupler, it allows some of the signal light to escape (reflectivity ~90%). FBGs have been used for providing feedback for both fiber lasers and fiber Raman lasers.

Using a series of gratings at appropriately chosen wavelength, a cascaded Raman laser (Figure 4.33(b)) has been demonstrated [59, 60]. In this design, four nested pairs of gratings (with reflectivity close to 100%) at appropriate Stokes wavelengths are used to build up the pump light for each Stokes scattering. The pump light is at 1117 nm. The first Stokes scattering produces stimulated emission at 1239 nm, which acts as a pump for the second Stokes scattering producing stimulated emission at 1390 nm and this acts as a pump for the third Stokes scattering which produces stimulated emission at 1480 nm. The 1480 nm light could be used as a pump source for Er-doped fibers. Thus, the cascaded Raman

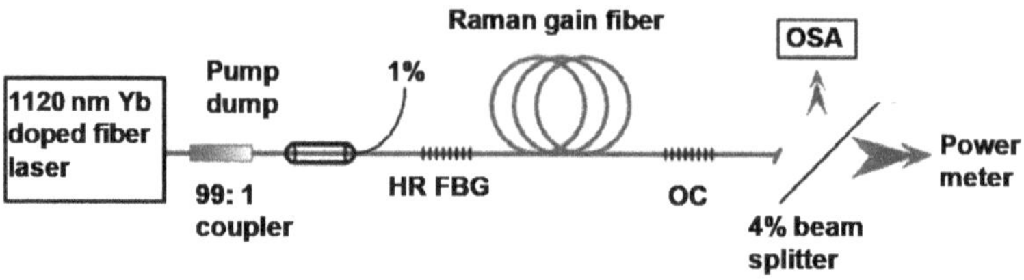

Figure 4.34. Schematic of a high-power Raman laser pumped by a Yb-doped double-clad fiber laser (DCFL) emitting at 1120 nm. HR FBG: high reflection fiber Bragg grating, OC: output coupler, OSA: optical spectrum analyzer [63].

process utilizing FBGs produces stimulated emission at 1480 nm using 1117 nm pump light.

4.6.1 *Raman laser*

Raman lasers with higher output power have been studied and demonstrated [60–65]. The experimental setup (Figure 4.34) using Yb-doped double-clad pump lasers is similar to that shown in Figure 4.5.3. The Raman gain fiber is a 75 m long commercial grade fiber with 10 μm core diameter. The FBGs have a reflectivity of 99% (HR FBG) at the pump input end and have a reflectivity of 50% at the output coupler end. The 3 dB bandwidth of the HR FBG is 1.49 nm, and that of the output coupler is 0.52 nm. The high power from YDFL (at 1120 nm) is delivered by a regular fiber with a core diameter of 10 μm. A 99%/1% coupler is spliced before the pump to monitor the stability of the pump source.

The Raman laser output at 1173 nm as a function of the injected pump has been measured. A Raman laser power of 114 W has been obtained with a 144 W of pump power. The conversion efficiency is 82%. The Raman laser spectrum has a full width at half-maximum of ~2 nm. For shorter Raman gain fiber lengths, the output power is lower. One importance of high-power output near 1230 nm is for its use as a pump source for a fiber Raman amplifier or fiber Raman laser emitting near 1300 nm. The first Stokes shift of 1230 nm from a silica fiber produces a Raman output near 1300 nm. Thus, a Raman amplifier for a 1300 nm spectral region, which is an important wavelength for optical communication, can be built using this process. Such an amplifier has been demonstrated.

4.7 Photonic Crystal Fiber

Photonic crystal fibers (PCF) also known as microstructure fibers (PCF) have been studied since the mid-1990s [66–70]. Nonlinear optics studies such as supercontinuum generation and amplification using rare-earth doping have been reported. Although much of these studies have been carried out using silica-based fiber, the photonic crystal fiber fabrication (described in the following paragraph) is, in principle, applicable to a wide range of materials. Photonic crystal fibers have nonlinearity and have been used in ultrashort pulse generation.

4.7.1 *Photonic crystal fiber fabrication*

The fabrication of PCF involves a process by which an arranged macroscopic structure is thinned using high temperatures. In the first step, individual capillaries are created using standard glass capillary fabrication procedures. Then they are manually organized (arranged) to form a multicapillary structure of required symmetry. In the middle, there is a glass rod in which the light propagates or there could be a hole for fibers with a photonic band gap. Capillaries with different aspect ratios (ratio of inner to outer diameter) can be chosen. The rods are positioned in the structure. Extra glass rods may be added to create a structure of sufficient mechanical integrity and desired dimensions.

The structure is then fused in a furnace to create an intermediate perform, a few mm in diameter. The preform is fused and drawn using a fiber drawing tower (described previously) to create the microstructure fiber. Figure 4.35 shows the schematic of the process. The fiber cross-section for various types of microstructure fibers is shown in Figure 4.36. One of the capillaries could be rare-earth doped for laser or amplifier fiber fabrication. The process is very general and could be applicable to a wide range of material choices. In principle, many types of glasses (silica, fluorides, germinates, etc.) can be used. Photonic crystal fibers have high nonlinearity due to small waveguide size and hence it is useful for the study of nonlinear optical processes.

Rare-earth-doped fiber lasers are important for a range of applications. Low-power lasers are important for communication and sensor

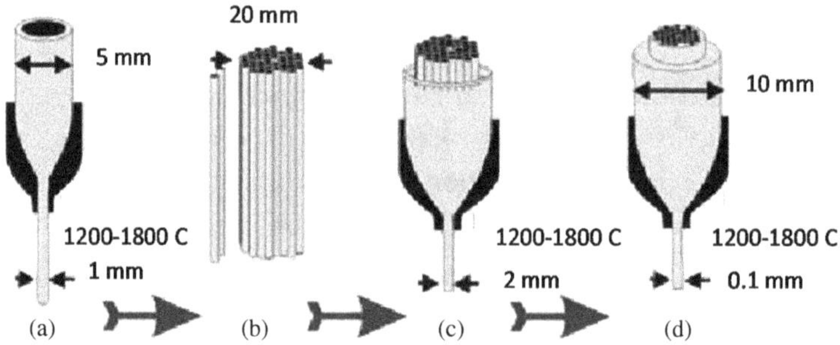

Figure 4.35. Schematic of the microstructure fiber fabrication. (a) Creation of individual capillaries, (b) perform creation, (c) drawing of preform with cladding — intermediate preform, (d) drawing of microstructure fiber [70].

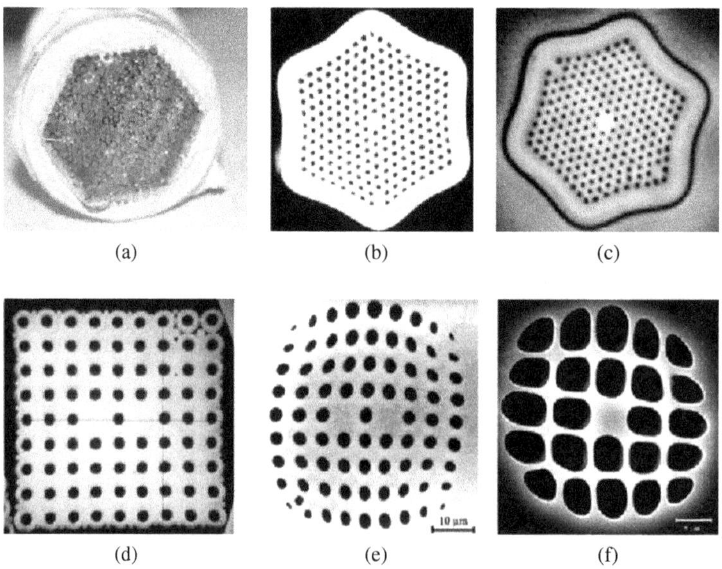

Figure 4.36. Fabrication of photonic crystal fibers: preforms, intermediate preforms, and final fibers: (a) the preform of fiber with holes spaced 1 mm with hexagonal lattice; (b) the intermediate preform with air holes spaced 250 μm with hexagonal lattice; (c) the fiber diameter of 120 μm, air holes diameter d = 3 μm; (d) the intermediate preform of double core fiber with square lattice; (e) a double core fiber with a square lattice, diameter of the 250 μm, air holes diameter $d = 2.5$ μm; (f) a multimode fiber with a square lattice; diameter of the fiber 160 μm, air holes diameter 3 μm [70].

systems. High-power lasers are important for industrial applications, such as metal cutting and shaping. A wide range of wavelengths in the near-infrared range can be obtained from these lasers. In addition to doping in SiO_2 glass, rare-earth-doped fluoride fibers have been fabricated. This is important for long wavelength (>2000 nm) operation where the SiO_2 glass absorption is high.

References

1. C. K. Kao and G. A. Hockmann, *Proc. IEEE*, **133**, (1966) 1151.
2. D. B. Keck, *IEEE Comm. Mag.*, **23**, (1985) 17.
3. J. B. MacChesney and P. B. O'Conner, US Patent 4,217,027 (1980).
4. S. R. Nagel, Chapter 4 in *Optical Fiber Telecommunications*, II (S. E. Miller and I. P. Kaminow Eds.) (Academic Press, NY, 1992).
5. K. J. Beales and C. R. Day, *Phys. Chem. Glasses*, **21**, (1980) 5.
6. K. J. Beales, S. F. Carter, P. W. France, and S. Partington, *J. Non-crystalline Solids*, **70**, (1985) 253.
7. K. L. Walker, G. M. Hornsy, and F. T. Geyling, *J. Colloidal Interface Sci.*, **69**, (1979) 138.
8. K. L. Walker, F. T. Geyling, and S. R. Nagel, *J. Am. Ceram. Soc.* **63**, (1980) 552.
9. N. K. Dutta, *Fiber Amplifiers and Fiber Lasers* (World Scientific Publishing Co., Singapore, 2014).
10. M. Shimizu, M. Yamada, M. Horiguchi, T. Takeshita, and M. Okayasu, *Electron. Lett.*, **20**, (1990) 1641.
11. Z. Burshtein, Y. Kalisky, S. Z. Levy, P. Le Boulanger, and S. Rotman, *IEEE JQE*, **36**, (2000) 1000.
12. M. J. F. Digonnet, *Rare Earth Doped Fiber Lasers and Amplifiers* (Marcel Dekker Inc., New York, 1993), Chapter 2.
13. T. Y. Fan and R. L. Byer, *J. Opt. Soc. Am.*, **B 3**, (1986) 1519.
14. E. Snitzer, *Phys. Rev. Lett.*, **7**, (1961) 444.
15. P. R. Watekar, S. Ju, and W. T. Han, In *Proc. CLEO, Paper OWF5*, 2005 (Baltimore, MD, USA).
16. J. Koenings, D. Kuppers, H. Lydtin, and H. Wilson, In *Proc. 5th Int. Conf. CVD*, 1975, p. 270 (San Jose, CA, USA).
17. P. Geittner, H. J. Hagemann, J. Warnier, and H. Wilson, *J. Lightwave Tech.*, **LT-4**, (1986) 818.

18. H. Lydtin, *J. Lightwave Tech.*, **LT-4**, (1986) 1034.
19. D. B. Keck, P. C. Shultz, and F. Zimar, US Patent 3,737,292, (1973).
20. D. B. Keck, R. D. Maurer, and P. C. Shultz, *Appl. Phys. Lett.*, **22**, (1973) 307.
21. P. C. Shultz, *Appl. Opt.*, **18**, (1979) 3684.
22. T. Izawa, T. Miyashita, and F. Hanawa, US Patent 4,062, 655 (1977).
23. T. Izawa and N. Inagaki, *Proc. IEEE*, **68**, (1980) 1184.
24. F. Hanawa, S. Sudo, M. Kawachi, and M. Nakahara, *Electron Lett.*, **16**, (1980) 299.
25. L. L. Blyler and F. V. DiMarcello, *Proc. IEEE*, **68** (1980) 120.
26. F. V. DiMarcello, C. R. Kurkjian, and J. C. Williams, Optical fiber communications Chapter 4 In *Optical Drawing and Strength Properties*, Vol. 1, (T. Li Ed.) (Academic Press, NY, 1985).
27. P. C. Becker, N. A. Olsson, and J. R. Simpson, *Erbium Doped Fiber Amplifiers* (Academic Press, NY, USA, 1997).
28. A. H. Gnauck, R. W. Tkach, A. R. Chraplyvy, and T. Li, *IEEE J. Lightwave Tech.*, **26**, 1032 (2008).
29. J. Simpson, *Proc. SPIE*, **1171**, (1989) 2; D. J. DiGiovanni, *Proc. SPIE*, **1373**, (1990) 2.
30. A. Kasukawa, Chapter 3, *WDM Technologies: Active Optical Components*, (A. K. Dutta, N. K. Dutta and M. Fujiwara Eds.) (Academic Press, NY, USA).
31. H. Asano, S. Takano, M. Kawaradani, M. Kitamura, and I. Mito, *IEEE Photonic Tech. Letts.*, **3**, 415 (1991) 415.
32. T. Tanbun-Ek, R. A. Logan, N. A. Olsson, H. Temkin, A. M. Sergent, and K. W. Wecht, *Appl. Phys. Lett.*, **57**, (1990) 224.
33. K. Shigihara, K. Kawasaki, Y. Yoshida, S. I. Yamamura, T. Yagi, and E. Omura, *IEEE JQE*, **38**, (2002) 1081.
34. G. Yang, G. M. Smith, M. K. Davis, A. Kussmaul, D. A. S. Loeber, M. H. Hu, H-K Nguyen, C. E. Zah, R. Bhat, *IEEE Photonic Tech. Letts.*, **16**, (2004) 981.
35. S. Pawlik, J. Muller, N. Lichtenstein, D Jaeggi, and B. Schmidt, In Proc. Optical Fiber Communication, Paper ThP3, (San Jose, CA, USA, 2000).
36. Y. He, H. An, J. Cai, C. Galstad, S. Macomber, and M. Kanskar, *Electron Letts.*, **45**, (2009) 163.
37. P. Crump, H. Wenzel, G. Erbert, P. Ressel, M. Zorn, F. Bugge, S. Einfeldt, R. Staske, U. Zeimer, A. Pietrzak, and G. Tränkle, *IEEE Photonic Tech. Letts.*, **20**, (2006) 1378.
38. G. P. Agrawal and N. K. Dutta, *Semiconductor Lasers* (Van Nostrand Reinhold, NY, 1992).

39. A. Tunnermann and H. Zellmer, Chapter B4.2, In C. Webb and J. Jones (Ed.), *Handbook of Laser Technology and Applications* (Institute of Physics, London, UK, 2004).
40. Y. Jeong, J. K. Sahu, D. N. Payne, J. Nilsson, *Optics Express*, **12**, (2004) 6088.
41. N. K. Dutta, *Optoelectronic Devices* (World Scientific, Singapore, 2018).
42. A. Othonos and K. Kalli, *Fiber Bragg Gratings* (Artech House, Boston, 1999), Chapter 5.
43. D. Hanna, Chapter B4.1, In C. Webb, and J. Jones (Ed.), *Handbook of Laser Technology and Applications* (Institute of Physics, London, UK, 2004).
44. A. Tunnermann and H. Zellmar Chapter B4.2, in C. Webb and J. Jones (Ed.), *Handbook of Laser Technology and Applications* (Institute of Physics, 2004).
45. V. Mizrahi, D. J. DiGiovanni, R. M. Atkins, S. G. Grubb, Y. K. Park, and J. M. Delavaux, *J. Lightwave Tech.*, **11**, (1993) 2021.
46. Y. Jeong, J. K. Sahu, D. N. Payne, and J. Nilsson, *Optics Express*, **12**, (2004) 6088.
47. A. Liem, J. Limpert, H. Zellmer, A. Tünnermann, V. Reichel, K. Mörl, S. Jetschke, S. Unger, H.-R. Müller, J. Kirchhof, T. Sandrock, and A. Harschak, In Conf. Lasers Electro-Optics, May 2004, Postdeadline Paper CPDD2, 2004 (Baltimore, MD, USA).
48. V. P. Gapontsev, D. V. Gapontsev, N. S. Platonov, O. Shkurikhin, V. Fomin, A. Mashkin, M. Abramov, and S. Ferin, Eur. Conf. Lasers Electro-Optics, July 2005, paper CJ1-1, 2005.
49. N. S. Platonov, D. V. Gapontsev, V. P. Gapontsev, and V. Shumilin, In Proc. CLEO 2002, May 2002, Postdeadline Paper CPDC3, 2002 (Baltimore, MD, USA).
50. J. Limpert, A. Liem, H. Zellmer, and A. Tunnermann, *Electron. Lett.*, **39**, (2003) 645.
51. Y. W. Lee, S. Sinha, M. J. F. Digonnet, and R. L. Byer, Proc. CLEO 2006, Paper CTuI5, 2006 (Baltimore, MD, USA).
52. Z. Chen, S. Ma, and N. K. Dutta, *Optics Express*, **17**, (2009) 1234.
53. G. P. Agrawal, *Nonlinear Fiber Optics* 4th (edn.) (Academic Press, NY, 2007).
54. R. G. Smith, *Appl. Opt.*, **11**, (1972) 2489.
55. R. H. Stolen, Chapter 10 In *Optical Fiber Telecommunications* (S. E. Miller and A. G. Chynoweth Eds.) (Academic, New York, 1979).
56. R. H. Stolen and J. E. Bjorkholm, *IEEE J. Quantum Electron.*, **QE-18**, (1982) 1062.
57. N. Bloembergen, *Am. J. Phys.*, **35**, (1967) 989.

58. S. Grubb, Proc. Optical Amplifiers and Their Applications, Paper PD3-1, 1994 (San Jose, CA, USA).
59. S. Grubb, Proc. Optical Amplifiers and Their Applications, Paper SA-4, 197, 1995 (San Jose, CA, USA).
60. Y. Feng, L. R. Taylor, and D. B. Calia, *Opt. Exp.*, **17**, (2009) 23678.
61. C. A. Codemard, J. Ji, J. K. Sahu, and J. Nilsson, In Proc. SPIE, San Francisco, CA, USA, 2010, Vol. 7580, p. 75801.
62. J. AuYeung and A. Yariv, *J. Opt. Soc. Am.*, **69**, (1979) 803.
63. H. Zhang, H. Xiao, P. Zhou, X. Wang, and X. Xu, *IEEE Photon. J.*, **5**, (2003) 1501706.
64. E. Bélanger, M. Bernier, D. Faucher, D. Côté, and R. Vallée, *J. Lightwave Tech.*, **26**, (2008) 1696.
65. A. J. Stentz, T. Nielsen, S. G. Grubb, T. A. Strasser, and J. R. Pedrazzani, In Optical Fiber Communication Conf. OSA Tech. Dig. Washington, DC, 1996.
66. J. C. Knight, T. A. Birks, P. St. J. Russell, D. M. Atkin, *Opt. Lett.*, **21**, (1996) 1547.
67. J. C. Knight, J. Broeng, T. A. Birks, and P. S. Russell, *Science* **282**, (1998) 1476.
68. P. S. Russell, *Science* **299**, (2003) 358.
69. R. F. Cregan, B. J. Mangan, J. C. Knight, T. A. Birks, P. S. Russell, P. J. Roberts, and D. C. Allan, *Science*, **285**, (1999) 1537.
70. R. Buczynski, *Acta Physica Polonica*, **106**, (2004) 141.

Chapter 5

Active Mode Locking of Fiber Lasers

5.1 Introduction

Mode-locked pulses have been reported using rare-earth-doped lasers using both active and passive mode-locking schemes. For active mode locking, the laser cavity generally has a modulator which modulates the optical loss or phase in the optical cavity [1–8]. When the modulation frequency equals the inverse of cavity round-trip time, mode-locked pulses at the modulation frequency are generated. Also, if the modulation frequency equals an integer multiple (n) of the inverse of the cavity round-trip time (f_c), the mode-locked pulses are generated at a frequency of f_m (= nf_c). This phenomenon is called harmonic mode locking. If the modulation frequency equals $(n + 1/p)f_c$ i.e., $f_m = (n + 1/p)f_c$, where p is an integer, mode-locked pulses are also generated at a frequency of pf_m i.e., at a multiple of the modulation frequency. This phenomenon is called rational harmonic mode locking and it is useful for generating optical pulses at a high data rate. For passive mode locking, a saturable absorber is generally used in the optical cavity. The saturation characteristics, particularly the time constant and the saturation power, of the saturation mechanism determines the mode-locking characteristics. In some cases, both active mode locking and passive mode locking are used in conjunction to produce stable mode-locked pulses.

The generation of ultra-short pulses with high repetition rate with wavelength near 1550 nm is important for high bit rate optical communication. Actively mode-locked Er-doped fiber lasers are a potential source of such pulses, however, the pulse repetition rate is usually limited by the bandwidth of the active mode locker used in the cavity unless a rational harmonic mode-locking scheme is used. Typically, a fiber laser in a ring cavity is used for active mode locking. The ring cavity has an optical loss modulator, typically a $LiNbO_3$ modulator. As mentioned earlier, when the modulation frequency is an integer multiple of the inverse of the cavity round-trip time, a series of pulses at the modulation frequency are generated. This process is called harmonic mode locking and is described in the following section. Other mode-clocking schemes such as rational harmonic mode locking are also discussed in this chapter.

5.2 Harmonic Mode Locking

The experimental setup for a harmonic mode-locking scheme is shown in Figure 5.1. The fiber laser has a 23-m long Er-doped fiber (EDF), a WDM coupler, a 10% output coupler, two polarization-independent isolators, a $LiNbO_3$ electro-optical modulator with a bandwidth of 11 GHz, and a

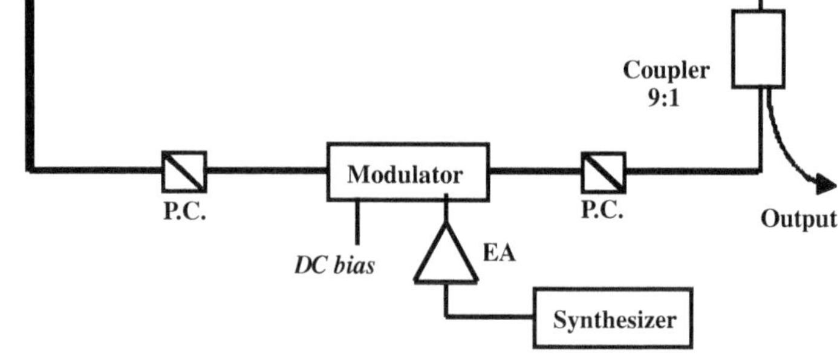

Figure 5.1. Schematic of the fiber laser.

Note: EDF: erbium-doped fiber, PC: polarization controller [5].

wavelength-tunable optical filter with a 3 dB bandwidth of 1.0 nm. A synthesizer is used to generate the RF signal which is amplified to 30 dBm to drive the modulator. All components are connected by a regular single-mode fiber. The fundamental cavity frequency f_c (= c/nL) is 2.58 MHz. The gain in EDF is created using a single-mode laser emitting at 980 nm as a pump.

The active harmonic mode locking in this fiber laser is obtained when the modulation signal at frequency $f_m = nf_c$ in the 1–10 GHz range is applied. The shortest and most stable optical pulses are obtained by careful tuning of the modulation frequency f_m and polarization of the circulating light through the polarization controller. The pulse width of mode-locked pulse trains is measured by a sampling oscilloscope or by an autocorrelator.

Figure 5.2 shows that for harmonic mode-locking case $f_m = nf_c$, the pulse width is proportional to $f_m^{-1/2}$. Figure 5.3 shows harmonic mode locking is possible for wavelengths from 1525 nm to 1565 nm for both $f_m \approx 1$ GHz and $f_m \approx 10$ GHz in this fiber laser. This is due to the broad gain spectrum of EDF. The measured pulse width as a function of wavelength is shown in Figure 5.3 for two different modulation frequencies.

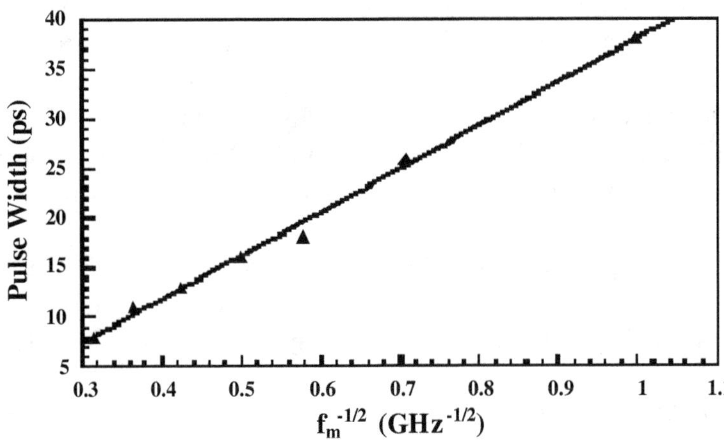

Figure 5.2. Pulse width versus inverse square root of the modulation frequency for harmonic mode-locking case. The corresponding frequency changes from 1 GHz to 10 GHz [5].

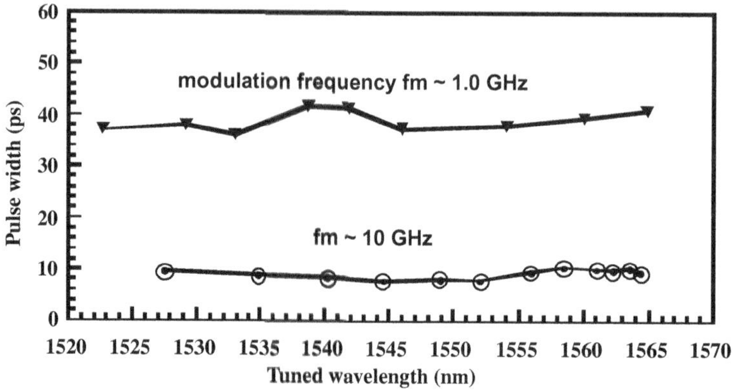

Figure 5.3. Pulse width as a function of wavelength for harmonic mode locking at 1 GHz and 10 GHz respectively for various laser wavelengths [5].

For the active mode-locking theory of fiber laser, the analysis of references [4, 9] is used. Assuming that the ring cavity has a length L, then the fundamental cavity frequency is $f_c = c/(Ln_{\text{eff}})$, where c is the velocity of light in vacuum and n_{eff} is the effective refractive index of the cavity. A modulation signal with frequency f_m is applied to the modulator to modulate the loss in the cavity. For harmonic mode locking, $f_m = nf_c$, or $\omega_m = n\omega_c$, where n is an integer and $\omega_c = 2\pi f_c$. As stated previously for a typical ring fiber laser, $f_c \sim$ 1–3 MHz and $f_m \sim$ 1–10 GHz. Thus, $n \sim$ 500–5000. However, the gain spectrum of the fiber laser is >1000 GHz i.e., ~10^6 cavity modes are present within the gain spectrum. This phenomenon allows harmonic mode locking i.e., although the modulator couples every ~1000th mode, the gain spectrum is broad enough that coupling of ~10^5–10^6 cavity modes can occur.

The envelope of mode-locked pulse $a(t)$ consists of its Fourier components a_q at frequency ω_q. There are various sources of losses in the fiber lasers; they are lumped into one single parameter l here so that after one round trip, the qth mode amplitude (a_q) is multiplied by $1 - l$ as it circulates once through the fiber laser. Similarly, it is assumed that due to gain, the qth mode amplitude (a_q) is multiplied by $G(q) = 1 + g(q)$

as it circulates once through the fiber laser, where $g(q)$ is the gain for the qth Fourier component at frequency ω_q. Assuming a Lorentzian spectral profile for the gain, and $\omega_q = \omega_0 + q\omega_c$, where ω_0 is the center frequency,

$$G(q) = 1 + \frac{g}{1 + \frac{(q\omega_c)^2}{\Omega_g^2}} = 1 + g\left[1 - \frac{(q\omega_c)^2}{\Omega_g^2}\right], \tag{5.1}$$

where Ω_g is the gain bandwidth and g is the gain at the center frequency. The modulator is assumed to pass the light at periodically timed sequences so that the Fourier component $a_q \exp(i(\omega_0 + q\omega_c)t)$ becomes after one pass (using $\omega_m = n\omega_c$)

$$\begin{aligned} &a_q \exp(i(\omega_0 + q\omega_c)t)M(1 - \cos(\omega_m t)) \\ &= -\left(\frac{M}{2}\right)a_q \exp(i\omega_0 t)\left[e^{i(q-n)\omega_c t} - 2e^{iqn\omega_c t} + e^{i(q+n)\omega_c t}\right]. \end{aligned} \tag{5.2}$$

In the above, the modulation factor is given by M $(1 - \cos(\omega_m t))$ at each pass of the components, where M is the modulation depth. So each Fourier component acquires sidebands that are injected into the nth adjacent modes.

Considering the above three elements, a recursion equation is obtained for the Fourier component $a_q^{(k+1)}(\omega)$ traveling around the cavity the $(k + 1)$th time in terms of its value $a_q^{(k)}$ upon the kth time:

$$a_q^{(k+1)} = a_q^{(k)} + g\left[1 - \frac{(q\,\omega_c)^2}{\Omega_g^2}\right]a_q^{(k)} - la_q^{(k)} + \frac{M}{2}\left(a_{q+n}^{(k)} - 2a_q^{(k)} + a_{q-n}^{(k)}\right). \tag{5.3}$$

Equation (5.3) is the basis of discussing harmonic mode locking. If q' is defined as $q' = q/n$, then Equation (5.3) becomes

$$a_{q'}^{(k+1)} = a_{q'}^{(k)} + g\left[1 - \frac{(nq'\omega_c)^2}{\Omega_q^2}\right]a_{q'}^{(k)} - la_{q'}^{(k)} + \frac{M}{2}\left(a_{q'+1}^{(k)} - 2a_{q'}^{(k)} + a_{q^1-1}^{(k)}\right). \tag{5.4}$$

Equation (5.4) shows that there are n pulses traveling around the cavity. In steady state, $a_{q'}^{(k+1)} = a_{q'}^{(k)}$, the superscript ($k$) can be dropped, and (5.3) and (5.4) reduce to a second-order difference equation. Moreover, when the number of oscillating modes is large, it is useful to replace the second-order difference by a second-order derivative, so Equation (5.4) is written as

$$\left[g\left(1-\frac{\omega^2}{\Omega_g^2}\right)-\ell+\frac{M\omega_m^2}{2}\frac{d^2}{d\omega^2}\right]a(\omega)=0, \tag{5.5}$$

where a_q has been replaced by the function $a(\omega)$ of the continuous frequency variable $\omega = q'\omega_c$. The solution of Equation (5.5) is the Hermite–Gaussian function and is given by

$$a(\omega)=H_v(\omega\tau)\exp\left(\frac{-\omega^2\tau^2}{2}\right), \tag{5.6}$$

where

$$\tau=\left(\frac{2g}{M}\right)^{1/4}\left(\frac{1}{\omega_m\Omega_g}\right)^{1/2} \quad \text{and} \quad g-l=M\omega_m^2\tau^2\left(v+\frac{1}{2}\right). \tag{5.7}$$

In general, the zeroth-order ($\nu = 0$) solution dominates. The following expressions for $a(\omega)$ and $a(t)$ follows:

$$a(\omega)=A\exp\left(\frac{-\omega^2\tau^2}{2}\right),$$

$$a(t)=\frac{\sqrt{2\pi}}{\tau}A\exp\left(\frac{-t^2}{2\tau^2}\right). \tag{5.8}$$

So, for harmonic mode-locking case $f_m = nf_c$, the solution is a Gaussian pulse, where the pulse width is proportional to τ and hence to $f_m^{-1/2}$. This agrees with the measurement shown in Figure 5.2. Furthermore, from Equation (5.7), lower pulse width is obtained by increasing the modulation frequency and modulation power while reducing the pump power for the cavity of the fiber laser.

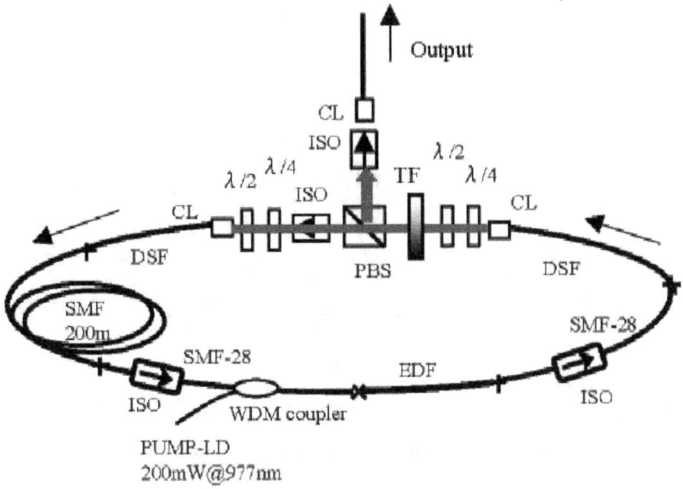

Figure 5.4. Experimental setup for low-frequency mode-locked pulses at cavity round-trip frequency.

Note: ISO — isolator, CL — collimating lens, TF — tunable filter, DSF — dispersion-shifted fiber, EDF — Er-doped fiber, WDM — wavelength division multiplexer for pump coupling, $\lambda/2$ — half-wave plate, $\lambda/4$ — quarter-wave plate [10].

So far, harmonic mode locking has been described. For low repetition rate pulses with short pulse width, mode locking at the cavity frequency (inverse of cavity round trip time) has been reported. Figure 5.4 shows the experimental setup. The Er-doped fiber ring laser has a 20 m long Er-doped fiber pumped by a 980 nm laser. The ring cavity has a dispersion-shifted fiber (DSF) and a 200 m long single-mode fiber (SMF) so that the cavity frequency is less than 1 MHz.

By adjusting the tunable filter to 1534 nm (bandwidth of 4.3 nm), four wave plates, and the pump power, stable mode-locked pulses at a repetition rate of 920 kHz are achieved at a pump power of 200 mW. The average laser output power is 4 mW. The autocorrelation trace of the pulse is shown in Figure 5.5. The Gaussian fit shows a pulse width of 815 fs. The time–bandwidth product is ~0.44.

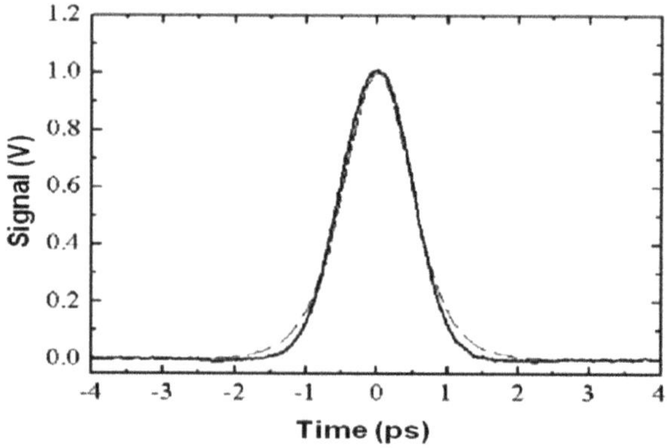

Figure 5.5. Autocorrelation trace of mode-locked pulse. The repetition rate is 920 kHz. The solid line is measured data and the dashed curve is a fit [10].

5.3 Rational Harmonic Mode Locking

A technique called rational harmonic mode locking has attracted considerable interest, where an optical pulse train with a repetition rate of $(np + 1)f_c$ can be produced when the modulation frequency is set at $(n + 1/\mathrm{p})f_c$, where f_c is the longitudinal mode spacing or fundamental cavity frequency of the fiber laser ring and n, p are integers. Several researchers have demonstrated their experimental results in actively rational harmonic mode-locked fiber lasers: Ahmed & Onodera [1] demonstrated the generation of 14 GHz and 21 GHz repetition rate optical pulses when the RF drive frequency of the modulator was ~7 GHz. Jeon *et al.* [2] observed a 15th rational harmonic mode locking when the modulation frequency f_m was ~1.5 GHz. Li *et al.* [3] showed a 19th rational harmonic active and passive mode locking in a figure-of-eight fiber laser when f_m was ~100 MHz. Das *et al.* [4] developed a theory of rational mode locking, and they also demonstrated 20th rational harmonic mode locking for the modulation frequency $f_m \approx 1.5$ MHz and 10th rational harmonic mode locking for $f_m \approx 15$ MHz. It is important to generate several GHz repetition rate pulses for optical communication studies.

Rational harmonic mode locking has been demonstrated using the laser schematic shown in Figure 5.1. The phenomena of rational harmonic mode

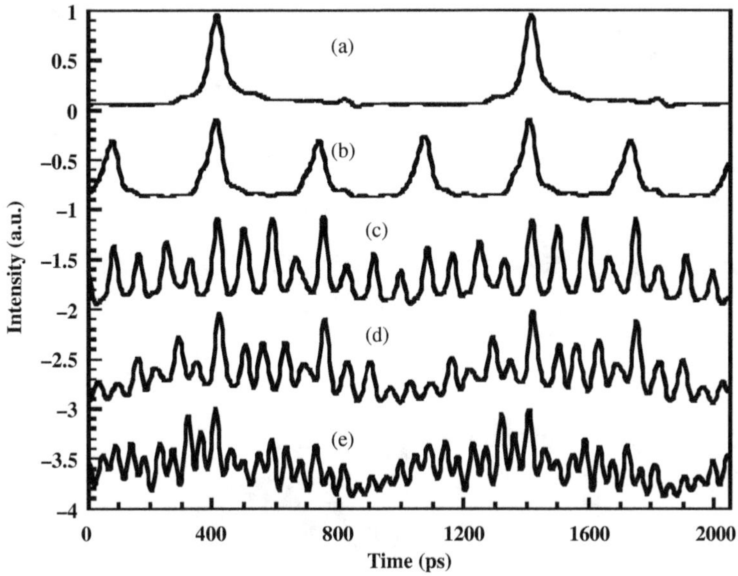

Figure 5.6. Generation of pulse train using rational harmonic mode locking around f_m = 1 GHz (observed on sampling oscilloscope). The observed pulse width is limited by the bandwidth of the oscilloscope. (a) f_m = 1001.7933 MHz = 388 f_c. Repetition rate of pulse train = 1 GHz. (b) $f_m = (388 + 1/3) f_c$. Repetition rate of pulse train = 3 GHz. (c) $f_m = (388 + 1/12) f_c$. Repetition rate of pulse train =12 GHz. (d) $f_m = (388 + 1/15) f_c$. Repetition rate of pulse train = 15 GHz. (e) $f_m = (388 + 1/22) f_c$. Repetition rate of pulse train = 22 GHz [5].

locking is observed when the modulation frequency $f_m = (n + 1/p)f_c$, where n and p are integers. Figure 5.6 shows the rational harmonic mode locking when $f_m \approx 1$ GHz, corresponding to the harmonic order $n = 388$. The 22nd order of rational harmonic mode locking was achieved in Figure 5.6.

The theoretical analysis of rational harmonic mode locking follows the analysis of harmonic mode locking. For the rational harmonic mode-locking case, the modulation frequency is $\omega_m = (n + 1/p)\omega_c$. So the modulation factor is given by $M\{1–\cos[(n + 1/p)t]\}$ and the recursion Equation (5.3) becomes

$$a_q^{(k+1)} = a_q^{(k)} + g\left[1 - \frac{(q\omega_c)^2}{\Omega_g^2}\right]a_q^{(k)} - la_q^{(k)} + \frac{M}{2}\left(a^{(k)}_{q+n+\frac{1}{p}} - 2a_q^{(k)} + a^{(k)}_{q-n-\frac{1}{p}}\right). \quad (5.9)$$

For this case, as $(n + 1/p)$ is not an integer, the modes in the cavity do not couple. But if a cavity of equivalent length pL is considered, then the modes can couple. This means that the pulse must propagate p times through the cavity before it can get through gain. Following the steps in Section 5.2, the above equations can be rewritten as

$$a_q^{(k+1)} = a_q^{(k)} + g\left[1 - \frac{\left(\frac{q}{p}\omega_c\right)^2}{\Omega_g^2}\right]a_q^{(k)} - la_q^k + \frac{M}{2}\left(a_{qp+np+1}^{(k)} - 2a_q^{(k)} + a_{qp-np-1}^{(k)}\right), \tag{5.10}$$

where $\omega_c = 2\pi f_c$. The quantity q' is redefined as

$$q' = \frac{q}{np+1}.$$

Then Equation (5.10) is written as

$$a_{q'}^{(k+1)} = a_{q'}^{(k)} + g\left[1 - \frac{\left[\left(n + \frac{1}{p}\right)q'\omega_c\right]^2}{\Omega_g^2}\right]a_{q'}^{(k)} - la_{q'}^{(k)} + \frac{M}{2}\left(a_{q'+1}^{(k)} - 2a_{q'}^{(k)} + a_{q'-1}^{(k)}\right). \tag{5.11}$$

Equation (5.11) is similar to Equation (5.4). Following the same process as in Section 5.2, it is possible to conclude that the solution for rational harmonic mode locking is also a Gaussian pulse given by Equation (5.8) and the pulse width τ is expressed as

$$\tau = \left(\frac{2g}{M}\right)^{1/4}\left(\frac{1}{\left(n + \frac{1}{p}\right)\omega_m\Omega_g}\right)^{1/2}. \tag{5.12}$$

Comparing Equation (5.11) and (5.4), it is possible to conclude that the repetition rate of the optical pulses traveling around the ring cavity with an effective cavity length pL is $(n + 1/p)\ \omega_c$, so the actual repetition

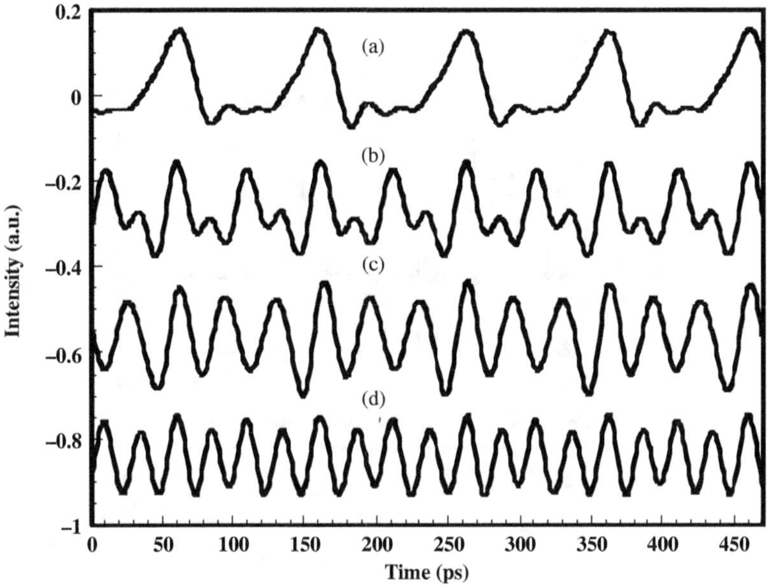

Figure 5.7. Rational harmonic mode-locking pulse trains using $f_m \approx 10$ GHz (observed on sampling oscilloscope). (a) $f_m = 9992.5129$MHz $= 3872\, f_c$. Repetition rate of pulse train = 10 GHz. (b) $f_m = (3872 + 1/2)\, f_c$. Repetition rate of pulse train = 20 GHz. (c) $f_m = (3872 + 1/3)\, f_c$. Repetition rate of pulse train = 30 GHz. (d) $f_m = (3872 + 1/4)\, f_c$. Repetition rate of pulse train = 40 GHz [5].

rate of the optical pulses traveling around the ring cavity with a real cavity length L is $(np + 1)\omega_c$, which is p times of the modulation frequency.

The pulse width parameter τ is proportional to $(n + 1/p)^{-1/2}$. Since n is much larger than 1 for the modulation frequencies used here, it is possible to conclude that the pulse width for rational harmonic mode locking is almost the same as that for harmonic mode locking. Rational harmonic pulse trains have been produced with a modulation frequency (f_m) of 10 GHz using the laser shown in Figure 5.1. Oscilloscope traces of the result (for $p = 2, 3, 4$) are shown in Figure 5.7.

Radio frequency (RF) spectrum is an important measure of frequency components in the pulse train. The RF spectrum is measured using a high-speed photodiode. Figure 5.8 shows the RF spectra of rational harmonic mode locking using $f_m = 1$ GHz; there exist some harmonics of the modulation frequency f_m, but the peak component always appears at the frequency pf_m, which is the repetition rate of the pulse train due to the rational

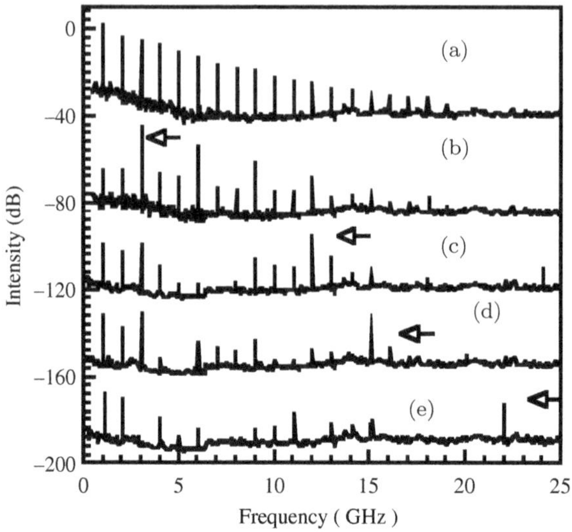

Figure 5.8. RF spectra of rational harmonic mode locking for $f_m \approx 1.0$ GHz. (a) $f_m =$ 1001.7933 MHz = 388 f_c. Repetition rate of pulse train = 1 GHz. (b) $f_m = (388 + 1/3)\, f_c$. Repetition rate of pulse train = 3 GHz. (c) $f_m = (388 + 1/12)\, f_c$. Repetition rate of pulse train = 12 GHz. (d) $f_m = (388 + 1/15)\, f_c$. Repetition rate of pulse train = 15 GHz. (e) $f_m =$ $(388 + 1/22)\, f_c$. Repetition rate of pulse train = 22 GHz [5].

harmonic mode locking. RF spectra for p values of 3, 12, 15, and 22 are shown in Figure 5.8. These p values correspond to mode locking at 3, 12, 15, and 22 GHz respectively using a modulation frequency of 1 GHz.

5.3.1 *Time domain analysis*

A theoretical analysis of harmonic and rational harmonic mode locking has been carried out in the time domain [11]. This is now discussed. The shape of the propagating pulse in the ring laser is assumed to be expressed by $f(t) = (1 + at + bt^2)\ \mathrm{Exp}(-\Gamma t^2)$, where a, b, Γ are coefficients that describe the pulse. The polynomial term represents an envelope function, and the pulse shape and pulse width are primarily determined by the exponential term. If it is assumed that before the first round trip, the pulse shape is Gaussian i.e., $f(t) = \mathrm{Exp}(-\Gamma_0 t^2)$, which implies that the initial values of a, b, Γ are $a = 0$, $b = 0$ and $\Gamma = \Gamma_0$, then, for the p-order rational harmonic mode locking, the final values of a, b, Γ are required to be the same as their initial values after p round trips. The analysis is carried out in two steps. The first step is to establish a recursion relationship between the new values and the old values of a, b, Γ after one round trip. The second step is to obtain a numerical solution by using the fact that the pulse returns to its original shape after p round trips.

Consider the modulated erbium-doped fiber laser (EDFL) as a ring composed of two components: one is the gain medium combined with the optical band pass filter and the other is the modulator. The gain curve of the gain medium is modeled as [9]

$$g(\omega) = \mathrm{Exp}\left[\alpha_m p_m \left(1 - \frac{4}{\Delta\omega_a^2}\omega^2\right)\right] = \mathrm{Exp}(\alpha_m p_m)\,\mathrm{Exp}\left(-\frac{4\alpha_m p_m}{\Delta\omega_a^2}\omega^2\right)$$
$$= g_0 \mathrm{Exp}(-\gamma\omega^2), \tag{5.13}$$

where

$$g_0 = \mathrm{Exp}(\alpha_m p_m), \tag{5.14}$$

$$\gamma = \frac{4\alpha_m p_m}{\Delta\omega_a^2}. \tag{5.15}$$

In the above, $\alpha_m \cdot p_m$ is the round-trip gain coefficient [9] and $\Delta\omega_a$ is the linewidth of the laser medium combined with the optical band pass filter inside the cavity. The pulse $f(t)$ whose shape is expressed by (5.13) travels through the gain medium:

$$f(t) = (1 + at + bt^2)\,\mathrm{Exp}(-\Gamma t^2). \tag{5.16}$$

The transmitted pulse $f_1(t)$ is obtained by Fourier transform of $f(t)$, multiplying it by $g(\omega)$, and then inverse Fourier transform back to time domain. The pulse shape $f_1(t)$ is given by [11]

$$f_1(t) = k(1 + a_1 t + b_1 t^2)\,\mathrm{Exp}(-\Gamma_1 t^2), \tag{5.17}$$

where

$$k = \frac{g_0}{\sqrt{1+4\Gamma\gamma}}\left(1 + \frac{2b\gamma}{1+4\Gamma\gamma}\right), \tag{5.18}$$

$$a_1 = \frac{a}{1+4\Gamma\gamma}\left(1+\frac{2b\gamma}{1+4\Gamma\gamma}\right)^{-1}, \tag{5.19}$$

$$b_1 = \frac{b}{(1+4\Gamma\gamma)^2}\left(1+\frac{2b\gamma}{1+4\Gamma\gamma}\right)^{-1}, \tag{5.20}$$

$$\Gamma_1 = \frac{\Gamma}{1+4\Gamma\gamma}, \tag{5.21}$$

where g_0 γ are defined by (5.14) and (5.15).

It follows from (5.15) that γ is inversely proportional to the square of the gain spectral width of the medium and the optical band pass filter and is generally ~(200 GHz)$^{-2}$. The quantity b is proportional to the square of the modulation frequency and is generally ~(10 GHz)^2, thus $b\gamma \ll 1$. Therefore, the quantities k, a_1, b_1, Γ_1 can be simplified as follows:

$$k = \frac{g_0}{\sqrt{1+4\Gamma\gamma}}, \tag{5.22}$$

$$a_1 = \frac{a}{1+4\Gamma\gamma}, \tag{5.23}$$

$$b_1 = \frac{b}{(1+4\Gamma\gamma)^2}, \tag{5.24}$$

$$\Gamma_1 = \frac{\Gamma}{1+4\Gamma\gamma}. \tag{5.25}$$

Thus after the pulse $f(t) = (1 + at + bt^2)\,\mathrm{Exp}(-\Gamma t^2)$ exits the gain medium, its mathematical form is unchanged. When the pulse arrives at the modulator, the transfer function of the modulator may be written as $T(\omega_m t + \varphi_i)$, where $T(t)$ is periodic over 2π, ω_m is the modulation frequency, and φ_i is the phase of the modulation when the pulse comes into the modulator for the ith time. Generally, the pulse width is much smaller than the modulation period, thus, it is possible to expand the transfer function $T_i(t) = T(\omega_m t + \varphi_i)$ as a Taylor series in time t around a time t_0 (with a modulator phase φ_i) when the pulse enter the modulator:

$$T_i(t) = T(\omega_m t + \varphi_i) = J_i \cdot (1 + u_i t + v_i t^2), \tag{5.26}$$

where

$$J_i = T(t)\Big|_{t=\varphi_i}, \tag{5.27}$$

$$u_i = \frac{T'\Big|_{t=\varphi_i}}{T\Big|_{t=\varphi_i}} \cdot \omega_m, \tag{5.27}$$

$$v_i = \frac{1}{2}\frac{T''\Big|_{t=\varphi_i}}{T\Big|_{t=\varphi_i}} \cdot \omega_m^2, \tag{5.28}$$

where T' and T'' denote the first-order and second-order derivatives of $T(t)$. For p-order rational harmonic mode locking, there exists p equally spaced optical pulses in one modulation period (2π). Therefore,

$$\varphi_i = \varphi_0 + \frac{2\pi}{p} i. \tag{5.29}$$

It is easy to see that after a pulse passes through the modulator, the output pulse shape is just the product of the input pulse shape and the transfer function of the modulator at phase φ_i. Now assume that for every round trip, the pulse passes through the modulator first and then the gain medium. Before it starts its ith round trip, the pulse shape has the form $f_i(t) = A_i\,(1 + a_i t + b_i t^2)\,\mathrm{Exp}(-\Gamma_i t^2)$, after this, the pulse exits the modulator, and it becomes

$$f_i'(t) = A_i(1 + a_i t + b_i t^2)\,\mathrm{Exp}(-\Gamma_i t^2) \cdot J_i \cdot (1 + u_i t + v_i t^2)$$

$$\approx A_i J_i \cdot [1 + (u_i + a_i)t + (a_i u_i + v_i + b_i)t^2]\,\mathrm{Exp}(-\Gamma_i t^2), \tag{5.30}$$

where the higher-order terms t^3 and t^4 have been neglected. After this pulse $f_i'(t)$ passes through the gain medium, its shape changes to

$$f_{i+1}(t) = A_{i+1}(1 + a_{i+1}\, t + b_{i+1} t^2)\,\mathrm{Exp}(-\Gamma_{i+1} t^2), \tag{5.31}$$

where A_{i+1}, a_{i+1}, b_{i+1} and Γ_{i+1} can be determined using Equations (5.19)–(5.22):

$$A_{i+1} = \frac{g_0}{\sqrt{1 + 4\Gamma_i \gamma}} A_i J_i, \tag{5.32}$$

$$a_{i+1} = \frac{u_i + a_i}{1 + 4\Gamma_i \gamma}, \tag{5.33}$$

$$b_{i+1} = \frac{a_i u_i + v_i + b_i}{(1 + 4\Gamma_i \gamma)^2}, \tag{5.34}$$

$$\Gamma_{i+1} = \frac{\Gamma_i}{1 + 4\Gamma_i \gamma}. \tag{5.35}$$

Thus, through Equations (5.29)–(5.32), the recursion relationship for the shape parameters after the pulse finishes one round trip is established. If it is assumed that the initial pulse takes the form $f_0(t) = \mathrm{Exp}(-\Gamma_0 t^2)$, and after p round trips, the pulse returns to its original shape for p-order rational harmonic mode locking, it implies that $f_p(t) = A_p(1 + a_p t + b_p t^2)$ $\mathrm{Exp}(-\Gamma_p t^2) = f_0(t) = \mathrm{Exp}(-\Gamma_0 t^2)$. Therefore, it follows

$$A_p = 1, \tag{5.36}$$

$$a_p = 0, \tag{5.37}$$

$$\Gamma_p - b_p = \Gamma_0. \tag{5.38}$$

Equation (5.35) makes the usual approximation that the envelope function is small compared to the exponential function. In order to calculate the pulse width, the initial phase of modulation φ_0 must be known, and this can be done by solving Equation (5.37). After obtaining φ_0, the pulse width can be calculated from Equation (5.38). If it is assumed that the gain spectral width is much larger than the pulse spectral width, then the product of Γ_i and γ is small. Under this assumption, the following expression for Γ_i a_i and b_i is obtained:

$$\Gamma_i = \Gamma_0(1 - 4i \cdot \Gamma_0 \gamma), \tag{5.39}$$

$$a_i = \sum_{j=0}^{i-1} u_j, \tag{5.40}$$

$$b_i = \sum_{j=0}^{i-1} a_j u_j + v_j, \tag{5.41}$$

where u_j and v_j are defined by (5.27) and (5.28).

Therefore, Equation (5.40) for the initial phase calculation becomes

$$a_p = \sum_{j=0}^{p-1} u_j = \omega_m \sum_{j=0}^{p-1} \frac{T'\big|_{t=\varphi_j}}{T\big|_{t=\varphi_j}} = 0. \tag{5.42}$$

And the equation for pulse width calculation (5.38) becomes

$$\Gamma_p - b_p = \Gamma_0(1 - 4p \cdot \Gamma_0 \gamma) - \sum_{j=0}^{p-1}(a_j u_j + v_j) = \Gamma_0. \tag{5.43}$$

Therefore, it follows

$$\Gamma_0 = \sqrt{\frac{-\sum_{j=0}^{p-1} a_j u_j + v_j}{4p\gamma}}. \tag{5.44}$$

The pulse width τ_p can now be determined as

$$\begin{aligned}\tau_p &= 2 \cdot \sqrt{\frac{\ln 2}{\Gamma_0}} \\ &= 4\sqrt{\ln 2} \cdot \frac{(p \cdot \alpha_m p_m)^{1/4}}{\sqrt{\omega_m \cdot \Delta\omega_a}} \left[-\sum_{i=0}^{p-1}\sum_{j=0}^{i-1} \left(\frac{T''\big|_{t=\varphi_i}}{2T\big|_{t=\varphi_i}} + \frac{T'\big|_{t=\varphi_i}}{T\big|_{t=\varphi_i}} \cdot \frac{T'\big|_{t=\varphi j}}{T\big|_{t=\varphi j}} \right) \right]^{-1/4},\end{aligned} \tag{5.45}$$

where p is the order of rational harmonic mode locking. After Γ_0 is determined from (5.44), the parameters for the pulse after i round trips can be calculated from the recursion Equations (5.32)–(5.35). Thus, the shape of the pulse train generated by a ring fiber laser can be simulated theoretically. This is important in determining the optimum bias level and modulation depth for generating equalized amplitude short pulses.

Calculation using this model shows that it is possible to get equal amplitude pulse train for $p < 5$. The equalization of the pulse train occurs only when the transmission coefficient of the modulator is the same every time the pulse arrives at the modulator. This can be illustrated by Figures 5.9 and 5.10 which show the simulated pulse trains (dashed line) together with the modulator's transfer function (solid line) for second-order and forth-order rational harmonic mode locking.

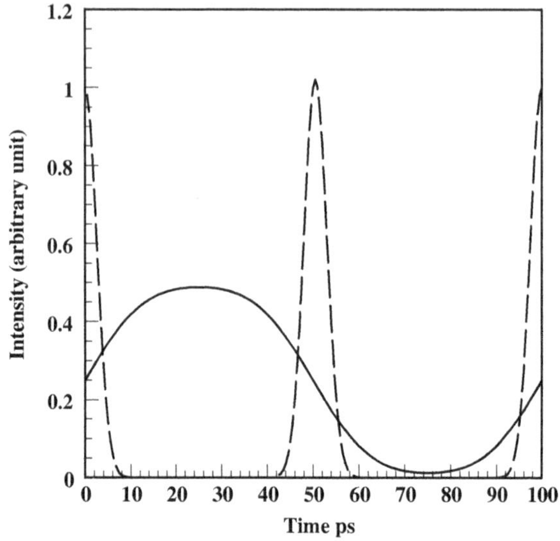

Figure 5.9. Simulated pulse train (dashed curve) for second-order ($p = 2$) rational harmonic mode locking. The RF is 10 GHz (period = 100 ps). The modulator is biased at $V_b = 0.5\ V_\pi$ and the modulation peak is $V_{\text{mod}} = 0.4\ V_\pi$. The solid curve shows the transfer function of the modulator [11].

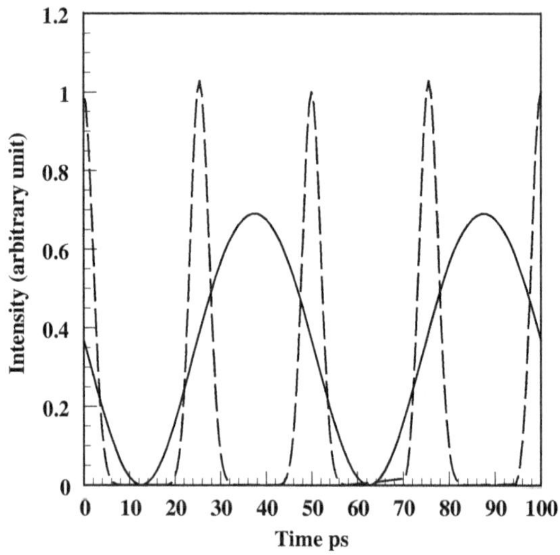

Figure 5.10. Simulated pulse train (dashed curve) for fourth-order ($p = 4$) rational harmonic mode locking. The RF is 10GHz (period = 100 ps). The modulator is biased at $V_b = V_\pi$. And the modulation peak is $V_{\text{mod}} = 0.4V_\pi$. The solid curve shows the transfer function of the modulator [11].

The experimental setup for the verification of theory is similar to Figure 5.1. A single pump EDFA is the gain medium and source of the ring laser cavity. A 90:10 splitter is placed at the output, so 10% of the total power is extracted out from the ring cavity. The center wavelength of the output was controlled by an optical filter with a 3.0 dB bandwidth of 1.0 nm. An $LiNbO_3$ electro-optic modulator with a bandwidth of 20 GHz is inserted in the cavity, which is modulated at 10 GHz by an electrical signal for mode locking.

The optical transfer function of the modulator can be expressed by

$$T(\omega_m t) = 1 + \mathrm{Cos}\left(\pi \frac{V_b + V_m \mathrm{Sin}(\omega_m t)}{V_\pi}\right), \tag{5.46}$$

where V_b is the modulator bias, V_m is the modulation amplitude, and V_π is the voltage needed to achieve π phase shift. By using the mechanism described earlier, the pulse width as a function of V_m as well as V_b has been calculated, and then by measuring the pulse width using an autocorrelator, a comparison between the theory and the experiment has been made. Figures 5.11 and 5.12 show the measured pulse width (dots) and the calculated pulse width (solid curve).

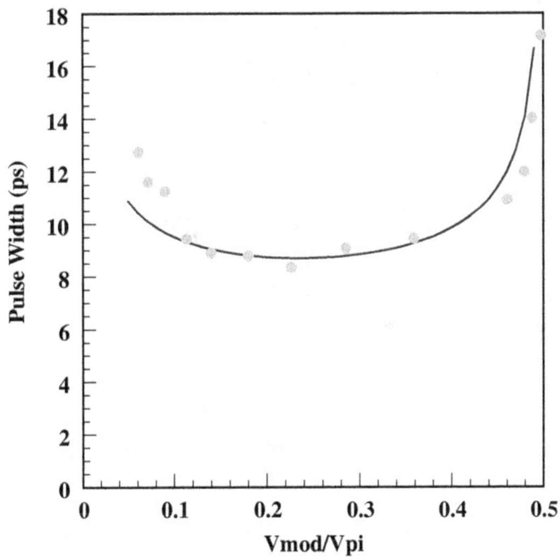

Figure 5.11. Pulse width for the case of harmonic mode locking ($p = 1$). The modulator is biased at $V_b = 0.5\ V_\pi$. The x-axis represents the modulation amplitude in a unit of V_π. The dots show the measured pulse width, and the solid curve shows the theoretical results [11].

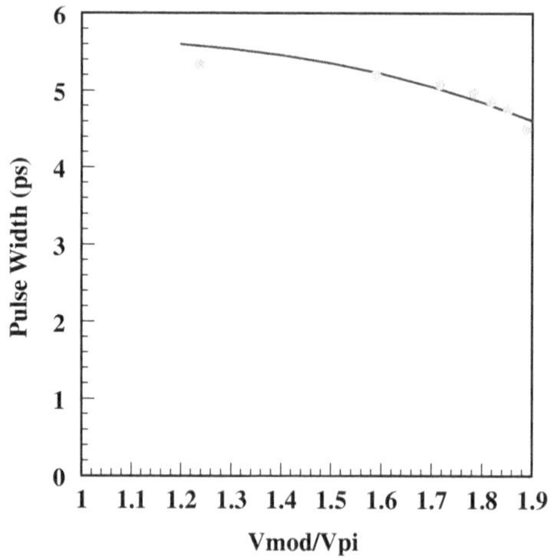

Figure 5.12. Pulse width for the case of fourth-order rational harmonic mode locking ($p = 4$) at 40 GHz. The modulator is biased at $V_b = V_\pi$. The x-axis represents the modulation amplitude in a unit of V_π. The dots show the measured pulse width, and the solid curve shows the theoretical results [11].

The pulse widths calculated using the above theory (as a function of the rational harmonic order and the optical transfer function of the modulator determined by bias and modulation voltage) agrees with the measured data. The theoretical work is based on a time domain analysis, which predicts that the pulse width decreases somewhat as the rational harmonic order goes up.

5.4 Stability of Mode-Locked Operation

The stability of mode-locked operation is important for many applications [12–18]. To commercialize the actively mode-locked fiber laser, one important thing to consider is the stabilization of the pulse train. Due to the cavity length drift caused by various environmental fluctuations (temperature and stress), without special treatment of the fiber laser, the mode-locked state cannot be sustained for very long. To overcome this problem, Shan *et al.* [12] have demonstrated a technique to stabilize the laser by dynamically adjusting the cavity length. Nakazawa *et al.* [13] have also stabilized the fiber laser with the help of a phase-locked loop based on the regenerative type of feedback. In their experiments, a portion of the output is extracted and converted to an electrical signal. This signal is transmitted through a narrow-band high Q band-pass filter, amplified, and then fed back to the RF port of the modulator and the loop is closed. The laser produced a pulse train at the center frequency of the band-pass filter. The output of the laser is at the same frequency as the RF driving frequency and therefore belongs to the harmonic mode-locking category.

Stable mode-locked operation has been demonstrated by cavity length control using a feedback loop. In the experiment, a piezoelectric controller (PZT) is used inside the optical cavity of a fiber ring laser. The ring laser has EDFA and an $LiNbO_3$ modulator for harmonic mode locking. The mode locking occurs even if there is a slight deviation between the multiples of cavity frequency and the frequency of the external modulator.

In the experiment (Figure 5.13), the cavity harmonic frequency is 10.005974 GHz, and the modulator frequency is 10.5999 GHz. So the difference between the two is 25 kHz. This difference frequency (~25 kHz) can be observed in the RF spectrum of the mode-locked pulse train.

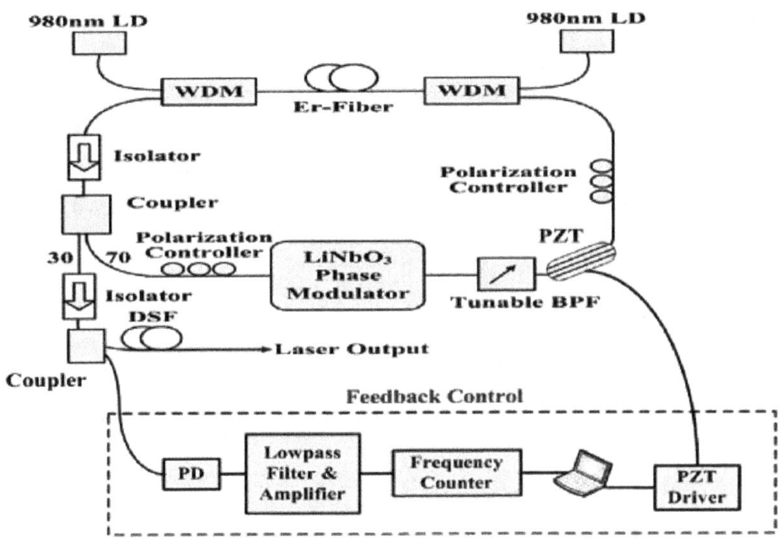

Figure 5.13. Schematic of a harmonic mode-locked laser with feedback control. Various components in the cavity are noted [15].

The difference in frequency results in a slow periodic modulation in the temporal position of the pulses. To stabilize the fiber laser, the feedback loop in the dashed region in Figure 5.13 is used. The deviation frequency is detected by the photodiode and the frequency counter and the computer sends a signal to the PZT driver to lock the deviation frequency. This is sometimes termed asynchronous mode locking.

5.4.1 *Stable mode-locked operation using PLL*

In order to extend the regenerative-type fiber laser to the rational harmonic region, a modified technique using a phase-locked loop (PLL) is used [19, 20]. When the laser pulse repetition rate is a multiple of the modulator's driving signal, the frequency of the optical pulse train is divided by the number of the rational harmonics (base-line extraction), and then the divided signal is fed back to the modulator.

This modified regenerative mode-locking technique has been used to stabilize the pulse train. In Figure 5.14, the schematics of the modified regenerative type of fiber laser is shown. A base-line extraction circuit is designed to extract the 10 GHz electrical signal from the input 40 Gb/s optical pulse train. This base-line extraction circuit is based on the mixed optical-electrical phase-locked loop. The principle can be briefly explained as follows. An offset signal Δf together with the output of the voltage-controlled oscillator (VCO) f_{vco} (~10 GHz) produces the sum frequency $\Delta f + f_{\text{vco}}$ through the single side-band modulator (SSM). This sum frequency is then sent to drive the Mach–Zehnder modulator which also takes the optical input at the rate of f_{train} (~40 GHz). Due to the high nonlinearity of the Mach–Zehnder modulator, a beat frequency at $4(\Delta f + f_{\text{vco}}) - f_{\text{train}} = 4\Delta f + (4f_{\text{vco}} - f_{\text{train}})$ is generated and is detected via the photodiode. The beat signal is then sent to the band-pass filter centered at $4\Delta f$. The phase detector takes the filtered beat signal and a reference signal oscillating at $4\Delta f$ and produces the error signal proportional to $4f_{\text{vco}} - f_{\text{train}}$. Therefore, when the loop is closed, the error is forced to be zero, and we get $4f_{\text{vco}} - f_{\text{train}} = 0$ which means the frequency of the optical pulse train is divided by 4. The operation of the modified regenerative fiber laser follows the following steps. First, the output of the integrator after the phase

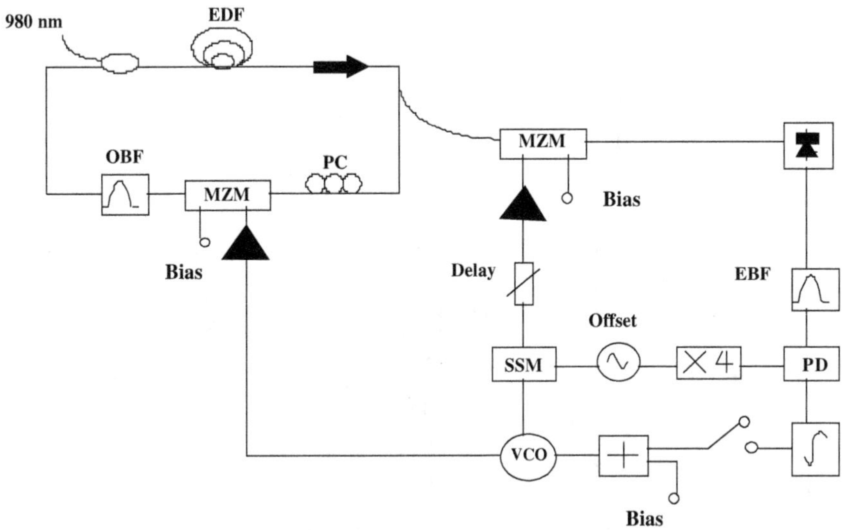

Figure 5.14. Configuration of the modified regenerative-type ring fiber laser employing the base-line extraction feedback technique. The offset frequency is Δf as described in the text.

Note: EDF: erbium-doped fiber. OBF: optical band-pass filter. MZM: Mach–Zehnder modulator. PC: polarization controller. SSM: single side-band modulator. VCO: voltage-controlled oscillator. EBF: electrical band-pass filter centered at 4 Δf. PD: phase detector. An integrator and an adder are also used in the circuits. The MZM inside the fiber laser cavity was driven at 10 GHz (VCO output) which is the fourth of the pulse train frequency at 40 Gb/s [20].

detector is disconnected from one of the input ports of the adder so that the loop is open. Then the DC bias voltage is carefully adjusted at the other input port of the adder which therefore changes the oscillating frequency of the VCO. At some VCO frequency which satisfies the requirement of the fourth-order rational harmonic mode locking, a temporally stable 40 Gb/s pulse train is obtained. At that point, the time delay inside the base-line extraction circuits is adjusted so that the output of the phase detector is minimum. Then the loop is closed by reconnecting the output of the integrator to one of the input ports of the adder. In Figure 5.15, the auto-correlator trace of the stabilized 40 Gb/s pulse train is shown. The pulse width is measured to be 2 ps (FWHM). A stable operation for many hours has been observed [20].

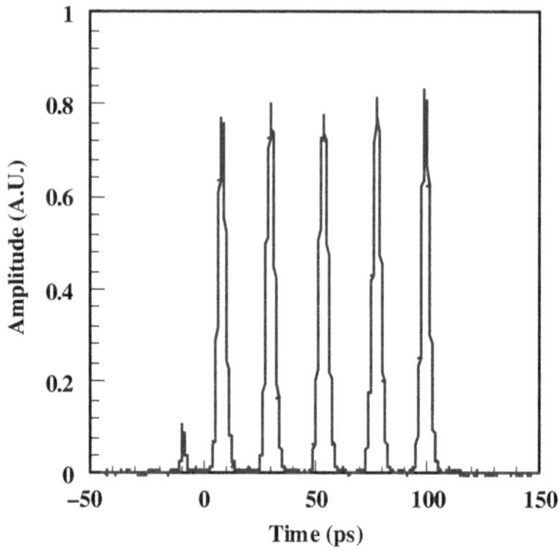

Figure 5.15. Autocorrelation trace and trace of the 40 Gb/s optical pulse train from the stabilized laser [20].

For high repetition rate pulse generation, a loop mirror consisting of several meters of polarization-maintaining fiber (PMF) can be used as a repetition rate doubler. The polarization-maintaining loop mirror has been previously used as the notch filter by Olsson, etc. [21] to filter out the original CW signal. The periodic notch effect in frequency space is caused by the phase difference between the fast and slow axes. The separation between two nearby notches is given by $2\pi c/(\Delta n\ L)$, where L is the length of the PMF, c is the velocity of the light, and Δn is the index difference between the fast and slow axes.

In Figure 5.16, the schematics and the principle of the repetition rate doubling are shown. As shown in Figure 5.17, if the notch separation is set to be the double of the mode spacing of the low-rate pulse train, some of the modes will be filtered out and thus produce a pulse train at the doubled repetition rate after passing through the loop mirror.

If the fiber length inside the loop mirror is ~9 m, the notch separation becomes 80 GHz, and therefore, for an input signal at 40 Gb/s, an output at 80 Gb/s is expected. Figure 5.17 shows the auto-correlator trace of the output 80 Gb/s pulse train for the input 40 Gb/s pulse train shown in Figure 5.15.

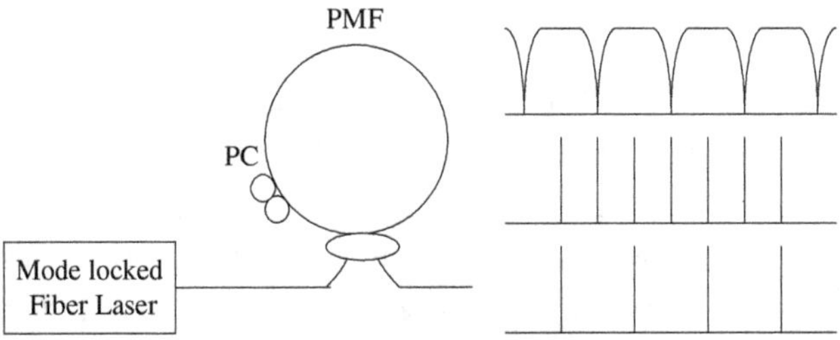

Figure 5.16. The left part is the schematic of the repetition rate doubler. The output of the mode-locked fiber laser is a pulse train at 40 Gb/s. After passing through the loop mirror, the repetition rate is doubled to 80 Gb/s. The upper trace of the right part shows the transmission of the loop mirror as a function of the wavelength. The middle trace of the right part shows the optical spectrum of the input low repetition rate pulse train. The bottom trace of the right part shows the optical spectrum of the output high repetition rate pulse train.

Note: PC: polarization controller. PMF: polarization maintained fiber [20].

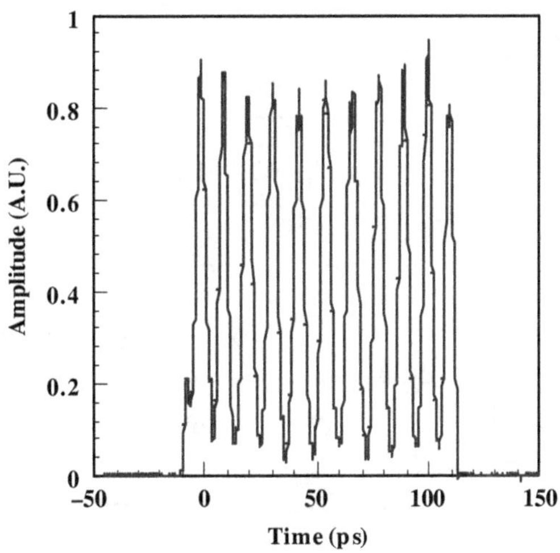

Figure 5.17. Autocorrelation trace of the 80 Gb/s optical pulse train from the stabilized laser followed by a repetition rate doubler [20].

5.4.2 *Dual wavelength mode locking*

The phase-locked loop (PLL) has been used to produce a stable dual-wavelength laser which may be important for some applications [22–28]. The schematic of the experimental setup is shown in Figure 5.18. The gain of the fiber laser is provided by an erbium-doped fiber amplifier (EDFA). The maximum output of the EDFA is 25 dBm. An $LiNbO_3$ Mach–Zehnder intensity modulator driven by the ouput of a voltage-controlled oscillator (VCO) @10 GHz is utilized for active mode locking. The VCO is part of a phase-locked loop circuit used to generate stable pulses at a precise repetition rate. The fiber laser is mode locked at 20 GHz. Tunable delay lines are used for adjusting the cavity length to recombine the separated pulses for mode-locked operation. A 70 m long dispersion compensating fiber (DCF) is used in the cavity. The dispersion parameter of the DCF is −89.6 ps/nm/km. The DCF produces a temporal separation between the pulses at two wavelengths which reduces gain competition in EDFA.

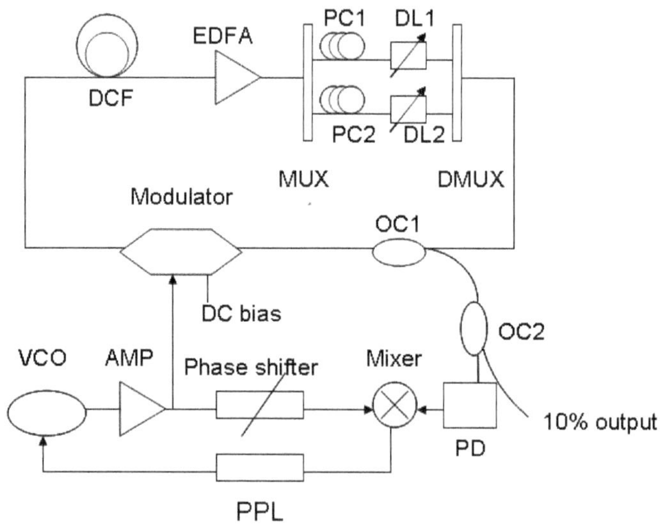

Figure 5.18. The experimental setup for the dual-wavelength fiber ring laser.

Note: EDFA: erbium-doped fiber amplifier. PC: polarization controller. DL: optical delay line. DCF: dispersion compensating fiber. MUX/DMUX: multiplexer/demultiplexer. OC: 90/10 optical coupler. VCO: voltage-controlled oscillator. AMP: 10 GHz RF amplifier. PD: photodetector. PC1, PC2: polarization controllers. PLL: phase-locked loop controller [28].

The harmonic mode locking happens at two different wavelengths when the modulation frequency f_m satisfies

$$f_{m1} = N \cdot \frac{c}{n_{\text{eff}} L_1}, \qquad f_{m2} = M \cdot \frac{c}{n_{\text{eff}} L_2}, \tag{5.47}$$

where N, M represent the harmonic number, which are integers, for two different wavelengths, c is the speed of light in vacuum, and n_{eff} is the effective refractive index of the ring cavity. Assuming two wavelengths have the same n_{eff}, L_1 and L_2 are the total lengths of the ring cavity in different wavelength channels. For simultaneous mode locking,

$$f_{m1} = f_{m2} \tag{5.48}$$

or $\Delta L = (M - N) \cdot \frac{L_1}{N}$, where $\Delta L = L_2 - L_1$. From Equation (5.47), the quantity $\frac{L_1}{N} = 50\,\text{ps}$ (the pulses are 50 ps apart), hence $\Delta L = (M - N) \cdot 50$ ps.

In high-speed optical transmission systems, the stability of the pulse trains is of major importance. Generally, for the EDF-based mode-locked multiple wavelength lasers, two main factors contribute to pulse instability: (1) the competition between different channels and (2) cavity length drift induced by the environmental temperature fluctuation.

The two pulse trains produced here at different wavelengths are separated by 25 ps after DCF, so they acquire gain at different times in the EDFA. After arrival at the MUX at different times, the two pulses combine (by adjusting the delay lines) at the output of DMUX. However, although the DCF separates the pulses at two different wavelengths, the cavity lengths may drift due to the thermal fluctuations. For this reason, a phase-locked loop (PLL) circuit is used in the cavity to maintain the mode-locking condtions. The phase-locked loop circuit includes a 10 GHz photodetector, a narrow band RF amplifier (9.0–10 GHz), a phase shifter, a 10 GHz mixer, a phase-locked loop controller, and a voltage-controlled oscillator (VCO). The VCO has a center frequency at 9.953 GHz, and the tuning range is from 5 MHz to 7 MHz. The tuning voltage is from 0 V to 12 V. The mixer detects the phase difference between the photodiode output and the VCO. The output of the mixer is sent to the PLL controller, which includes an integration circuit, a switch, and an analog adder circuit. When the switch is closed, the PLL controller will produce an output of $v_{\text{bias}} - \int v_{\text{mixer}}\, dt$. Thus the circuit continuously adjusts the modulation

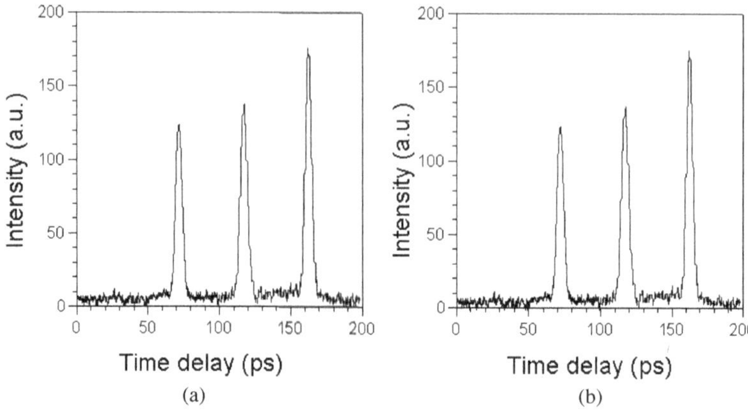

Figure 5.19. Autocorrelation trace (a) for 1546 nm wavelength and (b) for 1550 nm wavelength [28].

frequency to the mode-locking frequency and the long-term stabilization of harmonic mode locking is realized.

By carefully tuning the delay line 1, the harmonic mode locking in one channel is obtained, and then by adjusting the delay line 2, the cavity length is varied to combine the pulses and to achieve mode locking in the other channel. Then the PLL is closed to stabilize the pulses. The laser can be operated for various sets of mode-locked pulse wavelengths by using different channels of MUX/DMUX as working channels. Figure 5.19 shows the mode-locked dual wavelength laser operating at 1546 nm and 1550 nm. Since the modulator is polarization sensitive, by adjusting the polarization controller (PC) in either channel, it is possible to realize one-wavelength or two-wavelength operation. PLL is important for stable pulse output. Figures 5.20(a) and 5.20(b) show RF spectra of the harmonic mode-locked pulses measured by a spectrum analyzer with the switch of PLL open at 1546 nm and 1550 nm, respectively.

Without the PLL present, many sidebands are observed, and the separation between these sidebands is equal to the fundamental cavity frequency or the mode spacing of the fiber laser at 1.83 MHz, thus they are the super modes of the fiber laser. Figures 5.20(c) and 5.20(d) show the RF spectrum of the pulses after 1.5 hours with the switch of PLL closed at 1546 nm and 1550 nm, respectively. The super mode noise is

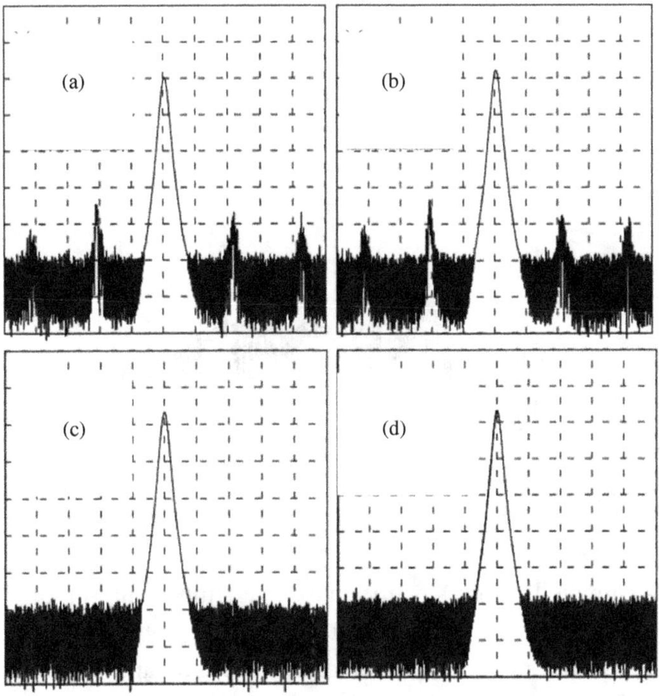

Figure 5.20. RF spectrum of pulses with and without PLL in the cavity at 1546 nm and 1550 nm. Peak center: 19.918006GHz. RES BW: 200 KHz. Super mode spacing ~1.83 MHz (a, c) — 1546 nm; (b, d) — 1550 nm [28].

suppressed. The RF spectrum is shown in Figure 5.21. The timing jitter of the pulse is calculated using the following equation [28]:

$$\text{Jitter(RMS)} = \frac{1}{2\pi f_c}\sqrt{\frac{P_n \cdot \Delta f_{\text{BW}}}{P_{\text{clk}} \cdot Rs}}, \tag{5.49}$$

where f_c is the center clock frequency, P_{clk} is the carrier frequency power, P_n is the 3 dB noise power of the shoulder, Δf_{BW} is the noise shoulder bandwidth, and R_s is the resolution of the spectrum analyzer. In this experiment, $P_{\text{clk}} = -20$ dBm = 0.01 mW, $P_n = -78$ dBm = 15.85 pW, $f_c =$ 19.9180046 GHz, $R_s = 1$ kHz, and $\Delta f_{\text{BW}} = 150$ kHz, so the estimated RMS jitter of the clock is 123 fs. The measured RMS jitter from the optical oscilloscope is 80 fs.

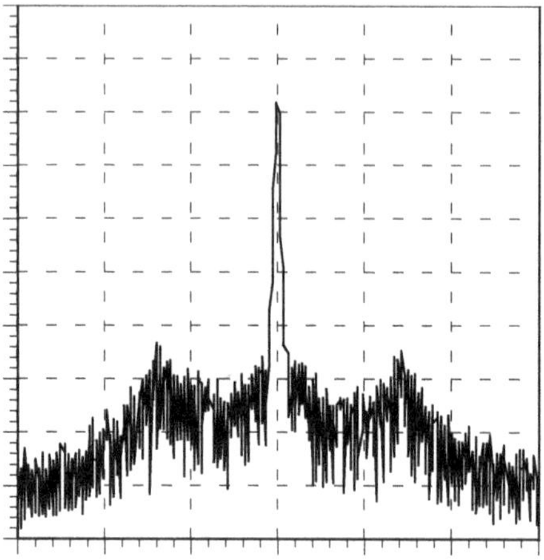

Figure 5.21. RF spectrum of pulses with PLL in the cavity at 1546 nm. Peak center: 19.9180046 GHz, RES BW 1 kHz. Span: 300 kHz [28].

Active mode locking of fiber lasers is important for applications where precise pulse rate is important. Among the applications are optical time division multiplexed communication systems and optical networking. Mode-locked diode-pumped solid-state lasers are now commercially available. They provide compact, reliable, turn-key, ultra-fast lasers. The pulse generation process is self-starting, and laser cavity adjustments are not critical. Such lasers are finding applications in materials studies, tomography studies, optical communication, and fiber Bragg grating-based sensors.

References

1. Z. Ahmed and N. Onodera, *Electron. Lett.*, **32**, (1996) 455.
2. M. Y. Jeon, H. K. Lee, J. T. Ahn, D. S. Lim, H. Y. Kim, K. H. Kim, and E.-H. Lee, *Electron. Lett.*, **34**, (1998) 182.
3. S. Li, C. Lou, and K. T. Chan, *Electron. Lett.*, **34**, (1998) 375.
4. P. Das, W. Kaechele, J. P. Theimer, and A. R. Pirich, *SPIE*, **3075**, (1997) 21.
5. C. Wu and N. K. Dutta, *IEEE J. Quantum Electron.*, **36**, (2000) 145.
6. T. Pfeiffer and G. Veith, *Electron. Lett.*, **29**, (1993) 1849.
7. M. Nakazawa, E. Yoshida, and Y. Kimury, *Electron. Lett.*, **30**, (1994) 1603.
8. E. Yoshida and M. Nakazawa, *Electron. Lett.*, **32**, 1370 (1996) 1370.

9. H. A. Haus, Short pulse generation. In I. N. Duling (Eds.), *Compact Sources of Ultrashort Pulses*, III (Cambridge University Press, Cambridge, UK, 1995).
10. M. S. Khan and N. Uehara, Conference on Lasers and Electro-optics, Pacific Rim, Seoul, Korea, 2007, Paper PDPA-4, Optica Publishing Group.
11. G. Zhu, H. Chen, and N. K. Dutta, *J. Appl. Phys.*, **90**, (2001) 2143.
12. X. Shan, D. Cleland, and A. Ellis, *Electron. Lett.*, **28**, (1992) 182.
13. M. Nakazawa and E. Yoshida, *IEEE Photon. Technol. Lett.*, **12**, (2000) 1613.
14. H. Takara, S. Kawanishi, and M. Saruwatari, *Electron. Lett.*, **31**, (1995) 292.
15. W. W. Hsiang, C. Y. Lin, N. K. Sooi, and Y. Lai, Proc. CLEO, paper OThC7 (Baltimore. MD, USA, 2005).
16. H. Chen, G. Zhu, N. K. Dutta, and K. Dreyer, *Appl. Opt.*, **41**, (2002) 3511.
17. G. Zhu, Q. Wang, H. Dong, and N. K. Dutta, *Opt. Express*, **12**, (2004) 3488.
18. H. Dong, G. Zhu, and N. K. Dutta, *Opt. Express*, **12**, (2004) 4297.
19. M. Nakazawa, E. Yoshida, and Y. Kimura, *Electron. Lett.*, **30**, (1994) 1603.
20. G. Zhu, H. Chen, and N. K. Dutta, *IEEE JQE*, **40**, (June 2004) 721; *Opt. Eng.*, **43**, (2004) 1056.
21. B. Olsson, P. Ohlen, L. Rau, and D. Blumenthal, *IEEE Photon. Tech. Lett.*, **12**, (2000) 846.
22. K. L. Lee and C. Shu, *IEEE Photon. Technol. Lett.*, **12**, (2000) 624.
23. H. Dong, G. Zhu, Q. Wang, H. Sun, and N. K. Dutta, *Opt. Express*, **12**, (2004) 4297.
24. D. Pudo and L. R. Chen, *Electron. Lett.*, **39**, (2003) 272.
25. S. Pan and C. Lou, *IEEE Photon. Technol. Lett.*, **18**, 604 (2006) 604.
26. F. Shih, C. Chen, and C. Lee, *Opt. Express*, **15**, (2007) 13844.
27. W. Zhang, J. Sun, J. Wang, and L. Liu, *IEEE Photon. Technol. Lett.*, **19**, (2007) 1418.
28. Z. Chen, S. Ma, and N. K. Dutta, *Opt. Express*, **17**, (2009) 1234.

Chapter 6

Pulse Compression Techniques and Mode Locking

6.1 Introduction

Short pulses are important for many applications including communications, materials studies, biology, and tomography. The pulses generated using a fiber laser can be compressed by propagation through various fiber types, loop mirrors, and gratings [1–14]. During the past two decades, a variety of pulse compression schemes have been proposed and demonstrated. The schemes include (i) adiabatic soliton compression techniques [2–4], (ii) multiple-step compression based on self-phase modulation in highly nonlinear fibers, noble gases, or semiconductor optical amplifiers (SOA) [5, 6], (iii) parametric gain in a fiber-optic parametric amplifier [2, 6], (iv) a comb-like profiled fiber (CPF) to emulate dispersion decreasing fiber [8, 9], and (v) saturable optical absorbers, such as nonlinear optical loop mirror (NOLM) or nonlinear amplifying loop mirror (NALM) [10–14]. The NOLM scheme can be adapted to various possible configurations of fiber types and fiber and laser designs. The pulse propagation in optical fibers is discussed in Section 6.2. Using various fiber types (i.e., with different chromatic dispersion and nonlinearity), pulse compression or pulse stretching can be done. Gratings have been used for pulse compression which is important for producing high-peak power pulses (Ref. [1], see also Section 8.6). They can also be used for pulse stretching.

6.2 Pulse Propagation in Optical Fibers

Pulse propagation in optical fibers has been extensively studied. For long lengths of transmission fiber, dispersion effects and nonlinearity play an important role in pulse propagation. For pulse compression, the appropriate choice of dispersion is important as stated in Chapter 3. Silica has inversion symmetry, thus the second-order nonlinearity $\chi^{(2)}$ in a silica fiber is 0. Thus, second harmonic generation and sum and difference frequency generation do not occur in silica fibers.

The basic equation for pulse propagation in fibers is derived in this subsection. For an analysis starting with Maxwell's equations, please see Ref. [15]. Consider the propagation of a plane light wave at frequency ω through a medium having third-order ($\chi^{(3)}$) nonlinearity. The polarization generated in the medium is given by

$$P = \varepsilon_0 \chi E + \varepsilon_0 \chi^{(3)} E^3. \tag{6.1}$$

For a light at frequency ω propagating along the z-direction with propagation constant k, the electric field is given by

$$E = E_0 \cos(\omega t - kz). \tag{6.2}$$

The polarization P is given by

$$P = \varepsilon_0 \chi E_0 \cos(\omega t - kz) + \varepsilon_0 \chi^{(3)} E_0^3 \cos^3(\omega t - kz). \tag{6.3}$$

For a plane wave given by Equation (6.2), the intensity is

$$I = \frac{1}{2} c \varepsilon_0 n_0 E_0^2, \tag{6.4}$$

where n_0 is the refractive index of the medium at low intensity. From Equation (6.3), the polarization at frequency ω is

$$P = \varepsilon_0 \left(\chi + \frac{3}{4} \chi^{(3)} E_0^2 \right) E_0 \cos(\omega t - kz). \tag{6.5}$$

Then,

$$P = \varepsilon_0 \left(\chi + \frac{3}{2} \frac{\chi^{(3)}}{c\varepsilon_0 n_0} I \right) E. \tag{6.6}$$

The polarization P and electric field are related by

$$P = \varepsilon_0 (n^2 - 1) E, \tag{6.7}$$

where n is the refractive index of the medium. From Equations (6.6) and (6.7), n^2 is given by (with $n_0^2 = 1 + \chi$)

$$n^2 = n_0^2 + \frac{3}{2} \frac{\chi^{(3)}}{c\varepsilon_0 n_0} I. \tag{6.8}$$

The last term in Equation (6.8) is usually very small, thus n can be written as

$$n \approx n_0 + n_2 I, \tag{6.9}$$

where

$$n_2 = \frac{3}{4} \frac{\chi^{(3)}}{c\varepsilon_0 n_0^2}. \tag{6.10}$$

The quantity n_2 is the nonlinear coefficient. The above equation shows that the refractive index of the medium becomes intensity dependent and the coefficient n_2 represents the strength of the intensity dependence.

For silica, $n_0 \approx 1.47$, $n_2 \approx 3.2 \times 10^{-20}$ m^2/W. For a fiber with a cross-section of 100 μm^2 and with a power of 100 mW, the change in refractive index is $\Delta n = n_2 I = 3.2 \times 10^{-11}$. For an EDFA, the cross-section is smaller by as much as a factor of 10, hence the nonlinear index is higher. For fibers, the nonlinear effect is expressed in terms of the propagation constant β. It is represented by

$$\beta_{NL} = \beta + \frac{k_0 n_2}{A_{\text{eff}}} P, \tag{6.11}$$

where $k_0 = 2\pi/\lambda_0$ and P is the power carried by the mode. The quantity A_{eff} represents the effective transverse cross-sectional area of the mode.

It is usual to describe the nonlinear characteristic of an optical fiber by the coefficient γ given by

$$\gamma = \frac{k_0 n_2}{A_{\text{eff}}}. \tag{6.12}$$

Thus, for the same input power and same wavelength, smaller values of A_{eff} lead to greater nonlinear effects in the fiber. Typically, $A_{\text{eff}} \approx 20\ \mu\text{m}^2$ for EDFA and $\gamma \approx 10\ \text{W}^{-1}\text{km}^{-1}$. The nonlinear effects are higher in high-gain EDFAs due to increase in power. The electric field in an optical pulse can be written as

$$E(x,y,z,t) = \frac{1}{2}[A(z,t)\psi(x,y)e^{i(\omega_0 t - \beta_0 z)} + cc], \tag{6.13}$$

where $A(z,t)$ represents the slowly varying complex envelope of the pulse, $\psi(x,y)$ represents the transverse electric field distribution of the mode, ω_0 represents center frequency, and β_0 represents the propagation constant at ω_0. In the presence of optical loss or gain, second-order dispersion, and third-order nonlinearity, (by using Maxwell's equations) the complex envelope $A(z,t)$ can be shown to satisfy the following equation [15]:

$$\frac{\partial A}{\partial z} + \beta_1 \frac{\partial A}{\partial t} + \frac{i\beta_2}{2}\frac{\partial^2 A}{\partial t^2} + \frac{\alpha}{2} A = i\gamma(\omega_0)|A|^2 A, \tag{6.14}$$

where

$$\beta_1 = \frac{d\beta}{d\omega}\Big|_{\omega=\omega_0} = \frac{1}{v_g}, \tag{6.15}$$

$$\beta_2 = \frac{d^2\beta}{d\omega^2}\Big|_{\omega=\omega_0} = -\frac{\lambda_0^2}{2\pi c} D, \tag{6.16}$$

where D represents the group velocity dispersion (measured in ps/nm-km). The quantity α is negative for gain. The various terms on the right-hand side of Equation (6.15) represent the attenuation/gain, group velocity, dispersion, and nonlinearity, respectively. Changing to the time frame defined by

$$T = t - z/v_g = t - \beta_1 z, \tag{6.17}$$

Equation (6.15) becomes

$$\frac{\partial A}{\partial z} + \frac{i}{2}\beta_2 \frac{\partial^2 A}{\partial T^2} + \frac{\alpha}{2} A = i\gamma |A|^2 A. \tag{6.18}$$

6.2.1 *Soliton equation*

In the absence of loss ($\alpha = 0$), the following equation is obtained from Equation (6.18):

$$\frac{\partial A}{\partial z} + \frac{i}{2}\beta_2 \frac{\partial^2 A}{\partial T^2} = i\gamma |A|^2 A. \tag{6.19}$$

The above equation has a solution (optical pulse) given by

$$A\ (z,t) = A_0 \sec h(\sigma T) e^{-igz}, \tag{6.20}$$

with

$$A_0^2 = -\frac{\beta_2}{\gamma}\sigma^2 \qquad g = -\frac{\sigma^2}{2}\beta_2. \tag{6.21}$$

In terms of pulse peak power $P_0 = A_0^2$, the pulse can be written as

$$A(z,t) = \sqrt{P_0}\ \sec h\left(\sqrt{\frac{P_0 \gamma}{|\beta_2|}} T\right) e^{-i\gamma P_0 z/2}. \tag{6.22}$$

Equation (6.22) represents a soliton and has the property that it propagates undispersed through the medium. The full width at half-maximum of the pulse envelope is given by $\tau_f = 2\tau_0$, where

$$\sec h^2 \sigma\tau_0 = \frac{1}{2},$$

which gives the FWHM τ_f:

$$\tau_f = 2\tau_0 = \frac{2}{\sigma}\ln\left(1+\sqrt{2}\right) \approx \frac{1.76}{\sigma}. \tag{6.23}$$

The peak power of the pulse is

$$P_0 = |A_0|^2 = \frac{|\beta_2|}{\gamma}\sigma^2. \tag{6.24}$$

The above equation gives the required peak power for a given pulse width (τ_f) for the formation of a soliton pulse. For a typical transmission fiber, $\tau_f = 10$ ps, $\gamma = 2.4\ \mathrm{W^{-1}km^{-1}}$ $\lambda_0 = 1550$ nm, $D = 2$ ps/nm-km, and the required peak power is $P_0 = 33$ mW. Soliton pulses are being extensively studied for application to ultra-long distance optical communication. Since optical loss is neglected for soliton formation, in actual systems, the pulses must be optically amplified at regular intervals to compensate for the loss suffered by the pulses.

Soliton transmission experiments over transoceanic distances have been carried out. It has been shown that the effects of soliton–soliton interaction and polarization mode dispersion can be reduced and error-free transmission over long distances can be obtained [16]. Figure 6.1 shows the measured waveforms before and after 10000 km transmission at 40 Gb/s. The experiment is carried out in a recirculating loop with erbium-doped fiber amplifiers for loss compensation. The result shows negligible pulse distortion.

6.3 Optical Fiber-Based Pulse Compressor

6.3.1 *Pulse compression theory*

From Equation (6.18), when the pulse width is >1 ps and the pulse peak power is not very high so that the nonlinear property of the fiber can be ignored, one can use the following simplified equation:

$$\frac{\partial A}{\partial z} + \frac{i}{2}\beta_2\frac{\partial^2 A}{\partial T^2} + \frac{\alpha}{2}A = 0. \tag{6.25}$$

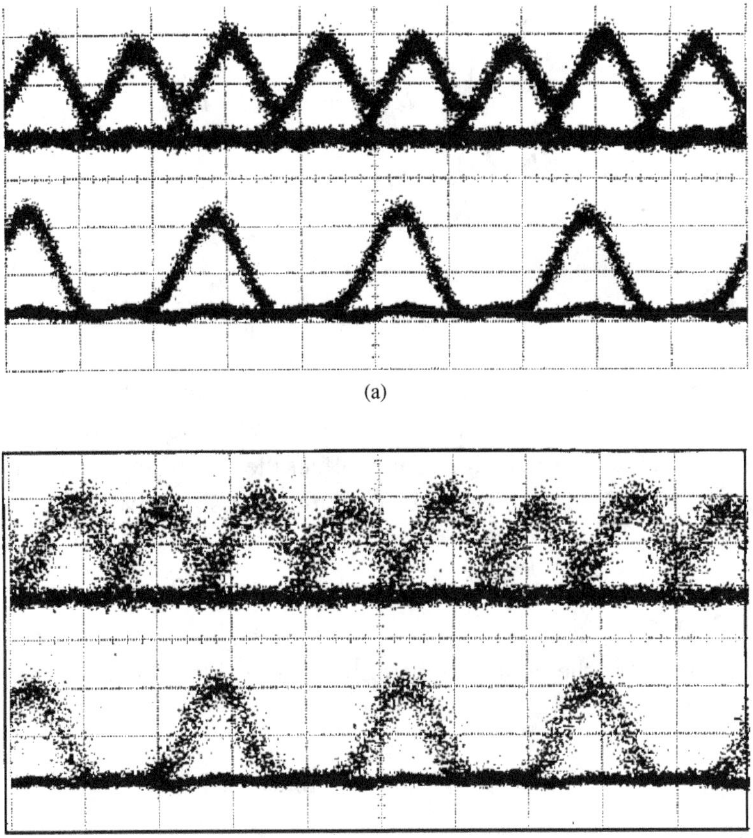

(a)

(b)

Figure 6.1. Measured waveforms at 40 Gb/s and demultiplexed waveforms at 20 Gb/s (a) before transmission i.e., at ~0 km, and (b) after 10,000 km transmission [16].

Introducing a normalized amplitude $U(z, T)$ by the following formula

$$A(z,T) = \sqrt{P_0}\exp(-\alpha z/2)U(z,T), \tag{6.26}$$

Equation (6.25) becomes

$$i\frac{\partial U}{\partial z} = \frac{1}{2}\beta_2\frac{\partial^2 U}{\partial T^2}. \tag{6.27}$$

The above equation is easily solved using Fourier transform. Using

$$U(z,T)=\frac{1}{2\pi}\int_{-\infty}^{\infty}\tilde{U}(z,\omega)\exp(-i\omega T)dT, \tag{6.28}$$

$$i\frac{\partial\tilde{U}}{\partial z}=-\frac{1}{2}\beta_2\omega^2\tilde{U}.$$

The solution is

$$\tilde{U}(z,\omega)=\tilde{U}(0,\omega)\exp\left(\frac{i}{2}\beta_2\omega^2 z\right).$$

The above equation shows that group velocity dispersion (GVD represented by β_2) changes the phase of each spectral component of the pulse. Although such a phase change cannot affect the spectrum, it can modify the pulse shape. Thus, the general solution of $U(z, T)$ is given by

$$U(z,T)=\frac{1}{2\pi}\int_{-\infty}^{\infty}\tilde{U}(0,\omega)\exp\left(\frac{i}{2}\beta_2\omega^2 z-i\omega T\right)d\omega,$$

where $U(0, \omega)$) is the Fourier transform of the incident field at $z = 0$ and is given by

$$\tilde{U}(0,\ \omega)=\int_{-\infty}^{\infty}U(0,T)\exp(i\omega T)dT.$$

For the linearly chirped Gaussian pulse, the incident field is given by [15]

$$U(0,T)=\exp\left(-\frac{1+iC}{2}\frac{T^2}{T_0^2}\right),$$

where C is a chirp parameter. $U(0, \omega)$ is obtained by Fourier transform and is given by

$$\tilde{U}(0,\ \omega)=\left(\frac{2\pi T_0^2}{1+iC}\right)^{1/2}\exp\left(-\frac{\omega^2 T_0^2}{2(1+iC)}\right).$$

The spectral width (at l/e-intensity point) from the above equation is $\Delta\omega = (1 + C^2)^{1/2}/T_0$.

If $C = 0$, the spectral width is Fourier-transform-limited and satisfies the relation $\Delta\omega T_0 = 1$. The solution for $U(z, T)$ is

$$U(z,T) = \frac{T_0}{[T_0^2 - i\beta_2 z(1+iC)]^{\frac{1}{2}}} \exp\left\{-\frac{(1+iC)T^2}{2[T_0^2 - i\beta_2 z(1+iC)]}\right\}. \tag{6.29}$$

From the above equation, the width T_1 after propagating a distance z in the fiber is related to the input pulse width (T_0) by the following equation:

$$\frac{T_1}{T_0} = \left[\left(1+\frac{C\beta_2 z}{T_0^2}\right)^2 + \left(\frac{\beta_2 z}{T_0^2}\right)^2\right]^{1/2}, \tag{6.30}$$

When $\beta_2 C > 0$, the pulse broadens monotonically with z. But if $\beta_2 C < 0$, the pulse goes through an initial narrowing. The pulse width becomes the minimum at $z_{\min}$ given by

$$z_{\min} = \frac{C}{1+C^2}\frac{T_0^2}{\beta_2},$$

and the minimum pulse width T_1, min is given by

$$T_{1,\min} = \frac{T_0}{(1+C^2)^{1/2}}.$$

The pulse width is Fourier transform limited i.e., $\Delta\omega T_{1,\min} = 1$ In general, the symbol D is used for dispersion instead of β_2 and Full Width at Half Maximum (FWHM) instead of $1/e$ widths. These two quantities are related by

$$\Delta\tau_{\text{FWHM}} = 2\sqrt{\ln 2}T_{1/e},$$
$$\Delta\omega_{\text{FWHM}} = 2\sqrt{\ln 2}\Delta\omega_{1/e},$$
$$\Delta\lambda_{\text{FWHM}} = 2\sqrt{\ln 2}\Delta\lambda_{1/e}.$$

For an experiment, the chirp is obtained from the measured spectral width ($\Delta\lambda$). The chirp parameter (C) is negative for a gain-switched laser.

Assuming $C >> 1$, the pulse width after propagation ($\Delta\tau_1$) is related to the pulse width before propagation ($\Delta\tau_0$) as follows (D replaces β_2):

$$\frac{\Delta\tau_1}{\Delta\tau_0} = \left[\left(1+\frac{D\Delta\lambda z}{\Delta\tau_0}\right)^2 + \left(2\ln 2\frac{D\lambda^2 z}{\pi c\Delta\tau_0^2}\right)^2\right]^{1/2}. \tag{6.31}$$

The above equation is stated in Section (3.3.3) and discussed for a gain-switched laser along with experimental data on pulse compression of gain-switched pulses.

6.3.2 *Pulse compression using multiple fiber types*

As stated above, the pulse compression using fibers depends on both the fiber dispersion (D) and the chirp parameter (C) of the input pulse. Often, multiple fiber types in series are used for pulse compression studies. This section describes a study.

The generation of ultrashort optical pulses at 1550 nm has attracted considerable attention for high data rate transmission systems. The combination of a gain-switched DFB laser diode and the pulse compression technique [17, 18] has been shown to be one of the promising ways to generate compact and stable femtosecond pulse sources. The pulse widths of the gain-switched laser pulses are around 10 ps after chirp compression with the dispersion compensating fiber. The compression in the dispersion-shifted fiber (DSF), which has a second-order dispersion zero wavelength around 1550 nm, requires lower energy to generate high-order soliton than that for the non-dispersion-shifted fiber. On the other hand, the compression in the standard single-mode fiber (SSF, $D = 17$ ps/nm/km at 1550 nm) produces a better pulse shape. Furthermore, the magnitude of the compression ratio, which is proportional to $1/D^{1/2}$, varies little with variation in input wavelength.

Figure 6.2 shows the schematic diagram of an experimental setup. The original pulses are generated from a 1540 nm gain-switched distributed feedback (DFB) laser at a repetition rate of 1 GHz. These pulses have a temporal width of around 40 ps and are launched into a 600 m dispersion compensated fiber (DCF) with dispersion $D = -90$ ps/km/nm. The pulse width is compressed down to 10 ps due to the linear chirp compensation

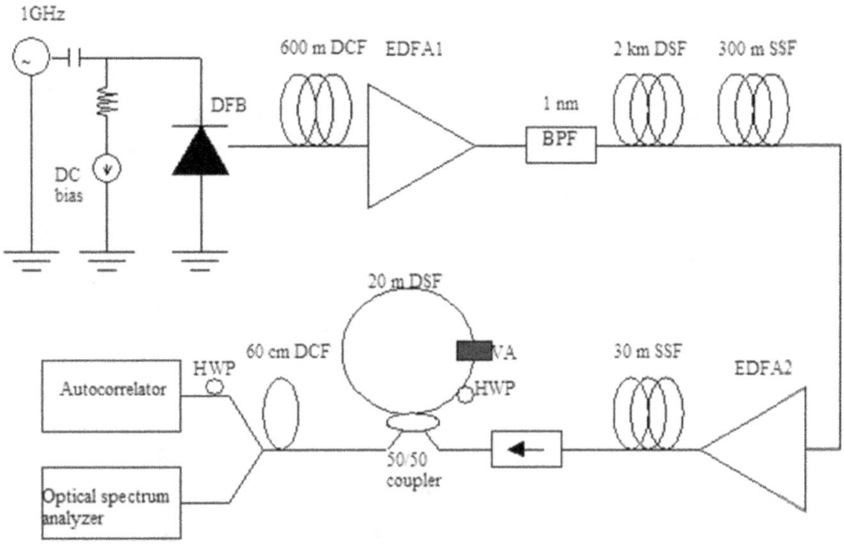

Figure 6.2. Experimental setup. DSF: dispersion-shifted fiber. SSF: standard single-mode fiber. DCF: dispersion compensated fiber. HWP: half-wave plate. VA: variable attenuator. BPF: band-pass filter.

and the time–bandwidth product of the output pulse is about 0.6. The chirp of the original pulse is controlled by varying the bias current and RF power to the DFB laser so that optimum pulse shape is obtained after the DCF.

The compressed pulses are linearly amplified in an erbium-doped fiber amplifier (EDFA) with a maximum output of 15 dBm and launched into a 2000 m dispersion-shifted fiber. The wavelength for zero dispersion and the dispersion slope for this fiber are 1552 nm and 0.09 ps/km/nm^2, respectively. Because of the self-phase modulation (SPM) in the dispersion-shifted fiber, the pulses are linearly chirped while the pulse width remains almost unchanged.

A 300 m long standard single mode fiber with dispersion of 17 ps/km/nm at 1550 nm was used to compensate for the chirp and compress the pulse. The relationship between the input pulse power and the output pulse width is shown in Figure 6.3, from which it can be seen that the pulse width is around 2 ps when the input power is 16 mW. Figure 6.4 shows the autocorrelation trace of the compressed pulse.

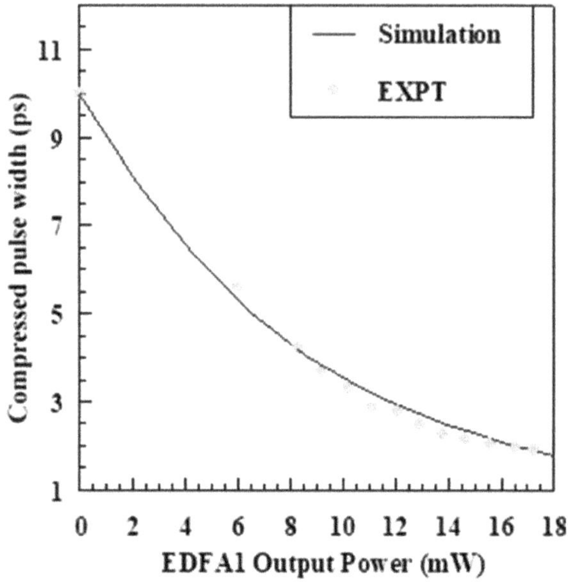

Figure 6.3. The relationship between the pulse width after compression stage 1 and the output power of EDFA1 (see Figure 6.2).

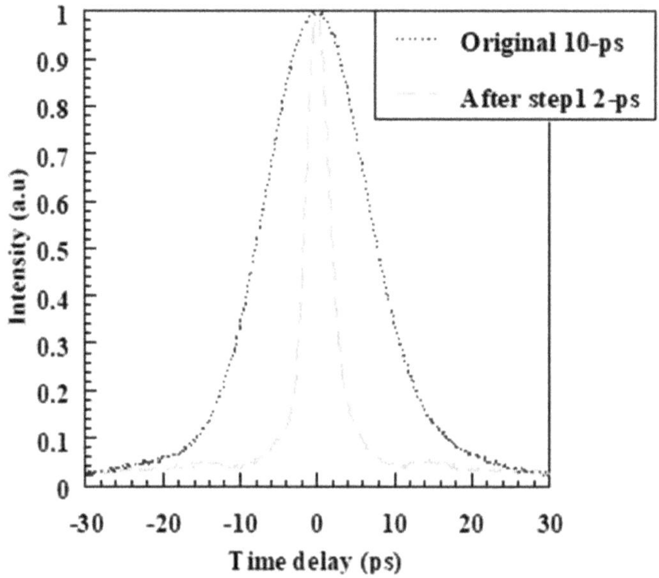

Figure 6.4. The autocorrelation traces of the initial gain-switched DFB laser pulses and the pulses after first step compression.

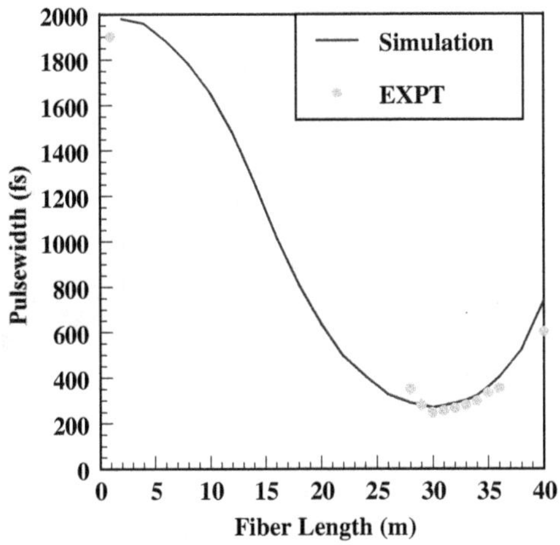

Figure 6.5. The relationship between the compressed pulse width after the second stage compression and the standard single-mode fiber length. The pulse width after the first compression stage is 1.9 ps.

These 2 ps laser pulses are used in further high-order soliton compression, which is achieved by transmission through standard single-mode fiber after the second EDFA. After amplification in the EDFA, the average output power is 22 dBm and the pulse width is not significantly changed. The measured pulse width after transmission through standard fiber (assuming sech^2 pulse shape) as a function of the fiber length is shown in Figure 6.5. The shortest-achieved pulse width is 250 fs for a fiber length of 30 m. From the initial pulse width of 2 ps, with an EDFA output power of 21.5 dBm and a repetition rate of 1 GHz, one can calculate a pulse peak power of 69 W before the compression.

The autocorrelation trace of the compressed 250 fs pulses is shown in Figure 6.6. It shows the typical wings associated with higher-order soliton compression, which degrades the pulse quality. A nonlinear optical loop mirror (NOLM) (described in detail in the following section) is designed to suppress the wings. The NOLM includes a 50/50 coupler, a polarization controller, a variable attenuator, and a 20-m-dispersion-shifted fiber. Because the pigtails of the coupler, polarization controller, and attenuator (about 5 m) are all standard single-mode fibers, the output pulses of the NOLM are much broader (FWHM: 700 fs) than the input pulses. A 60-cm-dispersion-compensated fiber is used to compensate for the

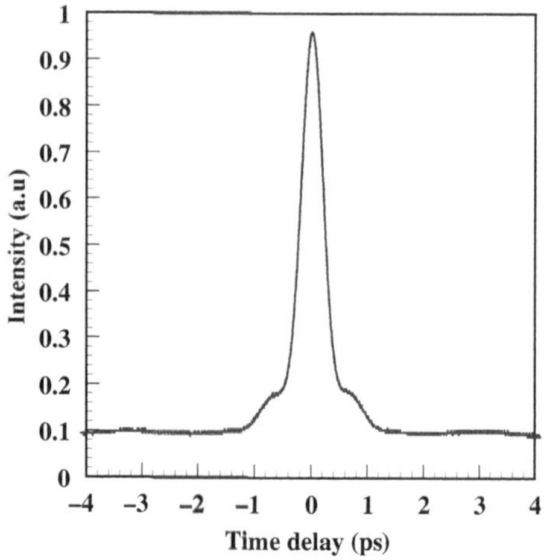

Figure 6.6. The autocorrelation trace of the 250 fs pulses after compression stage 2.

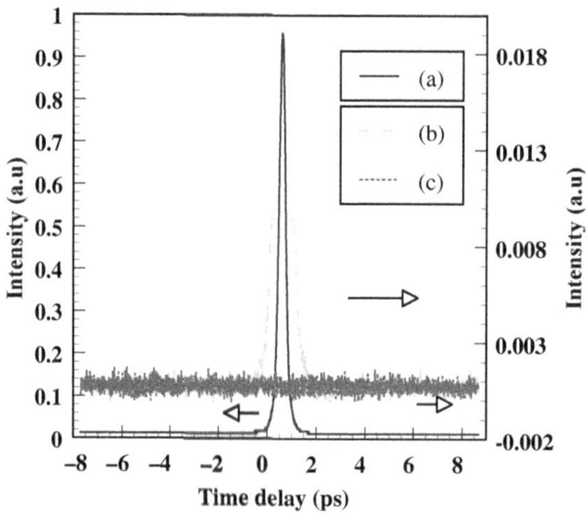

Figure 6.7. The autocorrelation trace of the compressed pulses after suppression of the pedestal with an NOLM. (a) The autocorrelation trace of the pulse, FWHM: 240 fs. (b) The autocorrelation trace of the pulse after NOLM. (c) The background of the autocorrelation trace without light.

linear chirp induced by those pigtails. Figure 6.7(b) shows the autocorrelator trace and optical spectrum of the NOLM output after chirp compensation when the attenuation in the loop is 3 dB and the output power is 11.5 dBm. These show clearly the removal of the pedestal and narrow spectral

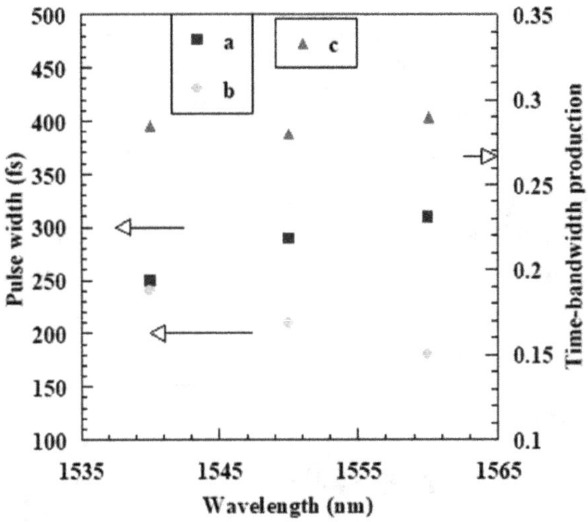

Figure 6.8. The minimum pulse width for different input wavelengths. (a) After stage 2 compression. (b) After the NOLM. (c) The time–bandwidth product of the pulse after the NOLM.

peaks, respectively. The peak-to-pedestal ratio is larger than 30 dB. The pulse width of the NOLM output after chirp compensation is 240 fs and the optical bandwidth is 9.9 nm, corresponding to a time–bandwidth of 0.285, which indicates the pulses are transform limited. Also with this same experimental setup, the compression of laser pulses with several input wavelengths around 1550 nm is examined. After the NOLM, transform-limited, pedestal-free pulses are generated. The shortest pulses (~180 fs) are obtained for an input wavelength of 1560 nm after NOLM (Figure 6.8).

The whole compression system including the NOLM is simulated by solving the nonlinear Schrodinger equation (Equation 6.14) (see also Ref. [15]) using the split-step method.

The parameters of the various fibers are listed in Table 6.1. The initial input pulses are transform-limited Gaussian pulses with an FWHM pulse width of 10 ps. The relationship between the compressed pulse width and the output power of EDFA1 in stage 1 and the relationship between the compressed pulse width and SSF length have been simulated and presented in Figures 6.3 and 6.5, respectively. Both of them show good agreement with the experimental observation. From this, it is clear that pulse widths as short as 200 fs can be achieved for laser input pulses with

Table 6.1. The parameters of fibers used in simulation.

Parameter	Definition	DSF	SSF	DDF
λ_0 (nm)	Dispersion zero wavelength	1552	1360	1550
D (ps/km/nm)	Second-order dispersion		$D(\lambda) = S(\lambda - \lambda_0)$	
S (ps/km/nm^2)	Dispersion slope at λ_0	0.09	0.09	0.0
α (dB/km)	Fiber loss	0.25	0.25	0.25
n_1	Group velocity index	1.4682	1.4682	1.4682
n_2 (m^2/W)	Nonlinear index	4×10^{-20}	2.6×10^{-20}	2.6×10^{-20}
A_{eff} (μm^2)	Effective mode area	75	75	75

wavelengths over the range of 1540–1560 nm just by varying the output power of EDFA1.

Pulses with ~200 fs pulse width were generated by using a two-stage pulse compression mechanism of gain-switched DFB laser pulses which were originally ~10 ps long. The pulse compression mechanism utilizes fiber nonlinearities and it involves propagation through specific lengths of various types of fibers and a nonlinear loop mirror. Simulation of the whole compression scheme by solving the nonlinear Schrodinger equation using the split-step method shows good agreement with the experiment results.

6.3.3 *Pulse compression using NOLM*

The following describes an optical pulse train compression scheme using a photonic crystal fiber nonlinear optical loop mirror (NOLM). A ring structure consisting of a photonic crystal fiber (PCF)-based nonlinear loop mirror (NOLM) and an EDFA is used for optical pulse compression. The structure of the loop mirror is schematically shown in Figure 6.9. Light entering the system from port 1 of the coupler is split into two ports 3 and 4 by the ratio of α and $1 - \alpha$. If the coupling ratio is close to 50%, a $\pi/2$ phase lag is applied to the light each time it is coupled across the waveguide in the coupler; in this way, the two counter-propagating lights will arrive at port 2 with a similar amplitude but π phase difference, resulting in their destructive interference and very low transmittance from port 2.

Calculations reported in [13] show that for α close to 50% and with proper polarization orientation, the NOLM reflectivity can reach higher

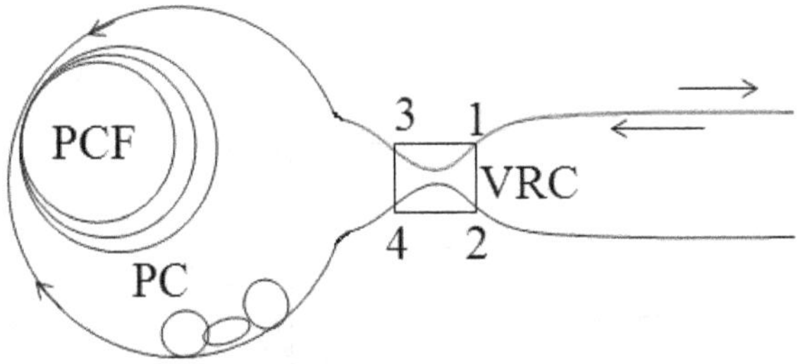

Figure 6.9. Schematic of PCF-based NOLM. VRC: variable ratio coupler (split ratio α).
Note: PCF: photonic crystal fiber. PC: polarization controller [14].

than 95%. The fiber used in the loop mirror is 25 m of highly nonlinear, dispersion-flattened photonic crystal fiber. This type of fiber has high nonlinearity and low second-order and third-order dispersions; it also has good polarization-maintaining characteristics, which is crucial for this scheme. The simulation results (obtained by solving the nonlinear Schrodinger equation [15] for pulse propagation using split-step Fourier method) show that using two cascaded NOLMs can significantly improve the result of pulse compression compared with that using a single NOLM. The impact of pulse energy on the output pulse width is shown in Figure 6.10(a). When the input pulse width is 1.8 ps, good compression results begin at input pulse energy ~1.5 pJ, for which the width of the reflected pulse from the first NOLM is 910 fs, and after the second NOLM, the pulse width reduces to 540 fs. The dependence of the output pulse width on the input pulse width is shown in Figure 6.10(b) with the input pulse energy fixed at 1.5 pJ. In both cases (Figures 6.10(a) and 6.10(b)), the 2-NOLM scheme has advantage over using the 1-NOLM scheme. It increases the amount of pulse width compression by >30%.

The schematic of the experimental setup is shown in Figure 6.11. A rational harmonic mode-locked ring laser incorporating a Mach–Zehnder-type modulator driven by a 10 GHz RF synthesizer is used to generate stable 40 GHz and 20 GHz RZ optical pulse trains.

In the cavity, a semiconductor optical amplifier (SOA) is used to serve as a fast saturable absorber in order to suppress self-pulsing behavior [16]. Without the SOA, a similar pulse is observed with less stability. The pulse autocorrelation trace of the generated pulse trains from the ring laser shows pulse widths of 4.4 ps and 4.1 ps for 20 GHz (second-order

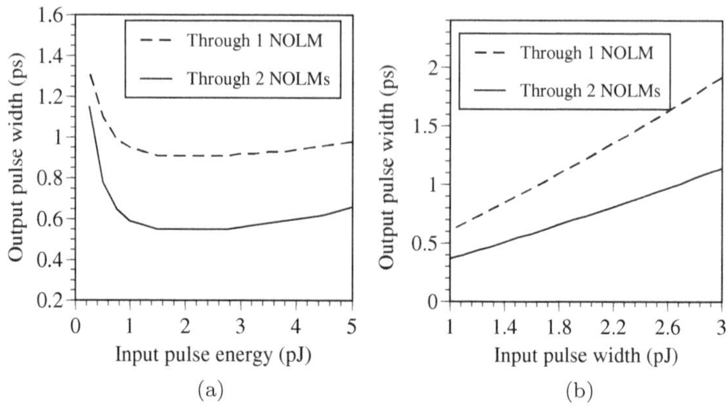

Figure 6.10. Simulated output pulse width through single or two cascaded NOLM (a) as a function of input pulse energy (with input pulse width 1.8 ps) and (b) as a function of input pulse width (with input pulse energy 1.5 pJ) [14].

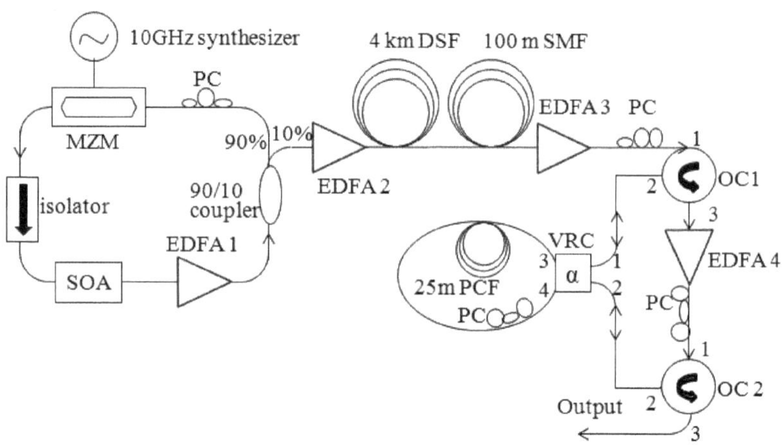

Figure 6.11. Experiment setup. MZM: Mach–Zehnder modulator, OC: optical circulator, PC: polarization controller, SOA: semiconductor optical amplifier, VRC: variable ratio coupler with coupling ratio α, DSF: dispersion-shifted fiber, PCF: photonic crystal fiber, SMF: single-mode fiber, EDFA: erbium-doped optical amplifier [14].

rational harmonic) and 40 GHz (fourth-order rational harmonic) pulses, respectively. These pulses are compressed after traversing the NOLM pulse compressor. The first external compression step is important as pulses shorter than 2 ps, in general, are needed for the NOLM to get good compression results. An EDFA is used to amplify the pulse train to an average power of ~17 dBm followed by a 4 km-long dispersion-shifted fiber (DSF), where the pulse spectrum is significantly broadened as a result of self-phase modulation. The DSF is followed by a segment of 100 m-long single-mode fiber in order to compensate for the chirp induced by the DSF. The output pulse width for 20GHz and 40 GHz pulses from the two types of fibers is measured to be around 1.8 ps. The pulse train is then guided through an optical circulator (OC) into port 1 of the variable ratio coupler. The coupling ratio and the polarization orientation in the loop mirror are carefully adjusted to make sure almost all the injected light emerges from port 1. The reflected light then goes into port 2 and out of port 3 of OC1 to enter another EDFA which amplifies the pulse train so that its average pulse energy is similar to that before it enters OC1. Through a second circulator (OC2), the light goes into the NOLM for a second time from port 2 of the variable coupler. The reflected light then exits from port 3 of OC2 as output. The autocorrelation traces of the output pulse train for 40 GHz pulses are shown in Figure 6.12. The results show good agreement with theoretical simulation, using input pulse width ~1.8 ps and input pulse energy ~1pJ. The cascaded NOLM

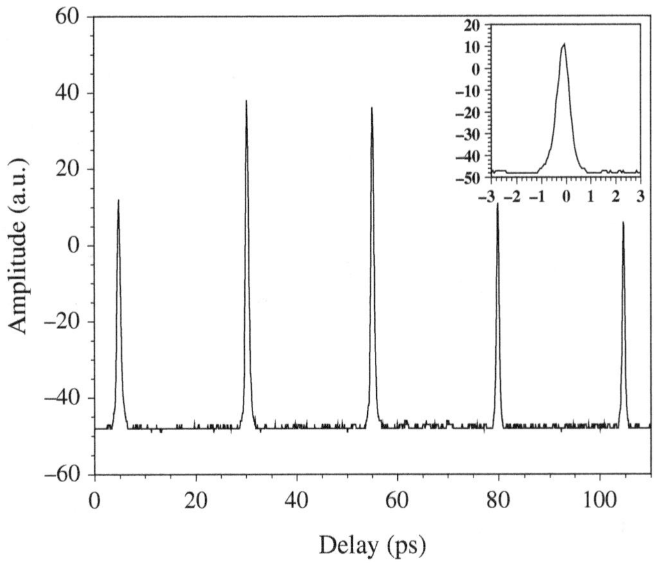

Figure 6.12. Autocorrelation trace of 40 GHz pulse train after the DSF-SMF and PCF-NOLM compressors. DSF length is ~4 km, SMF length is ~100 m, and pulse width is ~570 fs. The inset shows the autocorrelation trace of a single pulse [14].

pulse compression process shown above produces a pulse train operating at 20 GHz with full width at half maximum (FWHM) of ~610 fs and at 40 GHz with FWHM of ~570 fs.

6.4 Passive Mode-Locking Schemes

Although it is easier to generate stable mode-locked pulses at a specific repetition rate using active mode locking where the repetition rate is determined by the frequency of intracavity electrically driven optical modulators (as described in previous chapters), mode-locked pulses have been generated using other methods. These methods, generally known as passive mode-locking schemes, use saturable absorbers or comb filters along with highly nonlinear fibers [19–24]. Both saturable absorbers and highly nonlinear fibers are useful for generating shorter pulses in conjunction with the active mode-locking scheme. Examples of these are given in the following sections.

6.4.1 *Nonlinear fiber*

Generation of high repetition rate pulses using several types of intracavity optical filtering has been demonstrated. These include a Fabry–Perot filter [19], a sampled fiber Bragg grating [20], a programmable optical processor [21], and a micro-ring resonator [22, 23]. A mode-locking experiment using a silicon mirroring resonator (SMRR)-based comb filter along with a highly nonlinear fiber (HNLF) is now described. The schematic of the experimental setup is shown in Figure 6.13 [23]. The EDFA has a wide gain spectrum. The SMRR acts as a comb filter with a filter spacing of ~2 nm. The dispersive four-wave mixing in the HNLF locks the modes of the comb filter and causes passive mode locking of the laser.

The SMRR only allows the optical harmonic modes with 2 nm spacing to build up in the cavity by the filtering effect. The phases of these modes are locked by the FWM effect induced by the HNLF. The polarization-independent isolator forces the optical pulses to propagate in a single direction. A short (1 cm) section of polarization maintaining fiber (PMF) is inserted in the cavity, as shown in Figure 6.13. Silicon-on-insulator (SOI)-based SMRR with polarization-dependent grating coupler

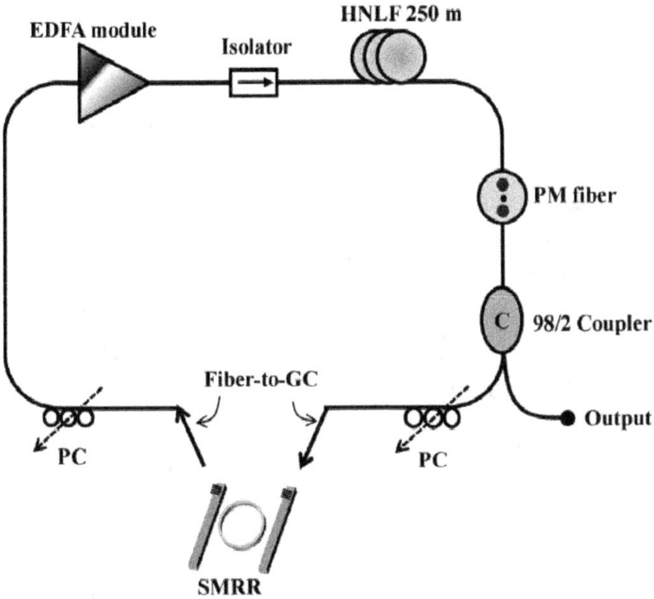

Figure 6.13. Schematic of the experimental setup [23].

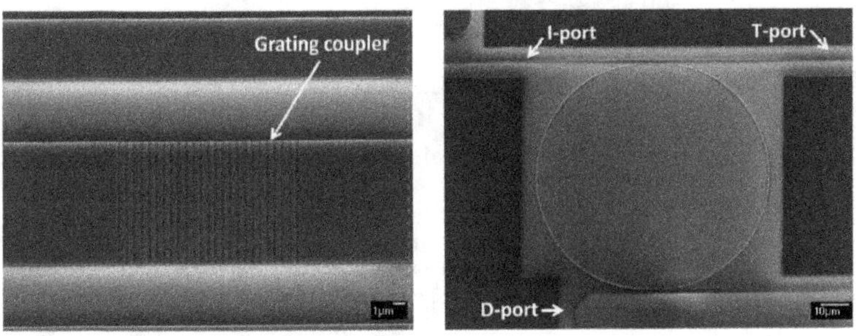

(a) (b)

Figure 6.14. SEM photograph of (a) grating coupler with 14 μm in length, 9 μm in width, and 580 nm in period; (b) silicon micro-ring resonator with two silicon nano-wires: the input port (I-port) and the drop port (D-port) [23].

(GC) is used for coupling to single-mode fiber. The scanning electron photomicrograph of the SMRR is shown in Figure 6.14.

The 250 m-long HNLF is used for enhancing the FWM effects. The group velocity dispersion of HNLF is ~0:86 ps/nm/km and the nonlinear coefficient is ~10.7 $W^{-1}km^{-1}$. To couple the light between the single mode fiber (SMF) and the SMRR effectively, the grating coupling (GC) technique is used. The GCs are 14 μm long and 9 μm wide with 580 nm period and 70 nm in depth. The birefringence introduced by the PMF with the polarization dependence of the GCs of the SMRR can act as a Lyot filter and help stabilize the laser through the combined effects of optical filtering and nonlinear spectral broadening. The EDFA output power can be measured without completing the ring cavity. When the output power is ~18 dBm, 10 comb filter modes are observed and when it is ~21 dBm, 12 comb filter modes at 2 nm spacing are observed. These modes lock due to FWM in HNLF. The laser modes and the autocorrelation trace for 12 modes are shown in Figure 6.15.

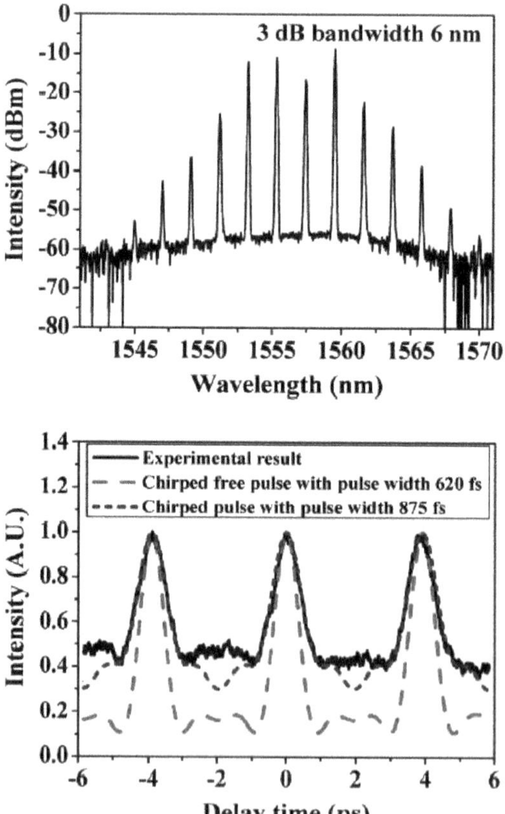

Figure 6.15. (Top) Optical spectrum of the laser output (bandwidth ~6 nm, 12 lasing modes above −65 dBm power level); (bottom) autocorrelation trace of the 250 GHz pulse train (pulse width ~875 fs) [23].

6.4.2 *Nonlinear fibers and active mode locking*

Stable optical sources that generate pulses at multiple wavelengths have attracted a lot of interest due to their potential applications in fiber-optical sensing, optical instrumentation, microwave photonic system, optical signal processing, and wavelength-division-multiplexing (WDM) transmission systems. Several researchers have demonstrated thw generation of multiwavelength lasing in an actively mode-locked ring laser (ML-RL) based on erbium-doped fiber (EDF) [25–27] and semiconductor optical amplifier (SOA) [28, 29]. However, it is difficult to achieve stable tunable pulses at a high repetition rate. The technique of nonlinear polarization rotation (NPR) as an intensity equalizer and generate stable pulses [28, 30, 31] is described in the following.

The schematic of the experimental setup is shown in Figure 6.16. The gain of the fiber laser is provided by an erbium-doped fiber amplifier (EDFA1), the length of EDFA is 13 m and the satuarated output is 20dBm. An optical isolator is inserted in the fiber ring to ensure unidirectional propagation of the laser mode in the cavity. An $LiNbO_3$ Mach–Zehnder intensity modulator driven by a synthesized microwave generator at ~10 GHz is utilized for active mode locking. Because the loss of an $LiNbO_3$ modulator is polarization sensitive, a polarization controller is used to achieve optimal modulation.

The fiber laser is mode locked at 20 GHz using the technique of rational harmonic mode locking [27]. A tunable delay line (tunability

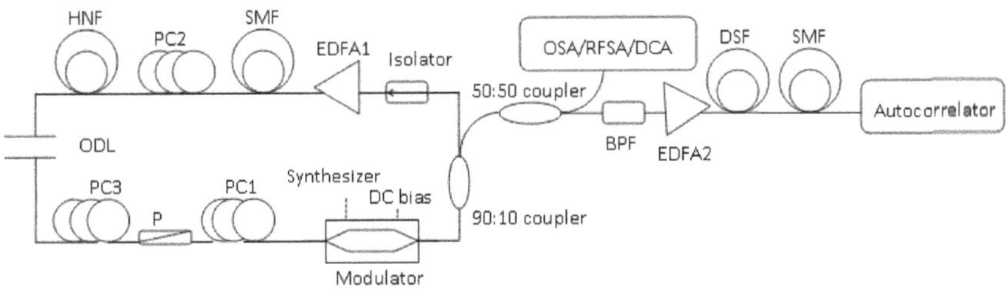

Figure 6.16. The experimental setup for the dual-wavelength laser. PC: polarization controller.

Note: P: polarizer. ODL: optical delay line. HNF: highly nonlinear fiber. SMF: single-mode fiber. DSF: dispersion shift fiber. OSA: optical spectrum analyzer. RFSA: radio frequency spectrum analyzer. DCA: digital communications analyzer [32].

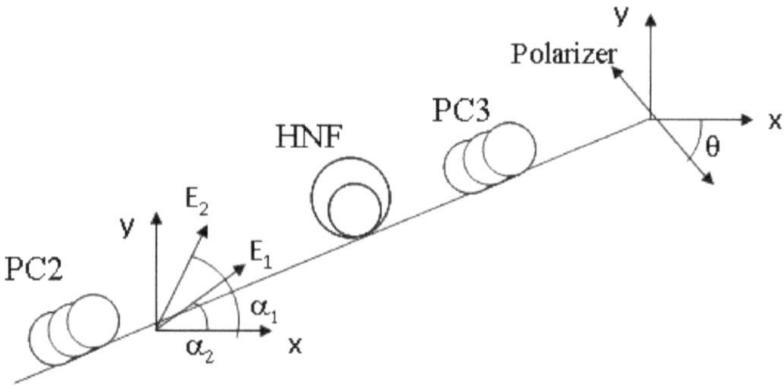

Figure 6.17. Operating principle of NPR.

Notes: *x*: fast axis of HNF, *y*: slow axis of HNF [32].

from 0.1 ns to 0.3 ns) is used for adjusting the cavity length and for wavelength tuning. The 500 m SMF introduces a dispersion in the cavity.

The 600 m HNF is used to compensate for the frequency chirp generated by the SMF. The laser output is coupled out with a 90:10 coupler. The tunable 3nm band-pass filter is used to filter out the pulse trains at different wavelengths. The DSF and SMF after the filter (outside the ring) are used to compress the pulses to a pulse width of ~1 ps. The chirp compensation mechanism is discussed in the following section.

Similar to the PC-SMF-PC-polarizer structure in [26], the PC-HNF-PC-polarizer structure can introduce the intensity-dependent loss and acts as an intensity equalizer for two wavelengths. Figure 6.17 illustrates the operating principle of NPR. The HNF dispersion is ~0 ps/nm/km at 1550nm and the dispersion is 0.019 ps/nm^{-2}/km. The birefringence Δn is 1.2×10^{-4}. The transmittivity T of this structure can be expressed as

$$T_j = \cos^2\alpha_j \cos^2\theta + \sin^2\alpha_j \sin^2\theta + \frac{1}{2}\sin 2\alpha_j \sin 2\theta \cos\left(\Delta\phi_{jL} + \Delta\phi_{jNL}\right), \tag{6.32}$$

$$\Delta\phi_{jL} = (n_{jy} - n_{jx})\beta_j L, \tag{6.33}$$

$$\Delta\phi_{jNL} = -\frac{\gamma_j}{3}[P_j L \cos 2\alpha_j + 4P_m L_c \cos 2\alpha_j], \tag{6.34}$$

where j, $m = 1$ or 2. $\beta_j = 2\pi/\lambda_j$ is the propagation constant. α_j is the angle between the fast axis of the HNF and the polarization direction of the input signal before HNF. Ej is the electric vector of the input signal. θ is the angle between the fast axis of HNF and the polarization direction of the polarizer. n_{jx} and n_{jy} are the linear birefringence coefficients. L is the length of HNF. γ_j is the nonlinear coefficient of HNF. P_j and P_m are the instantaneous power of the input signal at two wavelengths. L_c is the effective length of cross-phase modulation between the two wavelengths. The quantities α_1, α_2, and θ which determine the transmission through the PC-HNF-PC structure can be adjusted by changing the polarization controllers PC_2 and PC_3.

The γ (nonlinear coefficient) of SMF is $1 \times W^{-1}Km^{-1}$, and that of HNF is $10.5\ W^{-1}Km^{-1}$. The D (dispersion parameter) of SMF is 15 ps/km/nm at 1550 nm, and that of HNF is 0 ps/km/nm. So, it is reasonable to neglect the dispersion in HNF compared to that for SMF and neglect SPM and XPM in SMF compared to that in HNF. In HNF, the optical pulse propagation equation for the two pulses is given by [31, 33]

$$\frac{\partial A_j}{\partial z} = ie^{-\alpha_j z}\gamma_j\left(P_j\left|A_j\right|^2 + 2P_m\left|A_m\right|^2\right)A_j, \tag{6.35}$$

where $A_j(z,t)$ is the slowly varying pulse envelope, P_j is the peak power of pulses, α_j is the loss coefficient in the fiber, and z is the propagating distance. Supposing that $P_1 = P_2 = P$, $\alpha_1 = \alpha_2 = \alpha$, $\gamma_1 = \gamma_2 = \gamma$, and $|A_1| = |A_2| = |A|$, the general solution of (6.35) is

$$A(t) = A(0, t)\exp[i\phi_{NL}(t)], \tag{6.36}$$

$$\phi_{NL}(t) = 3\gamma P\,|A(0, t)|^2\,L_{eff}, \tag{6.37}$$

where $A(0, t)$ is the profile of the pulse before the pulses enter the nonlinear fiber, $L_{eff} = (1-\exp(-\alpha L))/\alpha$ is the effective length of HNF, and ϕ_{NL} is the nonlinear phase change. The SPM and XPM (cross-phase modulation)-induced spectral broadening is a consequence of the time dependence of ϕ_{NL}.

The optical frequency chirp in HNF is given by

$$\delta\omega_1(t) = -\frac{\partial \phi_{NL}}{\partial t} = -3\gamma \, \mathrm{PL}_{\mathrm{eff}} \frac{\partial |\mathrm{A}(0,t)|^2}{\partial t}. \tag{6.38}$$

For a Gaussian shape-like pulse, $A(0, t) = \exp(-t^2/(2T_0^2))$, where T_0 is the initial pulse width before the pulse enters the nonlinear fiber, Equation (6.38) can be written as

$$\delta\omega_1(t) = 6\gamma \, \mathrm{PL}_{\mathrm{eff}} \left(\frac{t}{T_0^2}\right) \exp\left(-\frac{t^2}{T_0^2}\right) \approx 6\gamma \, \mathrm{PL}_{\mathrm{eff}} \left(\frac{t}{T_0^2}\right) > 0. \tag{6.39}$$

The approximation has been made in the center of the pulse. The optical frequency chirp in the SMF is given by [30]

$$\delta\omega_2(t) = -\frac{(2L_{\mathrm{SM}}/L_D)}{1+(L_{\mathrm{SM}}/L_D)^2}\left(\frac{t}{T_0^2}\right) < 0, \quad \text{for } \beta < 0, \tag{6.40}$$

where L_{SM} is the length of SMF and $L_D = T_0^2/\beta$ is the dispersion length of SMF. When $\delta\omega_1(t) = -\delta\omega_2(t)$ after one round trip, the chirp induced by SMF is canceled by that induced by HNF, and it results in a stable pulse in the cavity. From $\delta\omega_1\,(t) = -\delta\omega_2(t)$, the stable condition can be simply written as

$$6\gamma \, \mathrm{PL}_{\mathrm{eff}} = \frac{(2L_{\mathrm{SM}}/L_D)}{1+(L_{\mathrm{SM}}/L_D)^2} \Rightarrow P \leq \frac{1}{6\gamma L_{\mathrm{eff}}}. \tag{6.41}$$

In Ref. [32], L_{SM} is 500 m, L_D is 2 km, α is ~ 0.9 dB/km, L is 0.6 km, and γ is 10.5 $\mathrm{W}^{-1}\mathrm{Km}^{-1}$, so the calculated P is 0.014 W, and for 20 GHz rate, the total average power before entering the HNF is 7 dBm. The high-order nonlinear part of HNF can also generate FWM effect and transfer energy between wavelengths. This FWM effect can make the system more stable, however its effect is small for the parameters (power and fiber lengths) used here [31].

In the experiment, the total cavity length of the fiber ring laser is 1.12 km. The laser operated at 1555.54 nm (λ_1) and 1559.94 nm (λ_2) with a longitudinal mode spacing of 118 kHz. The separation between the two wavelengths is given by [31, 33]

$$\Delta\lambda = 2/(f\, D_{\text{eff}}\, L), \tag{6.42}$$

where $D_{\text{eff}}L$ is the total dispersion in the cavity and f is the pulse repetition rate. For the parameter of this work, $\Delta\lambda \sim 4.4$ nm. The factor 2 in Equation (6.42) (which is not present in Ref. [33]) arises from second-order rational harmonic mode locking used here. The radio-frequency (RF) spectrum of the pulses shows that the side mode noise suppression ratio is higher than 45 dB. The laser is very stable and the power fluctuation is less than 1 dB for more than one hour. Simultaneous laser oscillations at more than two wavelengths (e.g., 1550, 1554, and 1559 nm) were also observed, but in such cases, the stability was not good. The dual wavelength operation was stable for the optical power levels present in the system.

Smooth wavelength tuning can be achieved by slightly varying either the frequency applied to the modulator or the cavity length without interrupting the stable pulse state. In the experiment, the repetition rate of the pulses is fixed at 20 GHz and the delay line is adjusted to achieve wavelength tuning. The two group wavelengths can be tuned from 1550.2 nm to 1563.8 nm as shown in Figure 6.18(a). The tuning range of the two wavelength groups is limited by the spectral range with high enough gain in EDFA and the transmission of the PC-HNF-PC structure. The PCs were adjusted to obtain stable pulsed operation.

A 3 nm tunable band-pass filter (BPF) is used to filter out these two wavelengths. Figure 6.18(b) shows the waveform of 20 GHz pulses right after the filter. The upper trace shows the waveform at λ_1, and the lower trace shows the waveform at λ_2. Figures 6.19(a) and 6.19(c) show the autocorrelation traces at the two wavelengths. For shorter pulses, the same experimental configuration is used. A 2 km DSF and a segment of SMF are used to compress the pulse trains after the output coupler using the nonlinear pulse compression technique described in Section 6.3.2. First, the optical bandwidth of the pulses is broadened with the SPM effect, which can be achieved by propagating the pulses through a nonlinear media such as a segment of DSF. Then the positively chirped pulses propagate in the anomalous dispersion regime of a segment of SMF. A proper length of SMF (~100 m) is chosen to achieve the anomalous dispersion compensation. The pulse widths out of the ring are ~7.1 ps and 6.5 ps (for the two wavelengths) and the 3 dB spectral bandwidth is

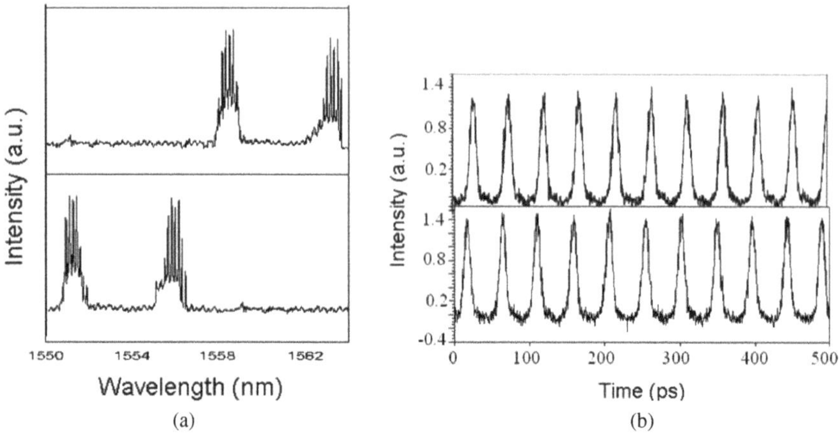

Figure 6.18. (a) An example of spectrum at two separate sets of wavelengths. (b) Trace of pulses (@20 GHz) at λ_1 and λ_2 after 3 nm band-pass filter after averaging. The two wavelength components are synchronized [32].

~0.6 nm, giving time-bandwidth products of ~0.5. This is slightly chirped for Gaussian-shaped pulses. The simulation has been carried out using a Gaussian input pulse with no linear frequency chirp. The pulse widths of the output pulses are 1.45 ps and 1.25 ps at the two wavelengths, respectively. The data are shown in Figures 6.19(b) and 6.19(d) and the data prior to compression are shown in Figures 6.19(a) and 6.19(c).

6.5 Saturable Absorber

A saturable absorber is a material whose absorption decreases as the intensity of the input light increases. The phenomenon of saturable absorption can be simply explained using a two-level system similar to that of a laser. From Equation (4.4), if the pump rate (P) is less than $N/2\tau$, the gain g is negative, and the gain saturation Equation (4.4) can be rewritten in terms of absorption ($\alpha = -g$) as

$$\alpha(I) = \alpha_0/(1 + I/Is), \tag{6.43}$$

where $\alpha_0 = (N - 2P\tau)\sigma$. Thus the simple two-level model can be used to explain the origin of saturable absorption. Two-dimensional materials

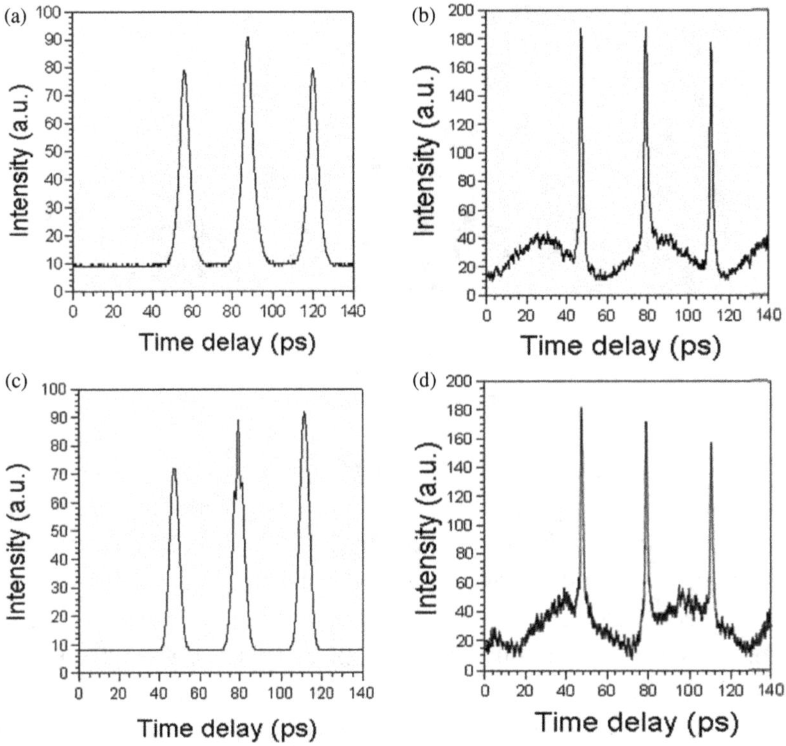

Figure 6.19. The pulses before (top left — 7.1 ps) (lower left — 6.5 ps) and after compression (top right — 1.45 ps) (lower right — 1.25 ps), at λ_1 and λ_2, respectively [32].

have been extensively studied as saturable absorbers for mode locking [34]. A general discussion on mode locking can be found in Ref. [24].

6.5.1 *Saturable absorption experiments*

Several types of saturable absorbers have been used for mode locking of fiber lasers. Over the last few years, carbon nanotubes (CNTs) have been used as saturable absorbers [35–42]. The CNT can be deposited at the end of a single-mode fiber. For film fabrication, 1.0 ± 0.2-nm diameter CNTs made by the high-pressure carbon monoxide (HiPCO) method are suspended in a UV-curable, commercial optical polymeric adhesive based on urethane. This adhesive is chosen for its transparency and good adherence

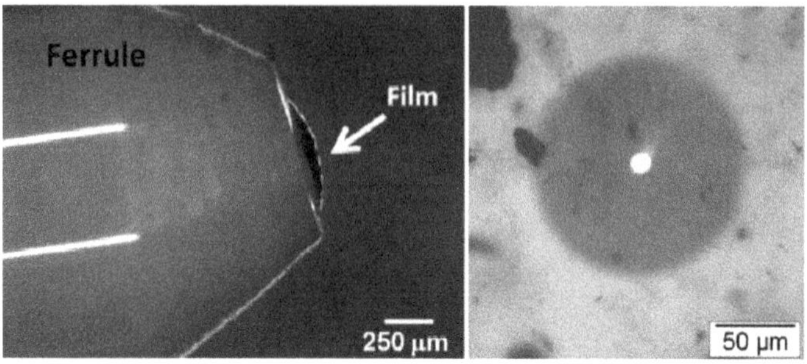

(a) (b)

Figure 6.20. Side view (a) and front view (b) of a droplet containing carbon nanotubes at the end of an FC/APC fiber connector [42].

to silica. It has a refractive index of 1.56 and a transparency of 90% at 1550 nm. For a homogeneous CNT/polymer dispersion, the samples are mixed in an ultrasonic bath for 90 minutes. The weight concentrations of CNT are in the 20–50 mg/ml range. A droplet of CNT/polymer suspension is collected with a micro-tip and transferred to the surface of the ferrule of an FC/APC fiber connector. The micro-tip has a 1 μm diameter tip and is fabricated by tapering a 1 mm diameter glass capillary. Sometimes, the CNT droplet is positioned using a micro-positioner. Figure 6.20 shows the microscope image of the side view and the front view of a CNT droplet at the end of a fiber connector (FC/APC type).

The connector with the CNT droplet is included in a standard ring fiber laser cavity. It has an Er-doped fiber, a polarization controller, an isolator, and a WDM for pump laser light coupling. The EDF is pumped by a pump laser emitting near 1480 nm to produce gain (Figure 6.21).

The pump laser power is 47 mW and the rest of the fiber laser has 8 m-long standard telecommunication fiber. The EDF length is 0.8 m. The shortest pulse width is obtained for a 20 μm thick CNT sample with a concentration of 40 mg/ml. The autocorrelation trace showing ~370 fs pulse width is shown in Figure 6.22.

Graphene is among the other saturable absorbers (SA) that have been studied [43–49]. In the experiment described in the following, a few layers of graphene deposited on a fused silica substrate are used as the saturable absorber. The performance of the SA is determined using a power-dependent transmission measurement. A mode-locked laser operating at

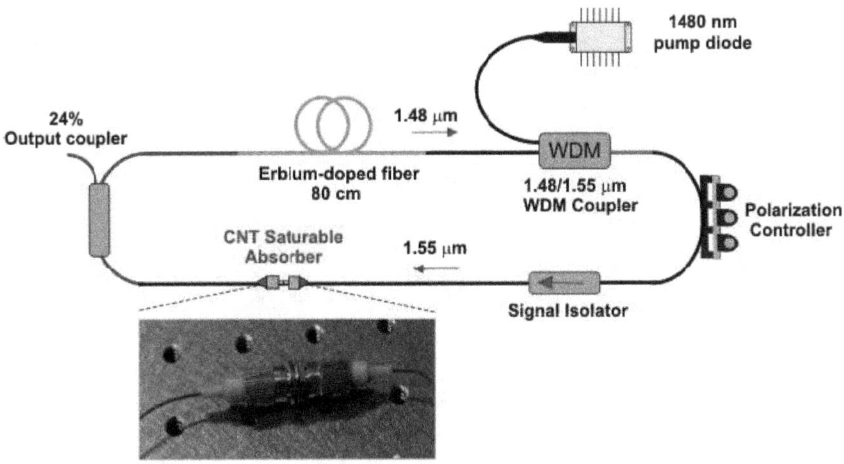

Figure 6.21. Schematic of a fiber laser setup with a photograph of the fiber connector containing CNT [42].

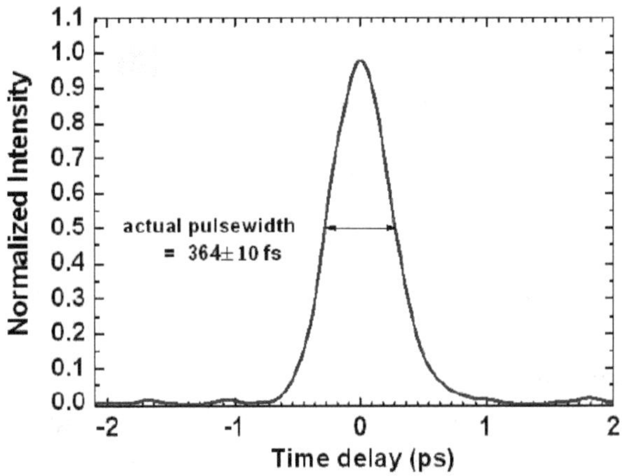

Figure 6.22. Autocorrelation trace of output pulses using carbon nanotube (CNT) saturable absorber [42].

1556 nm at a repetition rate of 169 MHz with 150 fs pulse width (1560 nm) and 3 mW average output power is used as a source for transmission experiment. The signal is transmitted through a variable optical attenuator and then to the SA. The results of the transmission measurement as a

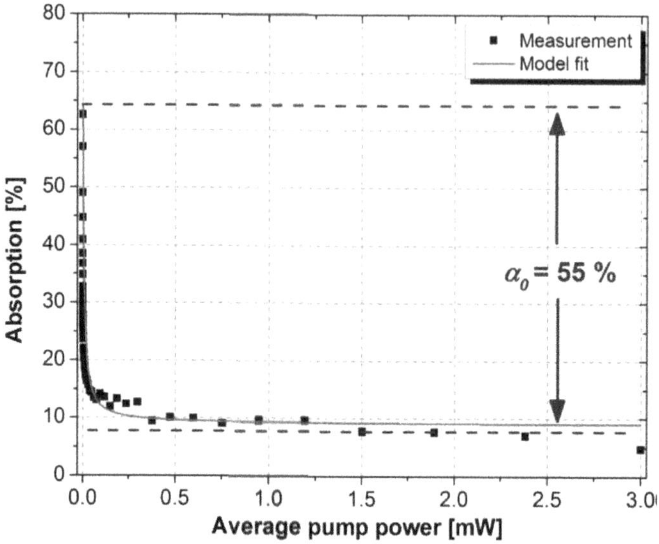

Figure 6.23. Optical transmission as a function of power of a film of grapheme [49].

function of average power are shown in Figure 6.23. The transmission data can be expressed by the equation

$$\alpha(I) = \alpha_0/(1 + I/Is) + \alpha_{ns}, \tag{6.44}$$

where $\alpha(I)$ is the power-dependent absorption coefficient, I is the intensity, I_s is the saturation intensity, and α_{ns} is the unsaturable absorption. Based on the results of Figure 6.23, a large modulation in transmission (55%) can be obtained with only 8% unsaturable absorption losses.

The mode-locked laser setup using graphene as a saturable absorber is shown in Figure 6.24. The Er fiber is 30 cm long and is pumped by a 980 nm pump laser. Two collimators on XYZ translation stages are used for transmitting the light through the grapheme film. The total optical loss between the collimators is 0.9 dB. The total length of the fiber laser loop is 3.1 m.

The threshold for laser action is 15 mW of pump power. At 60 mW of pump power, mode-locked operation is observed after adjusting the polarizer. The pulses are equally spaced by ~17 ns which corresponds to

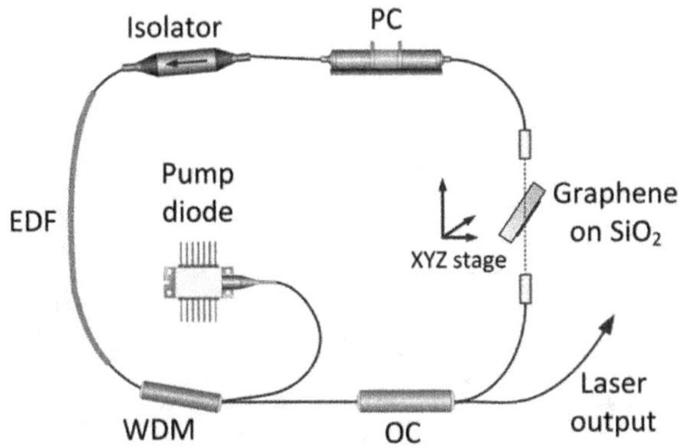

Figure 6.24. Schematic of the graphene mode-locked laser setup [49].

58 MHz repetition rate. The repetition rate agrees well with the inverse of cavity round-trip time. The autocorrelation trace of the pulses is measured and is found to be 315 fs. The pulses have 1.9 mW average power, 104 W of peak power, and pulse energy of 33 pJ. The time-bandwidth product of the pulses is 0.44.

6.5.2 *Saturable absorber theory*

In this section and in the previous section, several passive mode-locking schemes and laser systems have been described. The theory of passive mode locking using saturable absorbers has been extensively studied by Haus and others [50–54]. The passive mode locking can be separated into three categories: (i) slow saturable absorber mode locking [50, 51], (ii) fast saturable absorber mode locking [52], and (iii) soliton mode locking [53, 54]. The variation of gain, loss, and the light pulse as a function of time for the three categories are shown in Figure 6.25. For soliton mode locking, the pulse shaping is done by soliton formation i.e., the balance of group velocity dispersion and self-phase modulation produces a steady state for soliton pulse formation. In the first two cases, a short gain window is formed which defines the pulse width of a stable short pulse. For Figure 6.25(a), when the gain is higher than the loss, the pulse forms (rises), and the gain gets depleted due to upper state population depletion

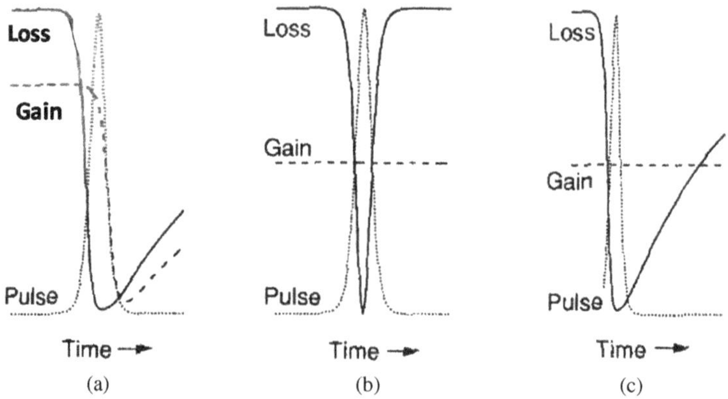

Figure 6.25. A schematic plot of pulse amplitude, loss, and gain as a function of time for three cases: (a) slow saturable absorber, (b) fast saturable absorber, and (c) soliton mode locking [54].

by the pulse, it results in the laser going below the threshold i.e., gain is less than loss, and the pulse intensity goes to zero. In Figure 6.25(a), a net-gain window is formed by the combined effect of saturation of the absorber and gain for which the absorber has to saturate and recover faster than the gain.

For rare-earth-doped lasers, the upper state lifetimes are long (~1–10 ms). This is much longer than the pulse duration which is typically less than 1 ns. As a result, there is no significant gain saturation (depletion as shown in Figure 6.25(b)). For fast saturable absorbers (absorber saturation time much shorter than the gain saturation time) shown in Figure 6.25(b), the gain is saturated to an average level determined by intracavity intensity. The short gain window is formed by the fast-recovering saturable absorber alone. The pulse formation by a fast saturable absorber is now described using the formalism developed by Haus and described in Ref. [54].

Consider the case when a light pulse of amplitude $A(t)$ travels through a gain medium, a saturable loss medium, and a constant loss medium. The cavity mirrors are included in the last term. In the frequency domain, $A(t)$ is written as $A(\omega)$, where $A(\omega)$ is the Fourier transform $A(t)$. For a homogeneously broadened gain medium described by a Lorentzian lineshape,

$$g(\omega) = \frac{g}{1+\left(\frac{\omega-\omega_0}{\Omega_g}\right)^2} \sim g\left(1-\frac{\Delta\omega^2}{\Omega_g^2}\right), \tag{6.45}$$

$$\text{for} \quad (\omega-\omega_0)^2/\Omega_g^2 \ll 1,$$

where $\Delta\omega = \omega - \omega_0$, g is the saturated gain coefficient for a cavity round trip, and Ω_g is the full width and half maximum (FWHM) of the gain bandwidth. Thus, in the frequency domain (for $g << 1$),

$$A_{\text{out}}(\omega) = e^{g(\omega)}A_{\text{in}}(\omega) \sim [1 + g(\omega)]A_{\text{in}}(\omega). \tag{6.46}$$

From the above equations,

$$A_{\text{out}}(\omega) = \left[1+g-\frac{g}{\Omega_g^2}\Delta\omega^2\right]A_{\text{in}}(\omega). \tag{6.47}$$

It can be shown that the above equation in the time domain becomes

$$A_{\text{out}}(t) = \left[1+g+\frac{g}{\Omega_g^2}\frac{\partial^2}{\partial t^2}\right]A_{\text{in}}(t). \tag{6.48}$$

Thus, the change in $A(t)$, in going through the gain medium, $\Delta A(t)$, is given by

$$\Delta A \sim \left[g + D_g\frac{\partial^2}{\partial t^2}\right]A \quad \text{with} \quad D_g = \frac{g}{\Omega_g^2}. \tag{6.49}$$

The quantity D_g is the gain dispersion. For the case of an ideal fast absorber, the loss (q) recovers instantaneously and thus it has the same time dependence as the pulse envelope. The loss q as a function of time is written as (with I as the pulse intensity and I_{sa} as the saturation intensity of the absorber)

$$q(t) = \frac{q_0}{1+I(t)/I_{sa}} \sim q_0 - \gamma I(t) \tag{6.50}$$

with $\gamma = q_0/\text{I}_s$.

Thus, the output pulse $A(t)$ from the loss modulator is given by

$$A_{\text{out}}(t) = e^{-q(t)} A_{\text{in}}(t) \sim [1 - q(t)]\, A_{\text{in}}(t). \tag{6.51}$$

The change in output $\Delta A(t)$ for a saturable absorber is

$$\Delta A(t) \sim \gamma |A|^2 A. \tag{6.52}$$

For an optical loss l, the change in $A(t)$ per pass is

$$\Delta A(t) = -lA. \tag{6.53}$$

Following Haus, for an ideal fast saturable absorber, the pulse is given by the requirement that

$$\sum \Delta A(t) = 0,$$

i.e.,

$$\left[g\left(1 + \frac{1}{\Omega_g^2}\frac{\partial^2}{\partial t^2}\right) - l + \gamma |A|^2 \right] A(t) = 0. \tag{6.54}$$

The above equation is obtained by substituting for $\Delta A(t)$ from previous equations for $\Delta A(t)$ for gain, saturable absorber, and optical loss. The solution of the above is

$$I(t) = I_0 \operatorname{sech}^2 (t/\tau) \tag{6.55}$$

with an FWHM pulse width of

$$\tau_p = 1.76\tau = 1.76 \frac{4D_g}{\gamma F_{\text{pa}}}, \tag{6.56}$$

where I_0 is the peak intensity and F_{pa} is the incident pulse energy (fluence) on the fast saturable absorber. An analytic solution is obtained, if it is assumed that the saturable absorber saturates linearly with pulse intensity over full modulation depth. In this case, $q_0 = \gamma I_0$ and for a soliton-shaped pulse, the fluence

$$F_{\mathrm{pa}} = \int I(t)\, dt = 2I_0\tau. \tag{6.57}$$

The minimum pulse width for a fully saturated absorber is then given by

$$\tau_{p,\min} = \frac{1.76}{\Omega_g}\sqrt{\frac{2g}{q_0}}. \tag{6.58}$$

6.5.3 *Saturable absorbers and active mode locking*

For high repetition rate mode locking, it is important to have a modulator at the appropriate wavelength, and all of the techniques (harmonic and rational harmonic mode locking and pulse stabilization) described earlier can be used to produce stable high repetition rate pulses. The availability of $LiNbO_3$ modulator for near 1500 nm wavelength allowed extensive study of Er-doped mode-locked lasers which is important for optical fiber communication studies. For other laser systems, short mode-locked pulses are generally produced using saturable absorbers as described previously. For Er-doped fiber lasers, the use of a saturable absorber in addition to an intracavity modulator (such as an $LiNbO_3$ modulator) can result in the generation of shorter optical pulses with high repetition rates than with intracavity modulation alone. These results are described in this section.

Since the initial report by Bao *et al.* [55], mode-locked fiber lasers based on graphene-saturable absorbers have attracted much attention due to their outstanding properties including ultrafast recovery time, wide range wavelength tunability, low saturation power, large modulation depth, and high damage threshold [55–63]. Despite the fact that those graphene-based lasers can produce ultrashort femtosecond pulses, the passive mode-locking scheme suffers from the drawback of low repetition rate, which is generally at a few MHz level.

Active mode-locking technique has been proven to be an effective way to acquire high-repetition-rate pulses by incorporating a modulator inside the laser cavity. In particular, the implementation of rational harmonic mode locking is able to overcome the modulator bandwidth limitation and further increase the repetition rate by detuning the modulation frequency a fraction of the cavity fundamental frequency from the harmonic frequency (Chapter 3). However, the pulse duration is usually larger than that of a passively mode-locked laser. A possible solution would be to build a hybrid mode-locked system incorporating both active and passive mode lockers.

A scheme of hybrid actively mode-locked erbium-doped fiber ring laser using graphene as a saturable absorber is discussed here. A 20-GHz optical pulse train is generated by applying the rational harmonic mode-locking technique, and with the insertion of a graphene-saturable absorber, the pulse width can be shortened by a compression factor of 0.53 from 5.28 ps to 2.82 ps.

In order to prepare the saturable absorber (SA), various methods have been described, such as chemical vapor deposition [55], mechanical exfoliation [56], and optical deposition [63]. A critical issue is that the properties of the SA rely on the thickness of the graphene layers, and self-aggregation of the particles would increase the absorption loss while decreasing the modulation depth [55]. A simple method has also been used to fabricate the graphene SA with desired features. The graphene material used is purchased from Graphene Laboratories Inc. The 12-nm-thick few-layer graphene flakes are placed on a piece of scotch tape and pressed repeatedly to adjust the thickness. Afterwards, the graphene nano-powder is directly deposited onto the previously cleaned end-face of a standard FC/APC fiber patch cord connector by pressing the ferrule to the tape. The patch cord is then tightly connected with another surface-cleaned patch cord using an FC/APC adaptor.

Figure 6.26 shows the schematic diagram of the fiber ring laser. A 23-m-long erbium-doped fiber (EDF) pumped by a 980-nm laser diode serves as the gain medium. An $LiNbO_3$ Mach–Zehnder (MZ) modulator driven by ~10 GHz radio frequency (RF) signal and a graphene SA are incorporated in the ring cavity simultaneously to realize hybrid mode locking. An isolator is employed to ensure unidirectional operation. With the aid of an intracavity polarization controller (PC), a stable pulse train can be generated and the laser output is extracted using a 90/10 optical

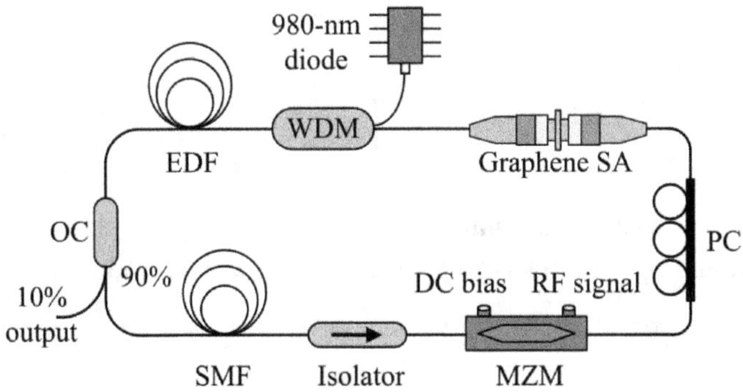

Figure 6.26. Schematic of the hybrid mode-locked fiber ring laser.

Notes: EDF: Er-doped fiber. MZM: Mach–Zehnder modulator. PC: polarization controller. SA: saturable absorber. SMF: single-mode fiber. OC: optical coupler. WDM: wavelength-division multiplexer.

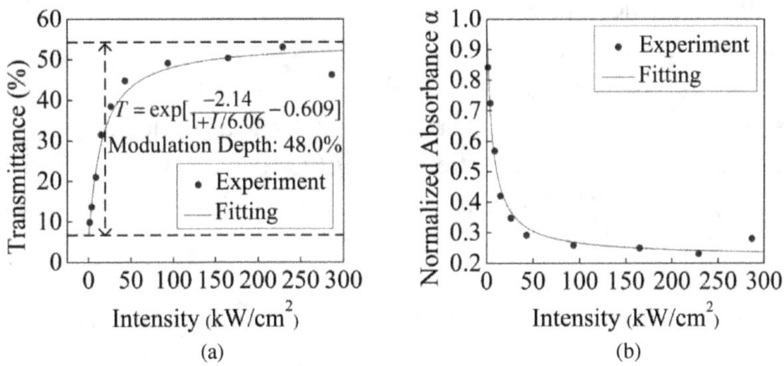

Figure 6.27. (a) Transmittance and (b) normalized absorption of the graphene SA as a function of input intensity [64].

coupler. All the components are connected by standard single-mode fibers (SMFs) with a total length of ~30 m.

To characterize the prepared graphene saturable absorber, the intensity-dependent absorption is measured as illustrated in Figure 6.27. The optical transmittance of the absorber is correlated with the linear absorption α_{lin} and nonlinear absorption $\alpha_{non}(I)$ expressed as

$$T(I) = \exp[-(\alpha_{non}(I) + \alpha_{lin})]. \tag{6.59}$$

The saturable absorbance $\alpha_{non}(I)$ is given by

$$\alpha_{non}(I) = \alpha_0 \left(1 + \frac{I}{I_{sat}}\right)^{-1}, \tag{6.60}$$

where I_{sat} is the saturation intensity. The fitting curve yields $\alpha_{lin} = 0.609$, $\alpha_0 = 2.14$, and $I_{sat} = 6.06$ kW/cm^2 respectively, with a modulation depth of 48.0%.

6.5.3.1 *Rational harmonic mode locking with graphene*

A 20-GHz pulse train is first obtained from the ring cavity shown in Figure 6.26 with the graphene-saturable absorber removed. The fiber laser with an MZ modulator incorporated is actively mode locked by carefully tuning the frequency of the RF signal and the polarization controller. When the fundamental frequency of the cavity is f_c and the modulation frequency f_m satisfies the condition $f_m = (n + 1/p) f_c$, where n and p are both integers, the laser resonator operates in the rational harmonic mode-locking (RHML) regime and pulses with a repetition rate of $(np + 1) f_c$ can be produced. In our case, f_c is 4.29 MHz, and f_m is set at $(2329 + 1/2) f_c \approx 9.99$ GHz to realize second-order RHML. Figure 6.28 shows the measured characteristics of the generated pulse train. The average output power is 1.7 dBm (1.5 mW). As can be observed from the oscilloscope trace, the spacing between adjacent pulses is 50 ps, which corresponds to 20 GHz repetition rate.

This is also confirmed by the recorded RF spectrum shown in Figure 6.28(b), where the peak with the largest amplitude is located at 19.98 GHz frequency. The result also shows low supermode noise with a signal-to-noise ratio of 25 dB. Figure 6.28(c) shows the optical spectrum centered at 1561 nm, with a 3-dB bandwidth of 0.80 nm. The autocorrelation trace of the pulse is presented in Figure 6.28(d). Assuming a hyperbolic secant pulse profile, the full width at half maximum (FWHM) is measured to be 5.28 ps. It turns out the time–bandwidth product (TBP) is equal to 0.520, not close to the transform-limit value for sech^2 pulses (0.315). This implies that the dispersion and nonlinearity of the fibers induce chirp in the pulses, which can be compensated by guiding the output through an external dispersion-compensation scheme as previously demonstrated in [65] or using grating-based compressors. Here we focus

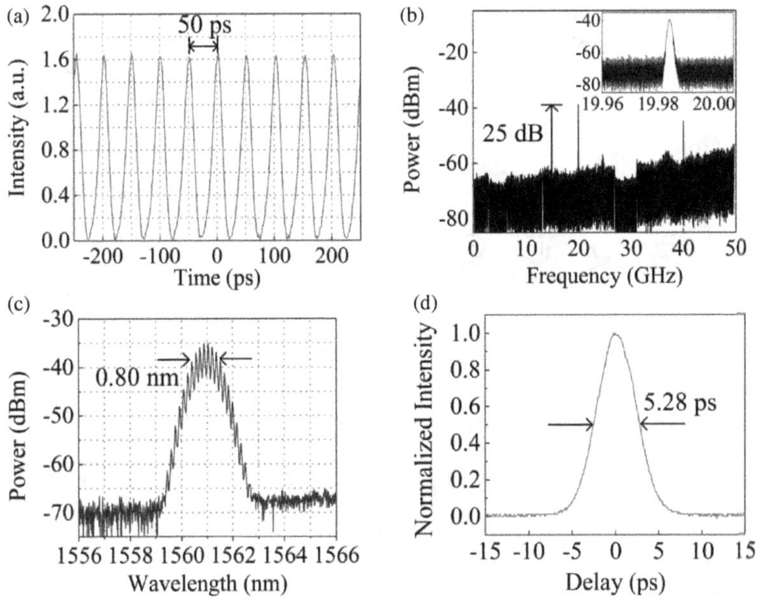

Figure 6.28. The measured (a) oscilloscope trace, (b) RF spectrum, (c) optical spectrum, and (d) autocorrelation trace of the 20-GHz pulse train from the rational harmonic mode-locked laser [64].

on the influence of a graphene SA inside the cavity on the generated pulse train.

The results of a hybrid mode-locked fiber laser with a graphene-saturable absorber in the ring cavity are now described. The length of the SMFs is carefully adjusted to ensure the insertion of the SA does not drift the cavity's fundamental frequency significantly. As a consequence, a 20-GHz pulse train can be obtained with the same experimental conditions. However, the combination of an active and passive mode locker results in the improvement of the laser performance by compressing the pulse duration, as seen in Figure 6.29. The FWHM of the pulse is shortened to 2.82 ps, indicating a compression factor of 0.53.

In this case, the role of the active mode locker is to produce a high-speed optical pulse train. Furthermore, by applying the RHML technique, an even higher repetition rate can be easily reached.

In Ref. [57], a passively harmonic mode-locked fiber laser based on a graphene-saturable absorber was described. The fundamental cavity

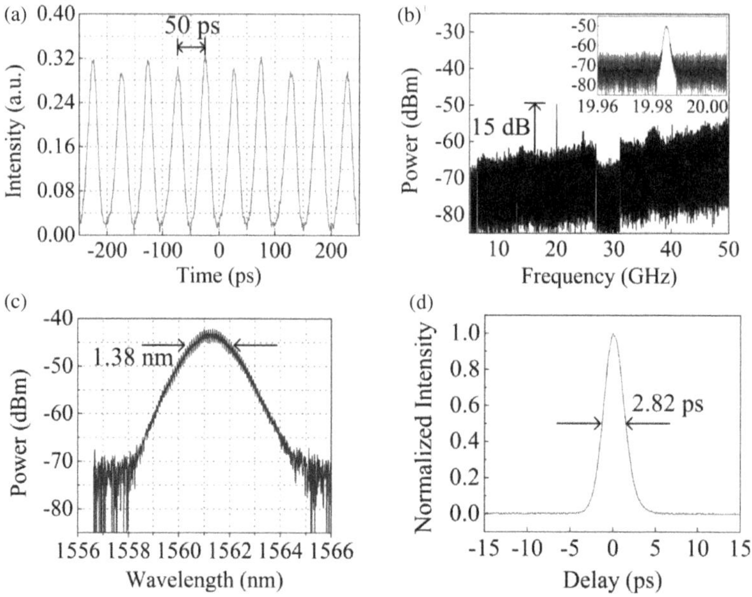

Figure 6.29. The measured (a) oscilloscope trace, (b) RF spectrum, (c) optical spectrum, and (d) autocorrelation trace of the 20-GHz pulse train generated by the hybrid mode-locked laser [64].

frequency (106 MHz) was relatively high and the repetition rate was 2.22 GHz corresponding to the 21st harmonic. Martinez *et al.* [58] reported a passively mode-locked fiber laser using graphene as a saturable absorber, which operated at a fundamental repetition rate of 9.67 GHz by reducing the cavity length to as short as 10 mm. On the other hand, the passive graphene mode locker acts as a pulse compressor. This is expected since the recovery time of graphene (~200 fs [59]) is much shorter than the pulse width, hence the transmittance reacts to the instantaneous intensity of the injected light. As can be seen in Equation (6.60), the central part of the pulse with higher intensity experiences less absorption than the leading and trailing edges, which gives rise to pulse narrowing. The optimal compression happens, however, within a limited range of the saturation intensity I_{sat}. These results are comparable to that reported in [66], where a hybrid mode-locked fiber laser with single-wall carbon nanotubes as saturable absorbers was built to obtain a 4-GHz pulse train.

6.5.3.2 *Numerical simulation*

A numerical simulation of the laser operation is as follows. The generalized nonlinear Schrödinger equation (GNLSE) governs light propagation along the fibers [15]:

$$\frac{\partial A(z,\tau)}{\partial z}+\frac{\alpha}{2}A(z,\tau)+\sum_{k\geq 2}\frac{i^{k-1}}{k!}\beta_k\frac{\partial^k A(z,\tau)}{\partial \tau^k}$$
$$=\frac{g}{2}A(z,\tau)+\frac{g}{2\Omega_g^2}\frac{\partial^2 A(z,\tau)}{\partial \tau^2}+i\gamma A(z,\tau)\left|A(z,\tau)\right|^2. \quad (6.61)$$

Here A(z, τ) is the pulse envelope, z is the propagation distance, τ is the time-delay parameter, α is the fiber loss, γ is the nonlinear parameter, and β_k represent the dispersion coefficients. The gain saturation effect of the erbium-doped fiber should be considered by expressing the gain factor as $g = g_0/(1+E/E_{sat})$, where g_0 is the small signal gain, E is the pulse energy, and E_{sat} is the gain saturation energy. Ω_g denotes the gain bandwidth. Pulse propagation in the passive SMF is simulated by setting $g = 0$.

Equation (6.60) models the graphene saturable absorber. The MZ modulator driven by an RF source can be described by the following single-pass transmission function [67]:

$$T=\cos^2\left(\pi\frac{V_b+V_m\sin(2\pi f_m\tau)}{V_\pi}\right) \quad (6.62)$$

in which V_π is the voltage required for a phase shift of π, V_b is the DC voltage bias, and V_m and f_m are the amplitude and frequency of the RF signal, respectively.

The parameters of the fiber laser components used for modeling are provided in Table 6.2, in which dispersion terms up to the third order are included. Simulation is initiated by launching into the system a seed pulse with small amplitude. The pulse evolution within the ring cavity is then iteratively modeled until a steady state is reached after many roundtrips. The split-step Fourier method is utilized to solve the GNLSE numerically.

Figure 6.30 shows the simulated steady-state solutions of the output pulses, respectively. The FWHM is 5.47 ps when operated solely in the

Table 6.2. Fiber laser parameters used for simulation.

Component	Parameter	Value
EDF	Length	23 m
	2nd dispersion $\beta 2$	-0.130×10^{-3} ps2/m
	3rd dispersion $\beta 3$	0.135×10^{-3} ps3/m
	Nonlinear parameter γ	3.69×10^{-3} /W/m
	Small signal gain g0	1.09 dB/m
	Gain saturation energy Esat	0.1 pJ
	Gain bandwidth	10 nm
SMF	Length	30 m
	2nd dispersion $\beta 2$	-21.7×10^{-3} ps2/m
	3rd dispersion $\beta 3$	0.182×10^{-3} ps3/m
	Nonlinear parameter γ	1.41×10^{-3} /W/m
MZM	Half-wave voltage $V\pi$	6 V
	DC bias Vb	0 V
	Modulation amplitude Vm	6 V

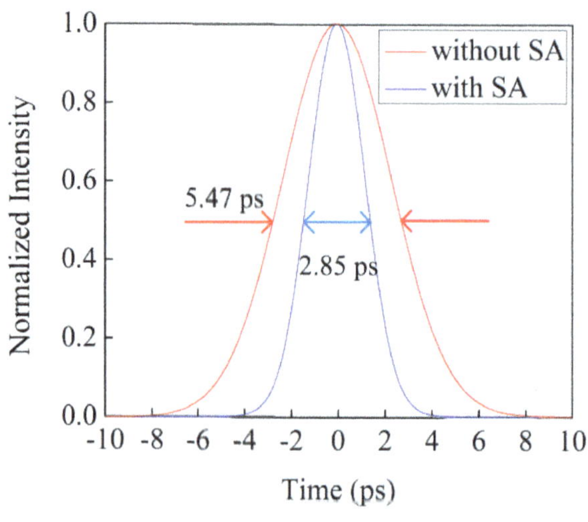

Figure 6.30. Comparison of the simulated output pulses generated from fiber lasers implementing rational harmonic active mode locking and hybrid mode locking with grapheme saturable absorber (SA) [64].

RHML regime and narrowed by a factor of 0.52–2.85 ps in the hybrid mode-locking regime, in good agreement with the experimental results. It should be mentioned that the asymptotic state is independent of the seed pulse. Pulses having the same characteristics can be obtained when pumped with white noise in the simulation as well. Saturable absorption in grapheme and other two-dimensional materials has been studied theoretically and experimentally [68–70].

6.6 Saturable Absorber Mirror (SAM)

A saturable absorber mirror (SAM) is a mirror with a saturable absorber in front of it. The incident light goes through the saturable absorber and is reflected back in the same direction by a mirror. The mirror generally consists of a multilayer dielectric or semiconductor material. With increase in incident intensity, the saturable absorber transmission increases which increases the light falling on the mirror and hence the reflected light increases. Thus the SAM has the property that its reflectivity increases if the incident power goes up. This is the principle of SAM and it is used as a mode-locking or Q-switching mechanism in lasers [71, 72]. A special case of SAM is a semiconductor saturable absorber mirror (SESAM). In this case, both the absorber and the mirror are made out of semiconductor material and the entire stack is grown on a semiconductor substrate by Metal Organic Chemical Vapor Deposition (MOCVD) technique. The mirror consists of a multilayer stack of two dissimilar semiconductors which can be grown lattice-matched on top of each other. A schematic of an SESAM is shown in Figure 6.31. The InGaAs quantum well is the saturable absorber and the multilayer stack of GaAs and AlAs serves as a Bragg reflector mirror. The absorption properties can be varied by increasing the number of quantum wells and the mirror reflectance can be varied by varying the number of layer pairs. The layers are generally quarter wave ($\lambda/4$ thickness) of GaAs and AlAs.

A schematic of an SESAM-based fiber laser is shown in Figure 6.32 [71]. A fiber loop is used as a second mirror for this case. The gain is provided by a 30 cm-long, high-concentration, Yb-doped single-mode fiber, with a peak absorption of 750 dB/m at 976 nm. The pump laser used

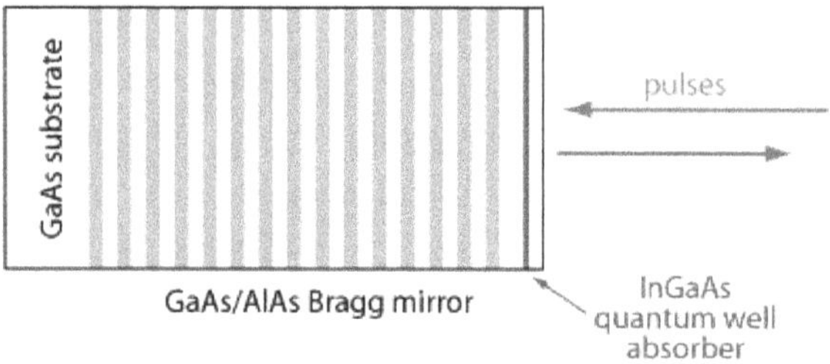

Figure 6.31. Schematic of a semiconductor saturable absorber mirror (SESAM). The InGaAs quantum well is the saturable absorber and the multilayer stack of GaAs and AlAs serves as a Bragg reflector mirror.

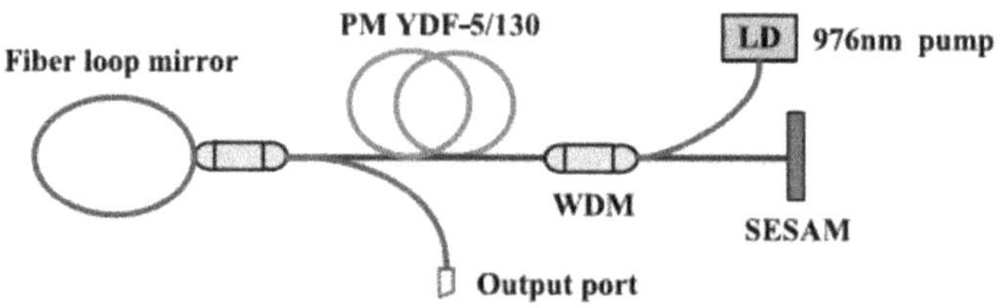

Figure 6.32. Schematic of the experimental setup. SESAM is the saturable absorber mirror. The fiber loop acts as the second mirror. Pump laser emits at 976 nm [71].

is a single-mode pigtailed laser diode emitting at 976 nm, capable of producing 400 mW. The pump laser is coupled to the gain fiber using a 976/1030 WDM multiplexer. An SESAM with 15 ps relaxation time and 37% modulation depth is used at the end of the laser cavity to permit reliable mode-locked operation. All components are polarization-maintaining (PM) components.

The all-fiber laser cavity with all the PM components can generate mode-locked pulses of 24 ps pulse width as shown in Figure 6.33. The output power measured over 2 h shows excellent environmental stability. The pulse energy is ~2 nJ.

SESAM offers a very versatile design for mode-locked operation. Both the absorption and the reflectivity can be easily varied through materials' growth. Thus optimized SESAM for high power or for a desired wavelength or very short pulses has been fabricated and used for producing mode-locked lasers.

Mode-locked diode-pumped solid-state lasers are now commercially available. They provide compact, reliable, turn-key, ultra-fast lasers. The

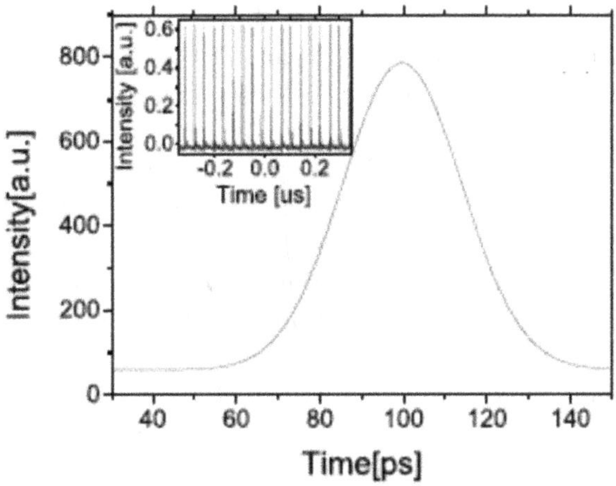

Figure 6.33. Autocorrelation trace of the mode-locked pulse; the inset shows the pulse train. The pulse width is 24 ps. The pulse repetition rate is 25.8 MHz [71].

pulse generation process is self-starting, and laser cavity adjustments are not critical. Such lasers are finding applications in materials' studies, tomography studies, optical communication, and fiber Bragg grating-based sensors.

So far, Er-doped mode-locked fiber laser results have been described. The methods described previously have been used for mode-locking studies of mode-locked lasers using dopants other than Er, such as Yb, Nd, and Tm. These laser systems emit at different wavelengths as described previously (Chapter 4). Mode locking using these dopants is discussed in Chapter 7.

References

1. D. Strickland and G. Mourou, *Opt. Commun.*, **56**, (1985) 219.
2. K. A. Ahmed, K. C. Chan, and H. F. Liu, *IEEE J. Sel. Topics Quant. Electron.*, **1**, (1995) 592.
3. M. N. Vinoj and V. C. Kuriakose, *J. Opt. A: Pure Appl. Opt.*, **6**, (1974) 63.
4. G. R. Lin, Y. C. Lin, K. C. Lin, W. Y. Lee, and C. L. Wu, *Opt. Express*, **18**, (2010) 9525.
5. K. Taira and K. Kikuchi, *IEEE Photon. Technol. Lett.*, **15**, (2003) 1288.
6. K. Taira and K. Kikuchi, *Electron. Lett.*, **40**, (2004) 15.
7. T. Torounidis, M. Westlund, H. Sunnerud, B. E. Olsson, and P. A. Andrekson, *IEEE Photon. Lett.*, **17**, (2005) 312.

8. Y. Ozeki, S. Takasaka, T. Inoue, K. Igarashi, J. Hiroishi, R. Sugizaki, M. Sakano, and S. Namiki, *IEEE Photon. Technol. Lett.*, **17**, (2005) 1698.
9. Y. Ozeki, S. Takasaka, J. Hiroishi, R. Sugizaki, T. Yagi, M. Sakano, and S. Namiki, *Electron. Lett.*, **41**, (2005) 1048.
10. C. De Dios and H. Lamela, *IEEE Photon. Technol. Lett.*, **22**, (2010) 377.
11. M. D. Pelusi, Y. Matsui, and A. Suzuki, *IEEE J. Quantum Electron.*, **35**, (1999) 867.
12. K. Smith, N. J. Doran, and P. G. J. Wigley, *Opt. Lett.*, **15**, (1990) 1294.
13. D. B. Mortimore, *J. Lightwave Technol.*, **6**, (1988) 1217.
14. S. Ma, W. Li, H. Hu, and N. K. Dutta, *Opt. Commun.*, **285**, (2012) 2832.
15. G. P. Agrawal, *Nonlinear Fiber Optics*, 4th Edn. (Elsevier, 2007).
16. I. Morita, K. Tanaka, M. Suzuki, and N. Edagawa, *IEEE J. Lightwave Technol.*, **17**, (1999) 2506.
17. Y. Matsui, M. Pelusi, and A. Suzuki, *IEEE Photon. Technol. Letts.*, **11**, (1999) 1217.
18. K. A, Ahmed, K.C. Chan, and H. F. Liu, *IEEE J. Sel. Top. Quantum Electron.*, **1**, (1995), 592.
19. E. Yoshida and M. Nakazawa, *Opt. Lett.*, **22**, (1997) 1409.
20. S. Zhang, F. Lu, X. Dong, P. Shum, X. Yang, X. Zhou, Y. Gong, and C. Lu, *Opt. Lett.*, **30**, (2005) 2852.
21. J. Schröder, T. D. Vo, and B. J. Eggleton, *Opt. Lett.*, **34**, (2009) 3902.
22. S.-S. Jyu, L.-G. Yang, C.-H. Yeh, C. W. Chow, H. K. Tsang, and Y. Lai, In Proc. CLEO 2013, Paper CM1I3, 2013 (Baltimore, MD, USA).
23. S. S. Jyu, L. G. Yang, C. Y. Wong, C. H. Yeh, C. W. Chow, H. K. Tsang, and Y. Lai, *IEEE J. Photon.*, **5**, (2013) 1502107.
24. H. A. Haus, Short pulse generation. In I. N. Duling (Eds.), *Compact Sources of Ultrashort Pulses*, III (Cambridge University Press, Cambridge, UK, 1995).
25. D. Pudo and L. R. Chen, *Electron. Lett.*, **39**, (2003) 272.
26. S. Pan and C. Lou, *IEEE Photon. Technol. Lett.*, **18**, (2006) 1451.
27. C. Wu and N. K. Dutta, *IEEE J. Quantum Electron.*, **36**, (2000) 145.
28. W. Zhang, J. Sun, J. Wang, and L. Liu, *IEEE Photon. Technol. Lett.*, **19**, (2007) 1418.
29. H. Dong, G. Zhu, Q. Wang, H. Sun, and N. K. Dutta, *Opt. Express*, **12**, (2004) 4297.
30. Y. D. Gong, M. Tang, P. Shum, C. Lu, J.Wu, and K. Xu, *IEEE Photon. Technol. Lett.*, **17**, (2005) 2547.
31. S. Pan and C. Lou, *IEEE J. Quantum. Electron.*, **44**, (2008) 245.
32. Z. Chen, H, Sun, S. Ma, and N. K. Dutta, *IEEE Photon. Technol. Letts.*, **20**, (2008) 2006.

33. S. Pan, X. Zhao, W. Yu, and C. Lou, *Opt. Laser Technol.*, **4**, (2008) 854.
34. J. Wang, X. Wang, J. Lei, M. Ma, C. Wang, Y. Ge, and Z. Wei, *Nanophtonics*, **9**, (2020) 2315.
35. J. W. Nicholson and D. J. DiGiovanni, *IEEE Photon. Technol. Lett.*, **20**, (2008) 2123.
36. J. W. Nicholson, R. S. Windeler, and D. J. DiGiovanni, *Opt. Express*, **15**, (2007) 9176.
37. Y. W. Song, S. Yamashita, C. S. Goh, and S. Y. Set, *Opt. Lett.*, **32**, (2007) 148.
38. K. Kieu and M. Mansuripur, *Opt. Lett.*, **32**, 2242 (2007) 2242.
39. N. Nishizawa, *et al.*, *Opt. Express*, **16**, (2008) 9429.
40. Z. Sun, *et al.*, *Appl. Phys. Lett.*, **93**, (2008) 061114-1.
41. A. G. Rozhin, Y. Sakakibara, S. Namiki, M. Tokumoto, H. Kataura, and Y. Achiba, *Appl. Phys. Lett.*, **88**, (2006) 051118-1.
42. R. M. Gerosa, D. Steinberg, H. G. Rosa, C. Barros, C. J. S. de Matos, and E. A. T. de Souza, *IEEE Photon. Technol. Lett.*, **25**, (2013) 1007.
43. Q. L. Bao, H. Zhang, Y. Wang, Z. H. Ni, Z. X. Shen, K. P. Loh, and D. Y. Tang, *Adv. Funct. Mater.*, **19**, (2009) 3077.
44. T. Hasan, Z. Sun, F. Wang, F. Bonaccorso, P. H. Tan, A. G. Rozhin, and A. C. Ferrari, *Adv. Mater.*, **21**, (2009) 3874.
45. H. Zhang, D. Y. Tang, L. M. Zhao, Q. L. Bao, and K. P. Loh, *Opt. Exp.*, **17**, (2009) 17630.
46. H. Zhang, D. Y. Tang, L. M. Zhao, Q. L. Bao, K. P. Loh, B. Lin, and S. C. Tjin, Laser, *Phys. Lett.*, **7**, (2010) 591.
47. H. Zhang, D. Tang, R. J. Knize, L. Zhao, Q. Bao, and K. P. Loh, *Appl. Phys. Lett.*, **96**, (2010) 111112.
48. P. L. Huang, S. Lin, C. Yeh, H. Kuo, S. Huang, G. Lin, L. Li, C. Su, and W. Cheng, *Opt. Exp.*, **20**, (2012) 2460.
49. G. Sobon, J. Sotor, I. Pasternak, K. Grodecki, P. Paletko, W. Strupinski, Z. Jankiewicz, and K. M. Abramski, *IEEE J. Lightwave Technol.*, **30**, (2012) 2770.
50. H. A. Haus, *IEEE JQE*, **QE-9**, (1975) 736.
51. G. H. C. *New IEEE, JQE*, **QE-10**, (1974) 115.
52. H. A. Haus, J. Appl. Phys. **46,** (1975) 3049.
53. F. X. Kartner and U. Keller, *Opt. Lett.*, **20**, (1995) 16.
54. U. Keller, Advances in lasers and applications. In D. M. Finlayson and B. D. Sinclair (Eds.), *Proceedings of fifty second Scottish Universities Summer School in Physics* (St. Andrews, 1998) (Institute of Physics (IOP) , London, UK).
55. Q. Bao, H. Zhang, Y. Wang, Z. Ni, Y. Yan, Z. X. Shen, K. P. Loh, and D. Y. Tang, *Adv. Funct. Mater.*, **19**, (2009) 3077.

56. Y. M. Chang, H. Kim, J. H. Lee, and Y. W. Song, *Appl. Phys. Lett.*, **97**, (2010) 211102.
57. G. Sobon, J. Sotor, and K. M. Abramski, *Appl. Phys. Lett.*, **100**, (2012) 161109.
58. A. Martinez and S. Yamashita, *Appl. Phys. Lett.*, **101**, (2012) 041118.
59. Y. W. Song, S. Y. Jang, W. S. Han, and M. K. Bae, *Appl. Phys. Lett.*, **96**, (2010) 051122.
60. H.-R. Chen, C.-Y. Tsai, H.-M. Cheng, K.-H. Lin, and W.-F. Hsieh, *Opt. Express*, **22**, (2014) 12880.
61. N. H. Park, H. Jeong, S. Y. Choi, M. H. Kim, F. Rotermund, and D.-I. Yeom, *Opt. Express*, **23**, (2015) 19806.
62. Y.-L. Qi, H. Liu, H. Cui, Y.-Q. Huang, Q.-Y. Ning, M. Liu, Z.-C. Luo, A.-P. Luo, and W.-C. Xu, *Opt. Express*, **23**, (2015) 17720.
63. A. Martinez, K. Fuse, B. Xu, and S. Yamashita, *Opt. Express*, **18**, (2010) 23054.
64. H. Hu, X. Zhang, W. Li, and N. K. Dutta, In Proc. SPIE, Micro- and Nanotechnol. Sens., Syst., and Appl., VIII, 983630, 2016 (San Jose, CA, USA).
65 S. Ma, W. Li, H. Hu, and N. K. Dutta, *Opt. Commun.*, **285**, (2012) 2832.
66. I. Hernandez-Romano, J. Davila-Rodriguez, D. Mandridis, J. J. Sanchez-Mondragon, D. A. May-Arrioja, and P. J. Delfyett, *IEEE J. Lightwave Technol.*, **29**, (2011) 3237.
67. H. Dong, H. Sun, G. Zhu, Q. Wang, and N. Dutta, *Opt. Express*, **12**, (2004) 4751.
68. G. Xing, H. Guo, X. Zhang, T. C. Sum, C. Hon, and A. Huan, *Optica Express*, **18**, (2010) 4564.
69. J. W. You, S. R. Bongu, Q. Bao, and N. C. Panoiu, *Nanophotonics*, **8**, (2019) 63.
70. J. Wang, X. Wang, J. Lei, M. Ma, C. Wang, Y. Ge, and Z.Wei, *Nanophotonics*, **9**, (2020) 2315.
71. L. Zhang, J. Zhou, Z. Wang, X. Gu, and Y. Feng, *IEEE Photon. Technol. Lett.*, **26**, (2014) 1314.
72. U. Keller, K. J. Weingarten, F. X. Kartner, D. Kopf, B. Braun, I. D. Juang, R. Fluck, C. Honninger, N. Matuschek, and J. Aus der Au, *IEEE J. Sel. Top. Quantum Electron.*, **2**, (1996) 435.

Chapter 7

Mode-Locked Fiber Laser for Various Rare-Earth Materials

7.1 Introduction

The methods described in the previous chapter have been used for mode-locking studies of mode-locked fiber lasers using Er dopants. Mode-locked fiber lasers have been studied using Yb, Nd, Tm, Ho, and other rare-earth dopants in the fiber. These laser systems emit at different wavelengths as described in Chapter 4. For high-repetition-rate mode locking, it is important to have a modulator at the appropriate wavelength, and all the techniques (harmonic and rational harmonic mode locking and pulse stabilization) described earlier can be used to produce stable high-repetition-rate pulses. The availability of the $LiNbO_3$ modulator for a near 1500 nm wavelength allowed extensive analysis of active mode locking of Er-doped mode-locked lasers, which is important for optical fiber communication studies. For other laser systems, short mode-locked pulses are generally produced using saturable absorbers.

7.2 Yb-Doped Mode-Locked Laser

Carbon nanotubes (CNTs) have been used for mode locking of Yb fiber lasers. The CNTs are immersed in an aqueous solution of polyvinyl alcohol (PVA), and PVA films with CNTs are thus created [1–4]. The films are cut into small pieces (1 mm × 1 mm) and are sandwiched between two

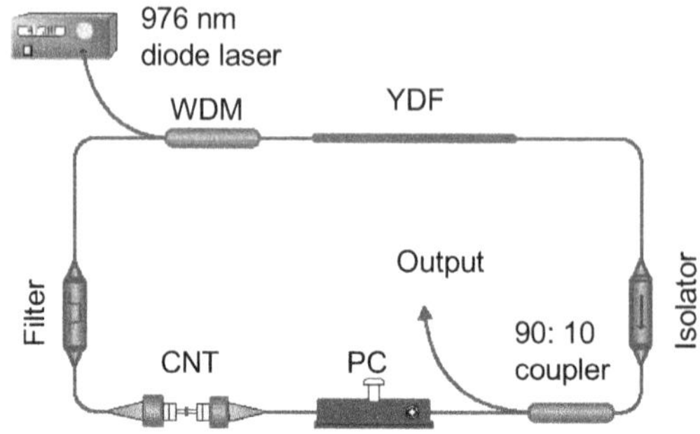

Figure 7.1. Schematic of a Yb-doped fiber laser. The laser uses CNTs for mode locking [3].

FC/PC fiber connectors. A schematic of the YDFL (Yb-doped fiber laser) in a ring cavity is shown in Figure 7.1. The Yb-doped gain medium is pumped by a diode laser emitting near 976 nm. The ring laser has an isolator for unidirectional light propagation and a polarization controller (PC) for optimizing the polarization state needed for optimum performance of CNTs. The YDF fiber is 0.8 m long and the total cavity length is 9.56 m. An optical filter is used to ensure mode locking at a certain wavelength.

Mode-locked pulses are observed at pump powers of ~47 mW. The pulse repetition rate is 21.5 MHz, which represents the inverse of the round-trip time for the 9.56-m-long cavity. The autocorrelation trace of the pulse is shown in Figure 7.2, which shows a pulse width of 310 ps. At higher pump powers, the mode locking changes to Q-switching.

Another type of saturable absorber is a semiconductor material whose bandgap is near the laser wavelength. It is configured in the form of a saturable absorber mirror (SAM). A special class of SAM is called SESAM. It stands for semiconductor saturable absorber mirror. It consists of a semiconductor waveguide material of appropriate band gap sandwiched between cladding layers (higher band gap layers) on both sides and is grown on a binary semiconductor substrate.

It can be attached to the end of a connector using transparent epoxy. This fiber-attached SESAM can be included in a fiber laser cavity using a circulator. A Yb fiber laser setup using SESAM is shown in Figure 7.3 [6, 7]. The Yb-doped fiber is 3 m long and is pumped by a diode laser emitting at 976 nm. The SESAM is mounted on an FC/PC connector and is

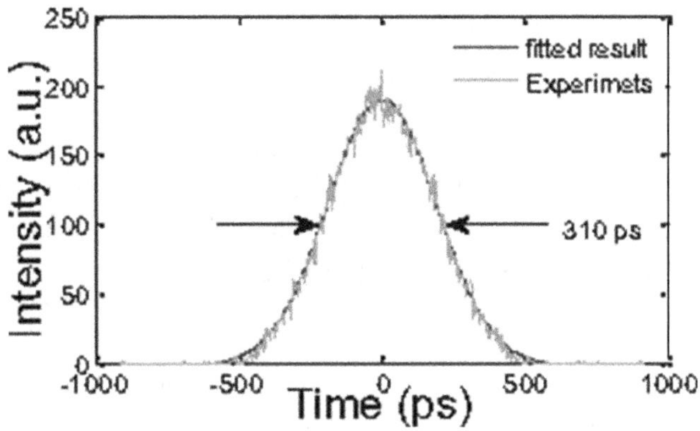

Figure 7.2. Autocorrelation trace of mode-locked pulses [3].

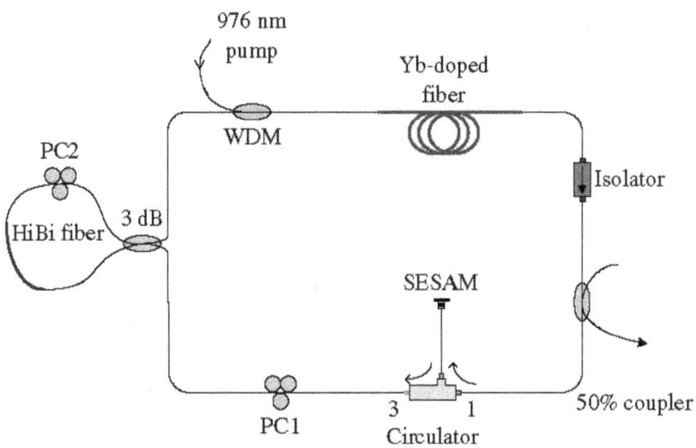

Figure 7.3. Schematic of a mode-locked tunable Yb fiber laser setup [4].

connected to the fiber laser using a three-port polarization independent circulator.

In the saturated mode, the SESAM has a modulation depth of 25% a saturation intensity of 21 μJ/cm^2, and a relaxation time constant of 5 ps. A fiber polarizer is used to control the polarization state in the cavity. A Sagnac loop mirror is added to the optical cavity. A highly birefringent (HiBi) fiber is used in the Sagnac loop mirror. The loop mirror has comb-like transmission spectrum. For a short length of HiBi fiber (~6 cm), only one transmission maximum is allowed, creating a filter. The peak wavelength of this filter can be tuned by varying PC2, and thus mode-locked pulses at different wavelengths can be obtained. By adjusting PC2, stable mode-locked pulses are observed at a pump power of ~110 mW. The pulse period repetition rate is 8.4 MHz, which corresponds to inverse round-trip time of the 24-m-long cavity. The pulse width measured using a scope is ~92 ps and is shown in Figure 7.4.

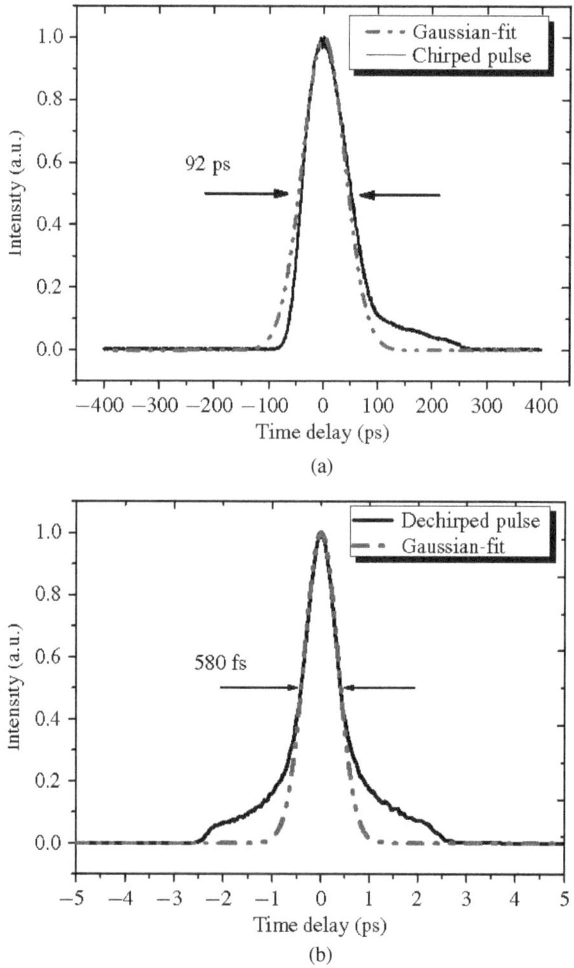

Figure 7.4. (a) Oscilloscope trace of a pulse before pulse compression, (b) autocorrelation trace after pulse compression using gratings [4].

The pulse is highly chirped (i.e., the wavelength of light varies significantly at different temporal positions in the pulse). The chirped pulse is compressed using a pair of gratings external to the laser cavity. The autocorrelation trace of the resulting pulse has a width of 580 fs (Figure 7.4). The pulse wavelength can be tuned from 1030 nm to 1060 nm.

Active mode locking using an $LiNbO_3$ modulator has been demonstrated for Yb-doped fibers. The pump laser is at 976 nm. The mode-locked frequency is ~10.3 GHz. Pulses with 2 ps pulse width with an average power of 38 mW have been obtained. A tuning range of 1022–1080 nm has been observed.

7.3 Nd-Doped Mode-Locked Laser

The Nd-doped fiber laser can be made to operate either around 1060 nm using a 4-level laser scheme or around 910 nm to 960 nm using a 3-level laser scheme [5]. Mode-locked fiber lasers have been demonstrated at both sets of wavelengths. A saturable absorber is generally used to obtain mode-locked pulses in these cases [6–12].

A semiconductor saturable absorber (SESAM) has been used to achieve mode locking at 910 nm [6]. The SESAM is specified at 920 nm (range 890–950 nm) with a saturation intensity of 15 $\mu J/cm^2$, modulation depth of 18%, and relaxation time of 3 ps. The experimental setup of the laser cavity is shown in Figure 7.5. The active gain medium is a single-mode Nd-doped silica fiber with a core diameter of 4.2 μm. The pump absorption at 808 nm (pump laser wavelength) is equal to 47 dB/m, which

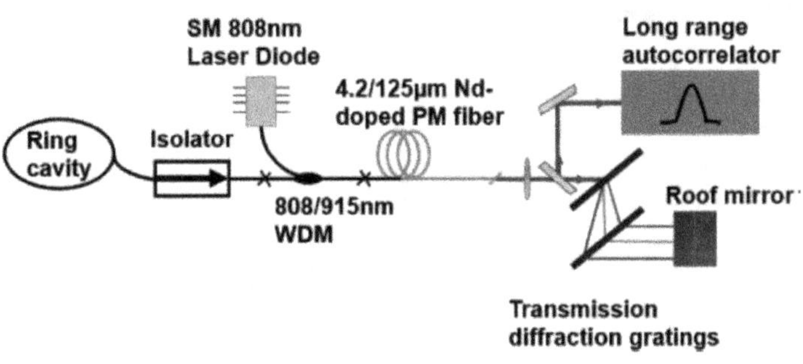

Figure 7.5. The schematic of the Nd-doped mode-locked laser and a grating compressor [6].

ensures a short fiber length. The Nd-doped fiber was pumped using a single-mode fiber-coupled laser diode (LD) delivering up to 250 mW at 808 nm. The pump LD is coupled to the active fiber with an 808 nm/915-nm-wavelength division multiplexer (WDM). For the experiment, the optimum length for the Nd-doped gain fiber is about 1 m. A circulator is used to insert the SESAM in the ring cavity, and two aspheric lenses (with focal lengths of 15 mm and 11 mm) are used to refocus the beam on the SESAM. Starting from a mode field diameter (MFD) of 4.5 μm at the fiber output, the focused beam diameter is estimated to be 3 μm on the saturable absorber.

The laser operates at a repetition rate of 28.2 MHz, which corresponds to the round-trip time of the 7.3-m-long ring cavity. The autocorrelation trace of the output pulse, measured using the second-harmonic generation, is shown in Figure 7.6. The autocorrelation trace fits perfectly with a Gaussian pulse shape, giving an estimated pulse duration of 8 ps (FWHM). The frequency chirp of the amplified pulses was studied by measuring the autocorrelation trace of the pulse after the grating compressor stage. The compressor consists of two identical transmission diffraction gratings (1850 lines/mm) in a double-pass Littrow Configuration (Figure 7.5). As shown in Figure 7.6, the minimum output pulse duration of 1.3 ps is experimentally obtained for a distance between gratings of ~5 cm, with the spectral FWHM bandwidth of 1.7 nm. Thus, the calculated time–bandwidth product (TBP) is <0.8. The compressed pulse pedestal represents the nonlinear part of the frequency chirp of the pulses. The shape

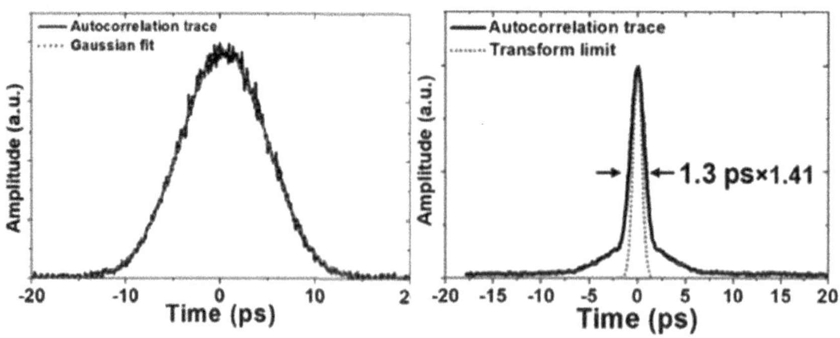

Figure 7.6. Autocorrelation traces before and after grating compressor [6].

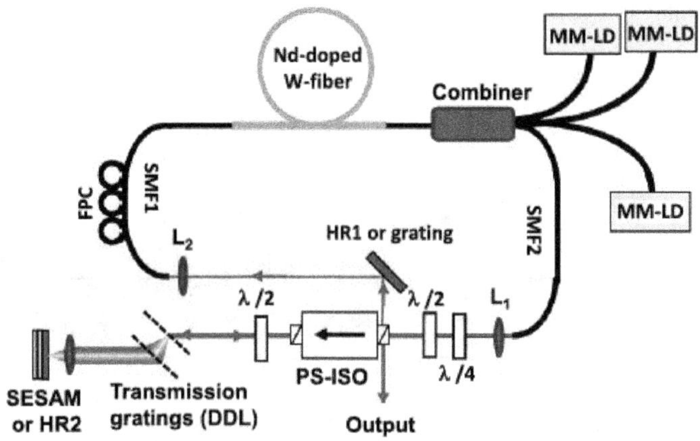

Figure 7.7. Experimental setup of the mode-locked Nd-doped W fiber laser. FPC, fiber polarization controller; PS-ISO, polarization sensitive isolator; HR, high-reflection mirror; SMF, single-mode fiber; DDL, dispersion delay line — Transmission grating pairs are used as dispersion delay lines as shown [13].

and estimated duration of the compressed pulse are independent of the pump power of the amplifier.

The experimental setup for an Nd-doped mode-locked laser near 930 nm is presented in Figure 7.7 [13]. The gain medium consists of a double-clad Neodymium-doped fiber with a 5 μm core diameter. This fiber is pumped through a multimode pump and signal combiner (MPC) by three LDs, delivering up to 6 W of total power around 808 nm. The specifically designed W-type Nd-doped fiber will induce an effective cutoff wavelength near 1060 nm.

The Nd-doped fiber length is 8 m. The passive fibers constituting the MPC and PC are 2 m and 3 m long, respectively. The total cavity length including the free space region is 15 m, leading to a repetition rate of 15 MHz. The laser design includes a pair of transmission gratings with 1250 lines/mm to partially compensate for the positive group-velocity dispersion (GVD) of the fibers. The mode-locking operation in this dispersion-managed (DM) laser is initiated and stabilized by the combined actions of a semiconductor saturable absorber mirror (SESAM) and nonlinear polarization evolution (NPE) polarizer, which acts as an effective saturable absorber (SA). A set of PCs are used for optimization and to minimize

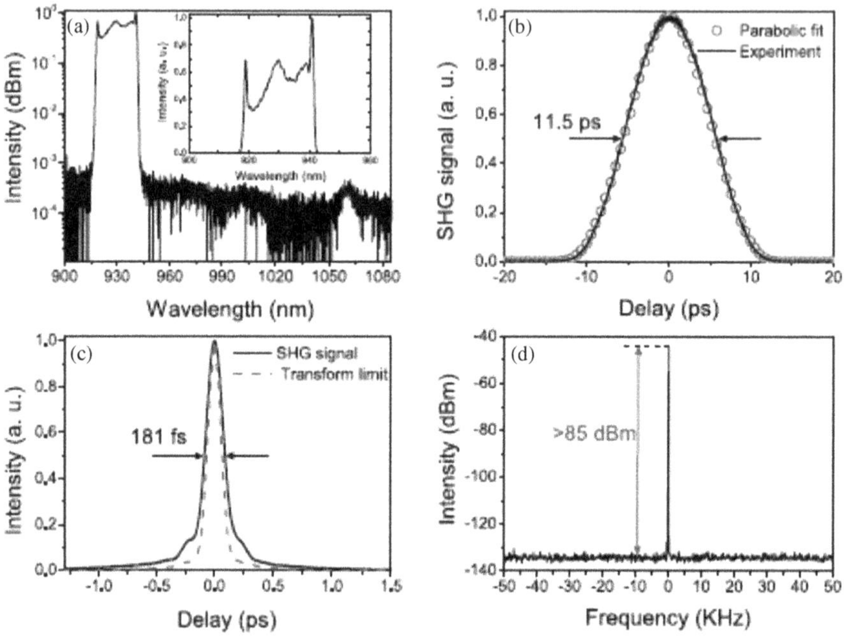

Figure 7.8. The fiber laser output of Figure 7.5 for 2.2 nJ pulse energy. (a) Optical spectrum on logarithmic and linear scales, (b) autocorrelation traces of output, (c) autocorrelation traces of output after chirp compression using a pair of external gratings, (d) radio frequency spectrum [13].

the cavity losses. The rejection port of the NPE polarizer is used as the output. The overall grating separation is varied in order to explore a large range of net GVD values. Single-pulse mode locking is self-starting with a repetition rate of ~15 MHz.

The experimental results are shown in Figure 7.8. These are obtained for net cavity dispersion of about +0.09 ps^2. The output optical spectrum is centered at 930 nm and presents a full width at half maximum (FWHM) of 23 nm. The very high contrast ratio (>30 dB) of laser emission at 930 nm compared with 1060 nm indicates a good filtering efficiency provided by the W-type fiber index profile.

Instead of a dispersion delay line (DDL) as shown in Figure 7.7, a regular single-mode fiber is also used in the experiment. In this case, 2 nJ pulses with 5.8 ps pulse width (prior to chirp compression) are obtained.

However, the pulse stability is not as good as with DDL. The details are described in Ref. [13].

7.4 Tm-Doped Mode-Locked Laser

Thulium (Tm) emits in the wavelength range of 1800–2000 nm. Tm-doped fiber has been used to make mode-locked fiber lasers in this wavelength range. Some of the experimental results are described here [14–18].

Figure 7.9 shows the experimental setup of a tunable Tm-doped mode-locked fiber laser [14]. The setup is similar to that described for other lasers. A 1.5-m Tm-doped fiber (core/cladding diameter: 9/125 μm) is bidirectionally pumped by two single-mode LDs with a center wavelength at 793 nm. The maximum output power of the two LDs are 170 mW and 200 mW. The 793-nm pump light is coupled into the laser cavity by two WDMs, and 40% of the light inside the cavity exits through one coupler. The coupler has a 40:60 splitting ratio. The optical cavity has a 70-m single-mode silica fiber, two PCs, and a polarization-dependent isolator (PD-ISO). Two PCs and one PD-ISO induce the NPE effect. The dispersion of Tm-doped fiber and single-mode fiber (SMF 28) at 1900 nm is −71 and −67 ps^2/km, respectively. The net total cavity dispersion is −5.37 ps^2. By adjusting the PCs, the emission wavelength can be tuned.

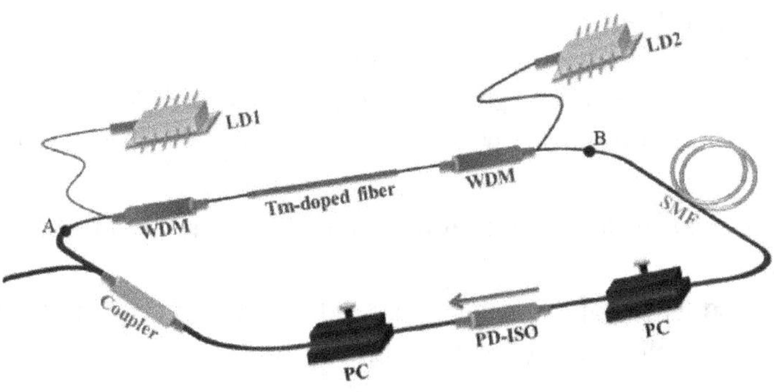

Figure 7.9. The experimental setup of the Tm-doped mode-locked fiber laser. LD, laser diode; WDM, wavelength division multiplexer; SMF, single-mode fiber; PC, polarization controller; PD-ISO, polarization-dependent isolator [14].

Tunable mode locking from 1839 nm to 1946 nm (107 nm range) is observed with the 40/60 coupler. Changing the coupler ratio changes the tuning range. In addition to the tunable single-wavelength mode locking, tunable dual-, tri-, and four-wavelength mode locking can also be achieved by tuning the PCs. The tuning range is 52 nm, (from 1864 to 1916 nm) for dual-wavelength mode locking, 49 nm, (from 1863 nm to 1912 nm) for tri-wavelength mode locking, and 55 nm (from 1860 nm to 1915 nm) for four-wavelength mode locking. Multi-wavelength emission is enabled when the net gain has multiple peaks, functioning as a periodic band-pass filter, and the optical power is equally distributed among two, three, or four peaks, to reduce the mutual mode competition. This multi-peak net gain is formed by the periodic modulation of cavity transmission, induced by NPE [14].

As previously described, graphene can be used as a satiable absorber in a fiber laser. Graphene has been used for Tm-based mode-locked fiber lasers [18]. A 12-layer graphene sample is grown by the CVD method on copper foil using the Aixtron Black Magic Pro deposition system. To prepare 12 graphene layers on PMMA foil, a graphene layer is detached from copper using the electrochemical delamination method [18]. In order to prepare a saturable absorber device for a fiber laser, two small pieces (about 1 × 1 mm) of the graphene/PMMA composite are sandwiched between two angle-polished fiber connectors. The saturable transmission of grapheme is shown in Figure 7.10.

The experimental setup of the all-PM femtosecond Tm-doped fiber laser is shown in Figure 7.11. The fiber laser has a 90-cm-long segment of Tm-doped fiber with 9 μm core diameter, a 1570/2000-nm-WDM, a polarization-sensitive isolator, a 10% PM output coupler, and the graphene-based saturable absorber. The laser is pumped by a 1568-nm LD, amplified in an Erbium–Ytterbium co-doped fiber amplifier (EYDFA) to the maximum level of about 340 mW.

The total length of the cavity is around 4.9 m. The laser starts to operate in the mode-locked regime by itself just after turning the pump on at around 215 mW of power. The pumping power might be further increased to 290 mW without losing the mode locking [18]. The Tm-doped fiber

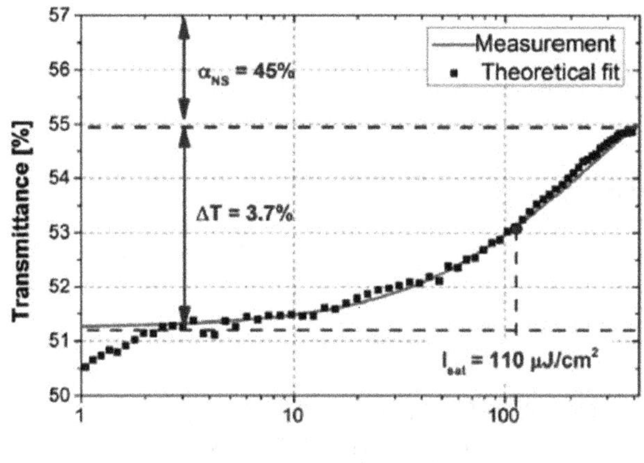

Figure 7.10. Transmission of the 12-layer graphene as a function of input power intensity measured using a 1560-nm laser [18].

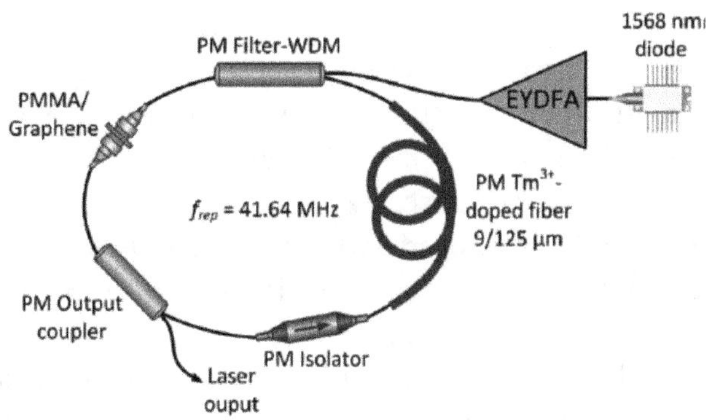

Figure 7.11. Experimental setup for a Tm mode-locked laser [18].

laser emits near 1876 nm. Figure 7.12 shows the measured autocorrelation trace of the pulses generated by the laser. Assuming a sech^2 pulse shape, typical for optical solitons, the pulse duration is equal to 603 fs. Using the measured 6.6 nm of spectral bandwidth, the calculated TBP is 0.341. This shows the pulses are nearly transform limited.

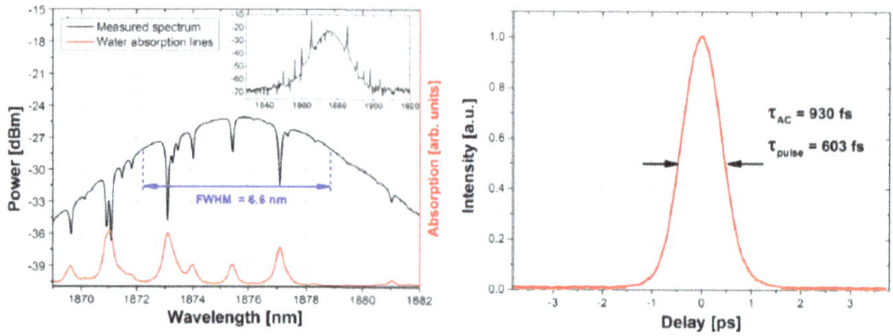

(a)

(b)

Figure 7.12. (a) Optical spectrum of the laser, together with the water absorption lines from the HITRAN database, and (b) the measured 603 fs pulse autocorrelation trace [18].

7.5 Ho-Doped Mode-Locked Laser

Ho emission is near 2080–2100 nm. Silica fibers can be doped with Ho to produce the gain medium using a suitable pump laser near 1950 nm. Graphene is used as a saturable absorber for the demonstration of mode-locked fiber lasers. The schematic of a mode-locked Ho-doped ring fiber laser is shown in Figure 7.13 [19]. The Ho fiber is 1 m long and is pumped by a 1950/2080-nm WDM device.

The graphene saturable absorber (GSA) is made using a composite of graphene and poly methyl methacrylate (PMMA) inserted between two fiber connectors. The GSAs consisted of several graphene layers stacked one by one onto the PMMA substrate (during this experiment, composites containing 2–12 graphene layers were investigated). The fabrication process and the optical properties of such composites containing more than 9 layers of graphene have been reported [20]. The transmittance of GSAs containing from 3 to 6 layers of graphene, as a function of incident power density, measured using the all-fiber setup operating at 1550 nm [21], is shown in Figure 7.14. The pump light source is a cascade of the two fiber lasers based on Er/Yb-doped double-clad and Tm-doped single-clad active fibers as shown in Figure 7.13. The system was pumped using a 9-W multimode semiconductor LD operating at 976 nm. The subsequent

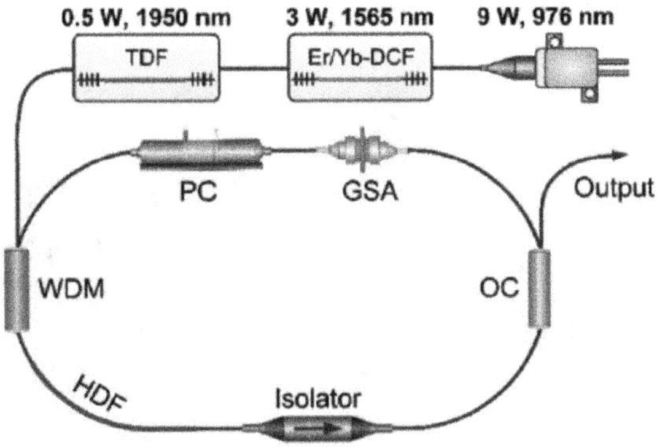

Figure 7.13. All-fiber Ho-doped mode-locked laser setup. HDF, Ho-doped fiber; OC, output coupler; GSA, graphene saturable absorber; PC, polarization controller; WDM, wavelength division multiplexer 1950/2080 nm [19].

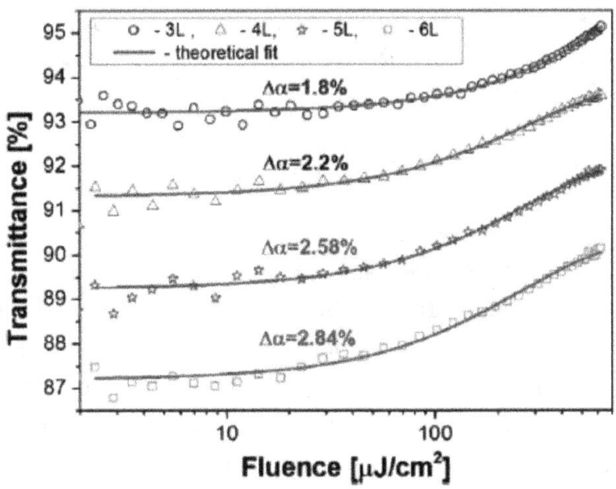

Figure 7.14. Measured energy-dependent transmittance of GSAs containing 3, 4, 5, and 6 layers of graphene ($\Delta\alpha$ is effective modulation depth) [19].

stages can produce up to 3 W at 1565 nm and 0.5 W at 1950 nm. The optical spectrum and the autocorrelation traces for two sets of graphene layers are shown in Figure 7.15.

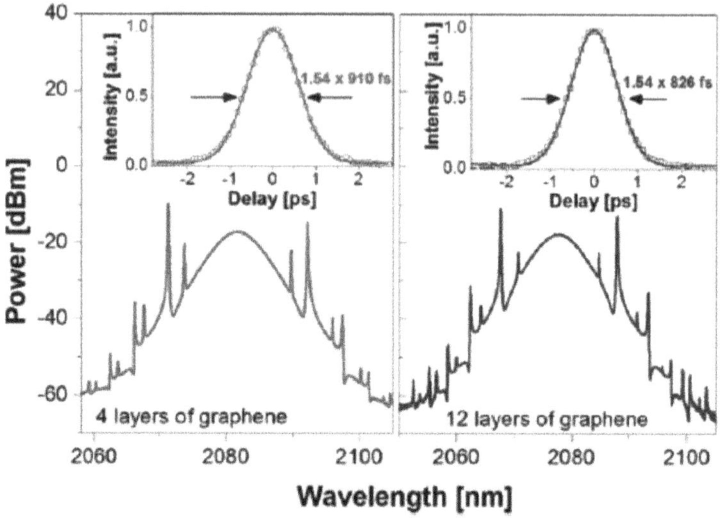

Figure 7.15. Optical spectra and the corresponding autocorrelation traces (inset graphs) measured for laser oscillators with GSAs containing 4 and 12 layers of grapheme [19].

7.6 Longer Wavelength Doped Fiber Results

Silica fibers have high loss at longer wavelengths (>2500 nm). With the development of fluoride fibers and their doping with rare earths, longer wavelength fiber lasers are now feasible [22]. Laser emission in this region is based primarily on optical transitions in Er-, Ho/Pr-, and Dy-doped gain materials. These long wavelength laser transitions are shown in Figure 7.16. Since SAM is a versatile technique for mode locking, many of these mode-locked laser systems are based on the SAM technique.

An example of a mode-locked fiber laser system operating near 3500 nm (3.5 μm) is shown in Figure 7.17. The gain medium is Er-doped fluoride fiber. A diode laser operating at 970 nm is used as a pump laser. Black phosphorus is used as SAM. The laser transition is shown in the inset. Since the autocorrelation materials at this wavelength are not well developed, the pulse width is estimated to be 10s of ps from the spectral width of 4.7 nm.

Silica fibers have high loss at long wavelengths (>2000 nm). Thus, the long wavelength mode-locked fiber lasers or CW fiber lasers are often

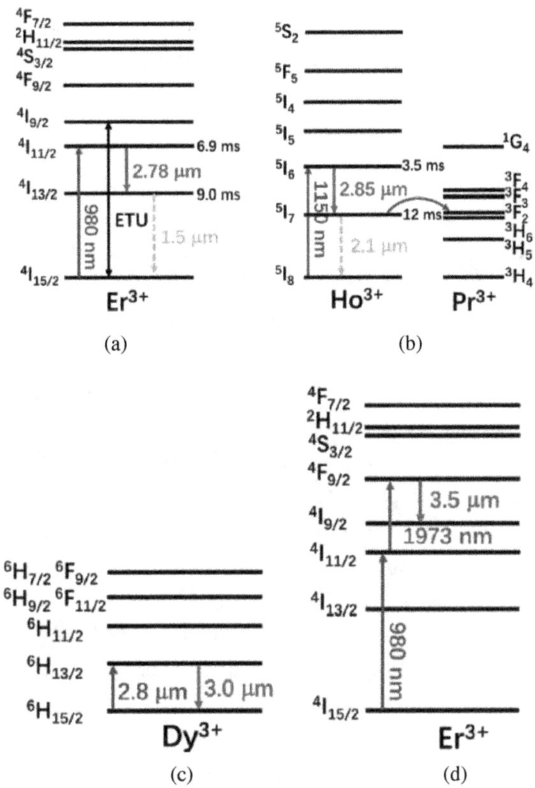

Figure 7.16. Energy level diagrams in Er, Ho/Pr, Dy, and Er (3.5 μm) showing the longer wavelength fiber laser transitions. ETU: Energy up conversion [22].

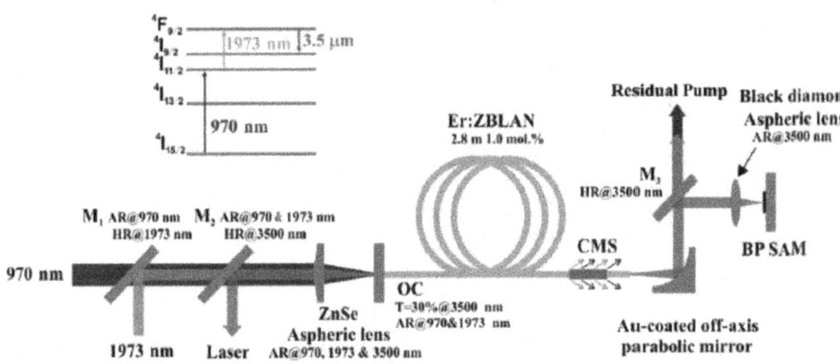

Figure 7.17. Schematic of the black phosphorus-based (BP-based) mode-locked Er-doped fluoride fiber laser at 3500 nm. OC: output coupler; CMS: cladding mode stripper; BP SAM: black phosphorus saturable absorber mirror [22].

based on rare-earth-doped fluoride fibers. Further development of fluoride fiber technology is likely to accompany the development of fiber lasers at long wavelengths.

References

1. S. M. Kobtsev, S. V. Kukarin, and Y. S. Fedotov, *Laser Phys.*, **21**, (2011) 283.
2. X. H. Li, Y. G. Wang, Y. S. Wang, X. H. Hu, W. Zhao, X. L. Liu, J. Yu, C. X. Gao, W. Zhang, Z. Yang, C. Li, and D. Y. Shen, *IEEE Photon. J.*, **4**, (2012) 234.
3. X. Li, Y. Wang, Y. Wang, Q. Wang, W. Zhao, Y. Zhang, X. Ya, and Y. Zhang, *Proc. IEEE Int. Nanoelectronics Conf.*, **139**, (2013) 55–57.
4. C. Ouyang, P. Shum, H. Wang, S. Fu, X. Cheng, J. H. Wong, and X. Tia, *IEEE JQE*, **QE-47**, (2011) 198.
5. N. K. Dutta, *Fiber Amplifiers and Fiber Lasers* (World Scientific Publishing Co., Singapore, 2014) and references therein.
6. K. Le Corre, T. Robin, B. Cadier, R. Becheker, T. Godin, A. Hideur, H. Gilles, S. Girard, and M. Laroche, *Optics Lett.*, **46**, (2021), 3564.
7. K. Kieu, B. Saar, G. Holtom, X. Xie, and F. Wise, *Opt. Lett.*, **34**, (2009) 2051.
8. C.-H. Hage, J.-T. Gomes, S. M. Bardet, G. Granger, M. Jossent, L. Lavoute, D. Gaponov, and S. Fevrier, *Opt. Lett.*, **43**, (2018) 5098.
9. P. Wang, X. Xu, Z. Guo, X. Jin, and G. Shi, *Appl. Phys. Express*, **12**, (2019) 032008.
10. A. Cook and H. Hendricks, *Appl. Opt.*, **37**, (1998) 3276.
11. D. B. S. Soh, S. Yoo, J. Nilsson, J. K. Sahu, K. Oh, S. Baek, Y. Jeong, C. Codemard, P. Dupriez, J. Kim, and V. Philippov, *IEEE J. Quantum Electron.*, **40**, (2004) 1275.
12. B. Leconte, B. Cadier, H. Gilles, S. Girard, T. Robin, and M. Laroche, *Opt. Lett.*, **40**, (2015) 4098.
13. K. Qian, H. Wang, M. Laroche, and A. Hideur, *Optics Lett.*, **39**, (2014) 267.
14. Z. Yan, B. Sun, X. Li, J. Luo, P. Ping Shum, X. Yu, Y. Zhang, and Q. Jie Wang, *Sci. Rep.*, **6**, (2016) 27245. DOI: 10.1038/srep27245.
15. Z. Sun, *Nano Res.*, **3**, (2010) 653.
16. Z. C. Luo, A. P. Luo, and W. C. Xu, *IEEE Photon. J.*, **3**, (2011) 64.
17. X. Liu, *IEEE Photon. Technol. Lett.*, **17**, (2005) 1626.
18. G. Sobon, J. Sotor, I. Pasternak, A. Krajewska, W. Strupinski, and K. M. Abramski, *Optics Express*, **23**, (2015) 9339.

19. J. Sotor, M. Pawliszewska, G. Sobon, P. Kaczmarek, A. Przewolka, I. Pasternak, J. Cajzl, P. Peterka, P. Honzatko, I. Kastik, W. Strupinski, and K. Abramski, *Optics Lett.*, **41**, (2016) 2592.
20. G. Sobon, J. Sotor, I. Pasternak, A. Krajewska, W. Strupinski, and K. M. Abramski, *Opt. Mater. Express*, **5**, (2015) 2884.
21. G. Sobon, *Photon. Res.*, **3**, (2015) A56.
22. J. Ma, Z. Qin, and G. Xie, *Appl. Phys. Rev.*, **6**, (2019) 021317.

Chapter 8

High-Power Pulse Generation

8.1 Introduction

High-power pulses are generally obtained by amplification of relatively wide pulses (several tens of ps to tens of ns) in a high-power fiber amplifier followed by pulse compression using, in many cases, a grating-based pulse compressor. The final pulse width is in the sub-ps regime. This allows the distribution of energy over a larger wavelength range on the input pulse to be amplified and hence it alleviates gain saturation and damage. Thus, the key to obtaining high-power short pulses is a high-power pulse amplifier and a pulse compressor.

For a regular single-mode amplifier, the pump co-propagates and/or counter-propagates with the signal in the rare-earth-doped core. The power output of such an amplifier is limited by the pump power that can be coupled in the small core which is typically a few microns (3–6 μm) in diameter. The pump power is generally produced by a single-mode pump laser. The output powers of single-mode lasers are generally limited by catastrophic facet degradation (for short wavelength 800 or 980 nm lasers) or by leakage current (for long wavelength 1480 nm lasers). The use of multiple pumps, such as using a co-propagating and counter-propagating configuration, increases the output power somewhat, but it is still limited to about 1 W. For higher power (~10 W to many kWs) applications, larger amounts of pump light need to be coupled into the fiber. The advantage of fiber amplifiers for such applications is that they produce a better-quality beam at the desired wavelength.

For free space communications, output powers of ~10 W may be needed, and for manufacturing applications, kW-level pulsed or CW powers may be needed. These fiber amplifiers therefore require coupling of 10–100 W and higher coupled pump power levels. One way of achieving higher coupled pump power is coupling the pump light into the large area (typically 100–200 μm diameter) cladding region [1–10]. A second cladding region (generally polymer-based) is used around this large area cladding to produce a multimode pump waveguide. The schematic of a double-clad fiber (DCF) is shown in Figure 8.1.

The core is doped with rare earth, such as Yb, Nd, or Tm. For single-mode DCF, the core is sufficiently small so that only the fundamental mode of the fiber is amplified. DCFs with single-mode cores are generally limited to lower power than that for large-area multimode (LMA) fibers. The LMA fibers typically have core regions 20–50 μm in diameter so that kW-level powers are produced.

Several methods of coupling pump light into a large core fiber have been studied [1–17]. In this type of design, the fraction of the pump power in the core interacts with the rare earth in the core and produces a gain for

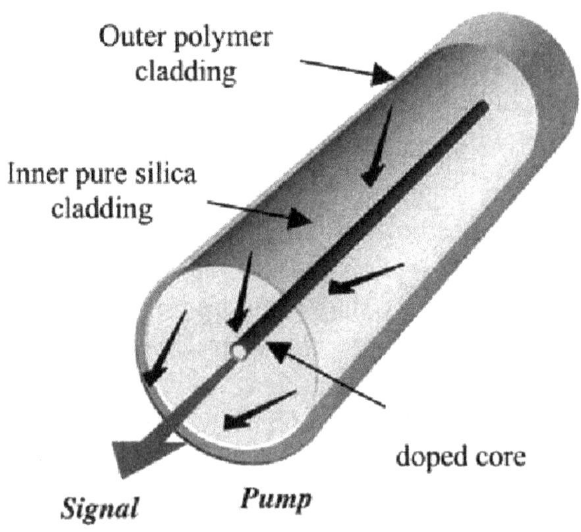

Figure 8.1. Schematic of a double-clad fiber (DCF) design. The coupled pump power in the multimode pump waveguide can reach kW levels [10].

the signal. The pump light transfers its power to the signal light (i.e., one pump photon creates one signal photon for 100% efficiency) as it propagates along the amplifier. The signal output can thus be much higher than ~1 W, generally feasible for single-mode amplifiers. A large cladding fiber is also suitable for coupling pump laser light from multiple semiconductor diode laser bars each of which has a typical emitting area size of 100 μm × 0.2 μm.

A 100 μm wide diode bar produces >10 W of power. The output power of many such laser diodes could be coupled into the cladding of a DCF fiber using suitable pump coupling mechanisms described later [1–5]. Various types of cladding-pumped high-power amplifiers have been used along with different pump coupling procedures. These are discussed in this chapter.

8.2 Fiber Waveguide Designs and Pumping for High Power

The pump travels along a large area multimode fiber waveguide. Several types of pump waveguide designs have been studied. These are shown in Figure 8.2. Enhanced pump coupling from diode laser bars is an important consideration for pump waveguide design. The simplest pump waveguide is circular in shape. It consists of a large area (~100 μm or more in diameter) of Ge-doped silica fiber, surrounded by another cladding layer typically 10–20 μm in thickness. For single-mode DCF used in somewhat low-power systems (10–20 W), the rare-earth-doped core design is similar to that described earlier but with the cladding diameter that matches the pump fiber for end pumping.

8.2.1 *Pump coupling*

Several methods have been developed for the coupling of pump light into a large-area double-clad fiber (DCF). Generally, they are classified as "end pumping" and "side pumping" schemes [1–5]. They are also discussed in detail in [11]. The tapered fiber bundle (TFB scheme) is an end pumping [2, 19] scheme discussed in the following.

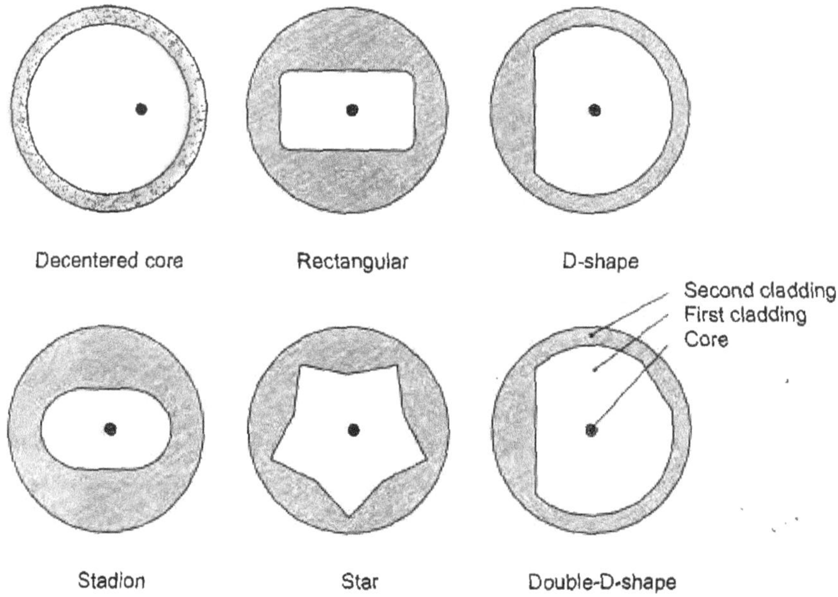

Figure 8.2. Various pump waveguide cross-sections. The central dot is the core [18].

The TFB allows the coupling of light from many semiconductor lasers into the large-area cladding (or large-area fiber). This type of end-pumping coupler is made by splicing a tapered fiber bundle (TFB) of N fibers (each of which carries pump light) into a single output fiber. By optimizing the length of the tapered region, high coupling efficiency from all N-input fibers to a single output fiber can be achieved.

A coupling experiment has been carried out using a TFB of the following design [19, Figure 8.3]. The pump fibers are seven multimode fibers with a core/clad diameter of 200/220 μm, and the NA (numerical aperture) of a core of 0.22. The output fiber is a DCF fiber with core/inner clad/outer clad diameter of 20/400/550 μm. The NA of core/inner clad is 0.06/0.46. The core in this case is not doped with rare earth, but for a DCF amplifier, it would be. The input fibers are coupled to a semiconductor laser each of which can be driven to produce a power of 430 W. The total coupled output power of TFB is plotted as a function of the total input

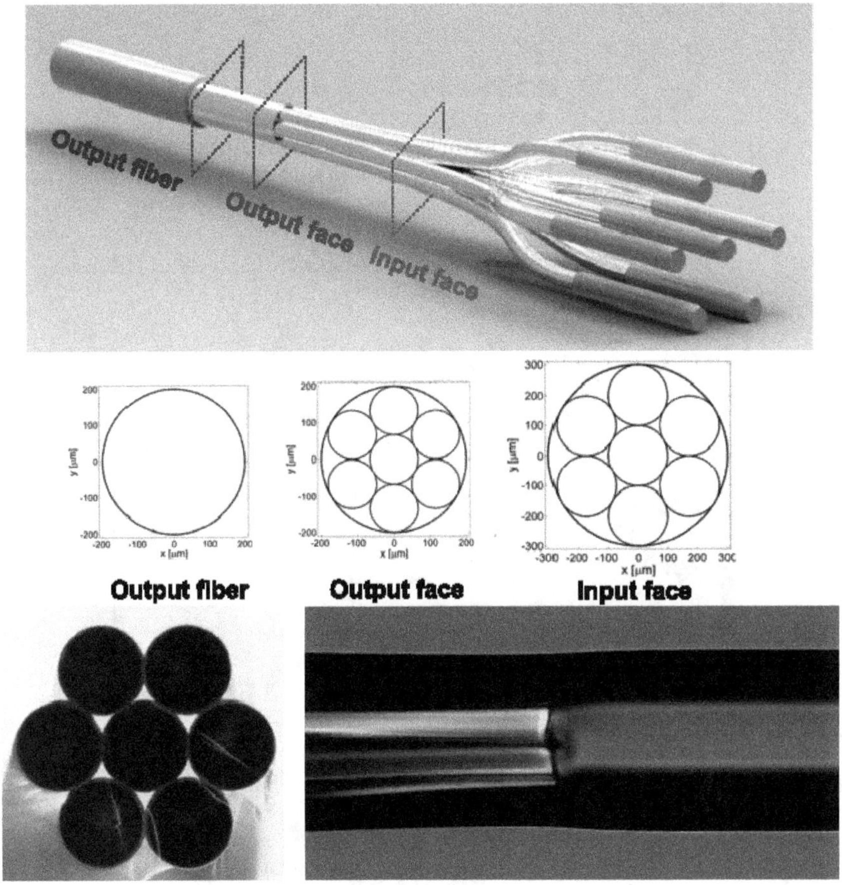

Figure 8.3. (a) Schematic of a TFB showing the fiber arrangement. (b) Photomicrograph of the cross-section of a fabricated TFB. (c) Side view of the coupling to a single fiber [19].

power of all seven lasers in Figure 8.4. The results show a 3 kW power output of TFB with >99% coupling efficiency.

In the particular TFB, and the experiment of Figure 8.4, the large area of the DCF allows very high coupling. High coupling efficiencies (>95%) have also been reported using 100/125 μm pump fibers and 10/125/100 μm DCF for 7X1 TFBs.

The 7X1 TFB has been used in a DCF amplifier configuration where the signal light is injected into the central fiber which has a single mode

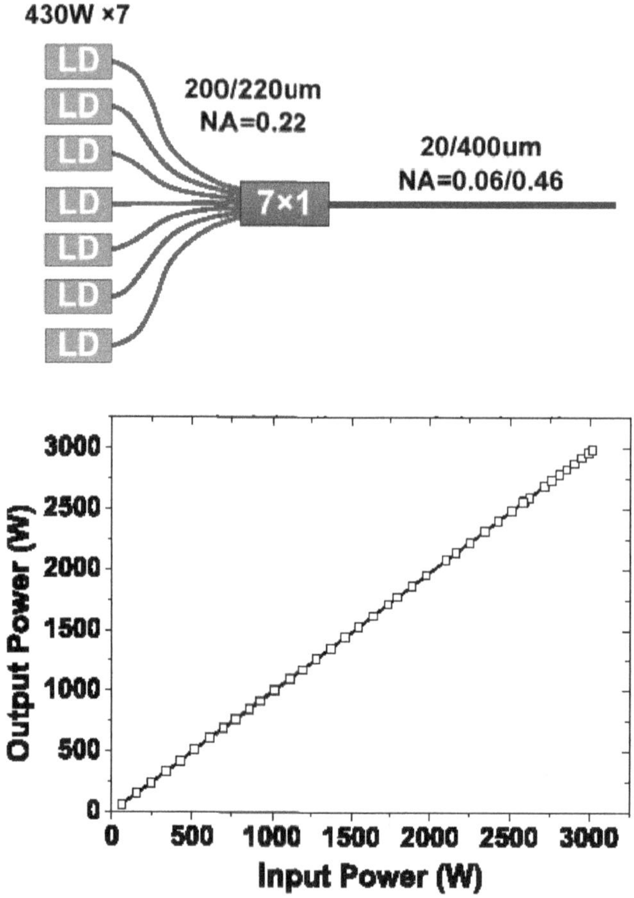

Figure 8.4. (a) Schematic of the experiment. (b) Output power as a function of total input power of a 7X1 TFB [19].

core (8 μm inner core/125 μm cladding) and the outer fibers are multi-mode fibers (100 μm/125 μm inner/outer cladding) for pump injection from a broad area semiconductor laser. The DCF in this case also has a core (8 μm diameter) to match the signal light. Another method for end pumping a DCF amplifier is shown in Figure 8.5 [20]. An important requirement here is the DC-fiber shape has to resemble the cross-section

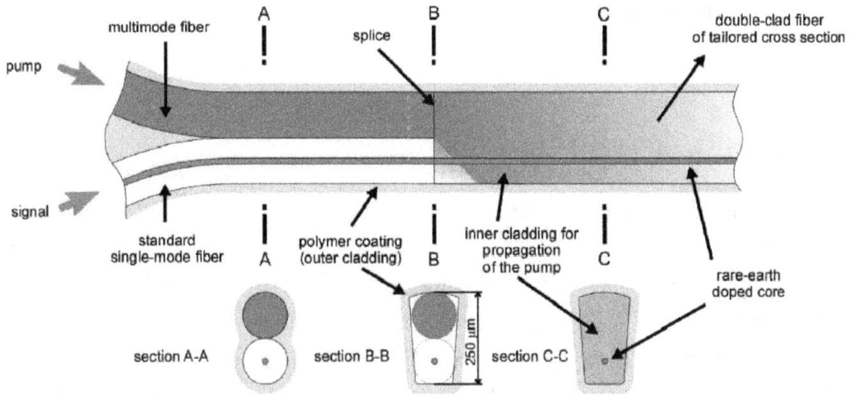

Figure 8.5. Schematic of an end pumping method [20].

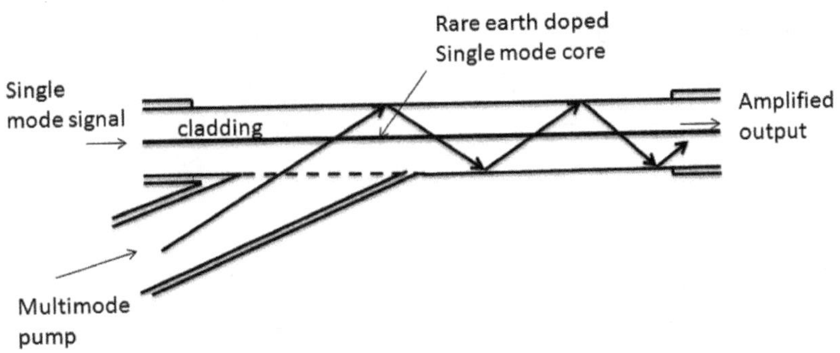

Figure 8.6. Fused fiber method for coupling pump light [4].

of the two input fibers as much as possible. This requirement is met by grinding the initial round preform to almost a rectangular shape, as shown in Figure 8.5.

The TFBs have been widely used in many relatively low-power (10–30 W) amplifiers for applications and for system demonstrations.

A commercially successful technology for side pumping is shown in Figure 8.6 [4]. In this scheme, the fiber carrying the multimode pump light is fused along the side of the doped fiber and the pump light then gets coupled to the DCF through internal reflection. Other methods of pump coupling have been demonstrated [21–36].

8.3 Double-Clad Fiber Results

High-power fiber laser or amplifier has been extensively studied using a DCF fiber with a Yb-doped core [37–42]. The amplifier used is a 12-m-long DCF with a rectangular inner cladding (100 μm × 200 μm) and a 10 μm diameter Yb-doped core. The absorption coefficients are 1.6 dB/m and 0.55 dB/m at 975 nm and 915 nm respectively for light propagating in the inner cladding. The amplifier is seeded with 100 mW light at 1064 nm, the output as a function of pump diode current is shown in Figure 8.7. Small signal gain for different wavelengths has been measured for a pump current of 3A i.e., ~1.4 W coupled power in the inner cladding (Figure 8.8). ~50 dB of small signal gain is observed at 1070 nm.

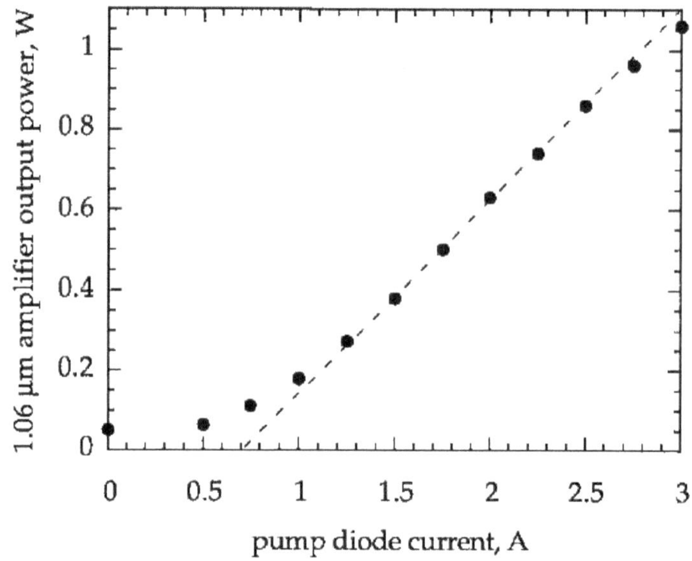

Figure 8.7. Amplified power output as a function of pump diode current. The maximum pump power coupled to the inner cladding is estimated to be 1.4 W. The seed power is 100 mW [41].

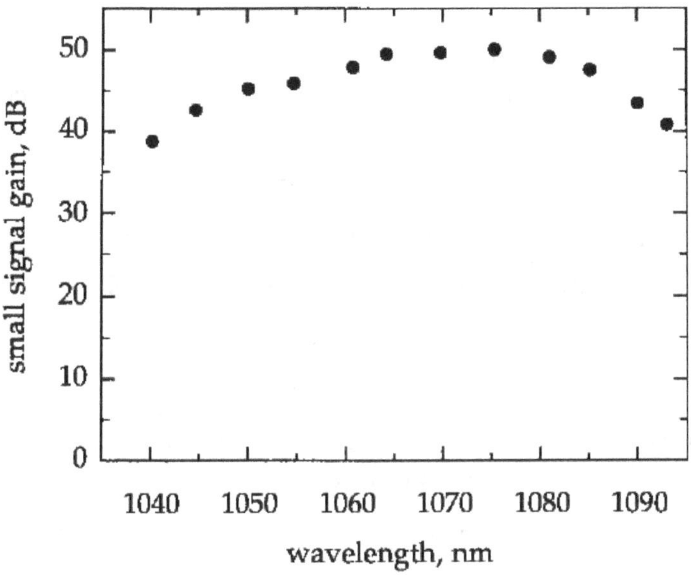

Figure 8.8. Small signal at various input signal wavelengths [41].

8.4 Double-Clad Fiber Model

A model of effective pump power absorption ratio in a double-clad fiber is discussed in this section. For this analysis, it is reasonable to divide the total power P propagating in a double-clad fiber into two mode ensembles. The first mode ensemble is the power in the inner core, P_i, which can be absorbed by the active rare earth ions. The other mode ensemble is the power between the inner core and the first cladding, P_f. Hence,

$$P = P_i + P_f. \tag{8.1}$$

When these two mode ensembles propagate along the double-clad fiber, they can be described by a system of two coupled differential equations:

$$\frac{dP_i}{dz} = -\alpha_a P_i - \gamma_1 P_i + \gamma_2 P_f, \tag{8.2}$$

$$\frac{dP_f}{dz} = -\alpha_s P_f - \gamma_2 P_f + \gamma_1 P_i, \tag{8.3}$$

where α_a is the absorption coefficient caused by the active rare earth ions in the inner core, γ_1 is the coupling constant from P_i to P_f, α_s is the propagation loss coefficient in the first cladding, and γ_2 is the coupling constant from P_f to P_i. By substituting the trial solution

$$P_i = A \exp(-\sigma z), \tag{8.4}$$

$$P_f = B \exp(-\sigma z) \tag{8.5}$$

to the coupled differential Equations (8.2) and (8.3), it follows

$$-\sigma A = -\alpha_a A - \gamma_1 A + \gamma_2 B, \tag{8.6}$$

$$-\sigma B = -\alpha_s B - \gamma_2 B + \gamma_1 A. \tag{8.7}$$

In order to have a nonvanish solution, the determinant of the coupled Equations (8.6) and (8.7) must be zero:

$$\begin{vmatrix} \sigma - \alpha_a - \gamma_1 & \gamma_2 \\ \gamma_1 & \sigma - \alpha_s - \gamma_2 \end{vmatrix} = 0, \tag{8.8}$$

which leads to the eigenvalue equation

$$(\sigma - \alpha_a - \gamma_1)(\sigma - \alpha_s - \gamma_2) - \gamma_1 \gamma_2 = 0 \tag{8.9}$$

with the solution

$$\sigma_{1,2} = \frac{(\alpha_a + \alpha_s + \gamma_1 + \gamma_2) \pm \sqrt{(\alpha_a + \alpha_s + \gamma_1 + \gamma_2)^2 - 4(\alpha_a \alpha_s + \alpha_a \gamma_2 + \alpha_s \gamma_1)}}{2}. \tag{8.10}$$

The coefficients $A_{1,2}$ and $B_{1,2}$ for the two solutions have the following relation:

$$B_1 = \frac{-\sigma_1 + \alpha_a + \gamma_1}{\gamma_2} A_1 = \frac{\gamma_1}{-\sigma_1 + \alpha_s + \gamma_2} A_1, \tag{8.11}$$

$$B_2 = \frac{-\sigma_2 + \alpha_a + \gamma_1}{\gamma_2} A_2 = \frac{\gamma_1}{-\sigma_2 + \alpha_s + \gamma_2} A_2. \tag{8.12}$$

The general solution of the coupled differential equations can be expressed as a superposition of the two solutions:

$$P_i(z) = A_1 \exp(-\sigma_1 z) + A_2 \exp(-\sigma_2 z), \tag{8.13}$$

$$P_f(z) = B_1 \exp(-\sigma_1 z) + B_2 \exp(-\sigma_2 z). \tag{8.14}$$

At the input of the fiber, the following boundary condition exists:

$$P_i(0) = \gamma P(0), \tag{8.15}$$

$$P_f(0) = (1 - \gamma) P(0), \tag{8.16}$$

where γ is the fraction of the input power coupled into the core. Using these boundary conditions, we obtain

$$A_1 = \left[\frac{\gamma_2 + (\sigma_2 - \alpha_a - \gamma_1 - \gamma_2)\gamma}{\sigma_2 - \sigma_1} \right] P(0), \tag{8.17}$$

$$A_2 = \gamma P(0) - A_1. \tag{8.18}$$

The total power propagating along the fiber can be expressed as

$$P(z) = P_i(z) + P_f(z) = (A_1 + B_1) \exp(-\sigma_1 z) + (A_2 + B_2) \exp(-\sigma_2 z), \tag{8.19}$$

where $A_{1,2}$, $B_{1,2}$, and $\sigma_{1,2}$ are given by (8.17) and (8.18), (8.10) and (8.11), and (8.12), respectively. Suppose the local effective attenuation coefficient is $\alpha(z)$, then the total power propagating at the position z is

$$P(z) = P(0) \exp[-\alpha(z) z]. \tag{8.20}$$

With Equations (8.19) and (8.20), the local effective attenuation coefficient is

$$\alpha(z) = -\frac{1}{z} \ln\left[(A_1 + B_1)\exp(-\sigma_1 z) + (A_2 + B_2)\exp(-\sigma_2 z) \right]. \tag{8.21}$$

The total power absorbed by the active rare earth ions is

$$\text{Absorbed Power} = \int_0^z \alpha_a P_i(z) dz. \tag{8.22}$$

Hence the effective absorbed power ratio is

$$\mu = \frac{\int_0^z \alpha_a P_i(z) dz}{P(0)}. \tag{8.23}$$

By using (8.17) and (8.18), the following expression for μ is obtained:

$$\mu = \int_0^z \alpha_a \left\{ \begin{array}{l} \dfrac{\gamma_2 + (\sigma_2 - \alpha_a - \gamma_1 - \gamma_2}{\sigma_2 - \sigma_1} \exp(-\sigma_1 z) \\ + \left[\gamma - \dfrac{\gamma_2 + (\sigma_2 - \alpha_a - \gamma_1 - \gamma_2)}{\sigma_2 - \sigma_1} \right] \exp(-\sigma_2 z) \end{array} \right\} dz. \quad (8.24)$$

Figure 8.9 shows the effective absorption ratio (μ) along the fiber for different coupling constants (γ_2). The quantity γ_2 is the power coupling from the first cladding to the core.

The results show that the coupling constants are the most important parameter to increase the effective pump power absorption coefficient. With all the other parameters the same, the effective absorption coefficient can be increased from 12% to 62% with a couple constant γ_2 from 0.1 m^{-1} to 0.8 m^{-1} for a 3 m-long fiber. Apart from the cladding/core ratio, there

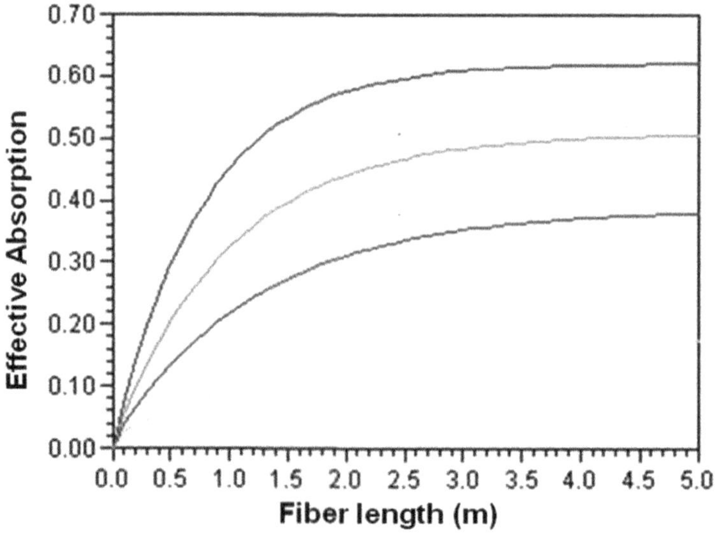

Figure 8.9. The effective pump power absorption along the fiber for different coupling constants for the power in the first cladding to the inner core. The values of γ_2 for the three curves are 0.3 m^{-1}, 0.5 m^{-1}, and 0.8 m^{-1}. Note the pump absorption increases rapidly with increasing γ_2. The pump propagation loss in the first cladding is 0.5 m^{-1} for the above figure.

are many other parameters that can change the coupling constant γ_2, such as the NA of the core, the shape of the first cladding, the launch angle of the pump power, and the bending of the fiber. The propagation loss of the pump light in the first cladding is not negligible. The loss in a fiber can be divided into two categories: material-related loss and structure-related loss. The material-related losses include absorption loss and scattering loss, which are small for silica fibers. The structure-related losses include bending loss, loss caused by cladding dimension, and loss caused by dimension perturbation. The propagation loss is another important parameter in determining the effective absorption ratio. When the propagation loss increases, more power is lost, and the effective pump absorption to the core decreases.

8.5 Pulse Propagation in Amplifiers

In this section, the basic rate equations for pulse propagation in fiber amplifiers are developed [43–50]. The amplitude change of an optical signal propagating through the amplifier is a result of gain and gain saturation. The propagation of an electromagnetic field inside the amplifier is governed by the wave equation:

$$\nabla^2 \vec{E} - \frac{\varepsilon}{c^2}\frac{\partial^2 \vec{E}}{\partial t^2} = 0, \tag{8.25}$$

where $\vec{E}(x, y, z, t)$ is the electric field vector of the wave, c is the velocity light, and ε is the dielectric constant of the amplifier medium. The dielectric constant is given by

$$\varepsilon = n_b^2 + \chi, \tag{8.26}$$

where n_b is the background refractive index which takes into dielectric waveguiding in the optical fiber medium. It is, in general, a function of x, y which represent the transverse coordinates of the waveguide medium. The susceptibility χ represents the effect of injected carriers on the dielectric constant. It depends on the injected carrier density n. The exact dependence of χ on n is quite complicated. However, a phenomenological model is found to be quite effective. In this model, χ is given by [46, 47]

$$\chi(n) = -\frac{\bar{n}c}{\omega_0}(\alpha + i)g(n), \tag{8.27}$$

where $\bar{n}$ is the effective mode index, ω_0 is the frequency, $g(n)$ is the optical gain, and α is the linewidth enhancement factor. The optical gain (g) approximately varies as

$$g = \sigma\,(n - n_0), \tag{8.28}$$

where σ is the net emission cross-section, n is the density of excited atoms, and n_0 is the density needed for transparency. It is useful to make some simplifying approximation of the wave equation. Assume a traveling wave amplifier with a single propagating optical mode whose electric field distribution is $\phi(x,y)$. Assume the light is linearly polarized and it remains linearly polarized in the amplifier. The electric field $\vec{E}(x,y,z,t)$ can then be written as

$$\vec{E}(x,y,z,t) = \hat{\varepsilon}\frac{1}{2}\{\phi(x,y)A(z,t)\exp[i(k_0 z - \omega_0 t)] + c.c.\}, \tag{8.29}$$

where $\hat{\varepsilon}$ is the polarization unit vector, $k_0 = \bar{n}\omega_0 / c$, ω_0 is the photon angular frequency, and $A(z,t)$ is the slowly varying amplitude of the propagating wave. Using Equations (8.25) and (8.29) and neglecting second derivatives of $A(z,t)$ with respect to t, and z and integrating over the transverse dimensions x,y, the following is obtained:

$$\frac{\partial^2\phi}{\partial x^2} + \frac{\partial^2\phi}{\partial y^2} + \left(n_b^2 - n^{-2}\right)\frac{\omega_0^2}{c^2}\phi = 0, \tag{8.30}$$

$$\frac{\partial A}{\partial z} + \left(\frac{1}{vg}\right)\frac{\partial A}{\partial t} = gA, \tag{8.31}$$

where the group velocity $v_g = c/n_g$, and the group index n_g is given by

$$n_g = \bar{n} + \omega_0\left(\frac{\partial \bar{n}}{\partial \omega}\right). \tag{8.32}$$

The group velocity dispersion has been neglected in the above Equation (8.31). The evolution of carrier density (n) can be described by the following Equation (8.33) [47]. The equation assumes that the carrier density is nearly uniform in the transverse direction and hence an average value can be used, and the upper and lower state lifetimes are long:

$$\frac{\partial}{\partial t}g(z,t) = -\sigma\, g(z,t)I(z,t)/(hv_s). \tag{8.33}$$

Using $I(z,t)$ as the intensity, Equation (8.31) may be rewritten as

$$\left[\frac{\partial}{\partial z} + \frac{1}{v_G}\frac{\partial}{\partial t}\right]I(z,t) = g(z,t)I(z,t). \tag{8.34}$$

For pulse propagation, Equations (8.31) and (8.33) can be further simplified by using the retarded time frame:

$$\tau = t - z/v_g, \tag{8.35}$$

$$\frac{\partial I}{\partial z} = g(z,t)I. \tag{8.36}$$

Define a quantity I_s (saturation power) = $h\nu_s/\sigma$, then

$$\frac{\partial}{\partial t}g(z,t) = -g(z,t)I(z,t)/(I_s). \tag{8.37}$$

The above equations have the following analytical solutions:

$$I(z,t) = \frac{I_{\text{in}}\left(t - z/v_g\right)}{1 - \left[1 - G_0^{-1}(z)\right]\exp\left(-I_s^{-1}\int_{-\infty}^{t-z/v_g} dt' I_{\text{in}}(t')\right)}, \tag{8.38}$$

$$g(z,t) = \frac{g_0(z)}{1 - \left[1 - \exp\left(I_s^{-1}\int_{-\infty}^{t-z/v_g} dt' I_{\text{in}}(t')\right)\right]G_0(z)}, \tag{8.39}$$

where $G_0 = \exp(g_0 L)$ is the small gain of the amplifier of length L and $G(z)$ is the integral of $g(z)$ from 0 to z i.e.,

$$G(z) = \int_0^z g(z')dz'. \tag{8.40}$$

For many applications, short-intensity modulated signals are injected into the amplifier. Amplification of such pulses of light is considered in this section. Equation (8.36) can be integrated over the amplifier length to provide the following expression for output power $I_{out}(\tau)$:

$$I_{out}(\tau) = I_{in}(\tau)\exp[h(\tau)], \tag{8.41}$$

where $I_{in}(\tau)$ is the input power and $h(\tau)$ is the total integrated net gain defined as

$$h(\tau) = \int_0^L g(z,\tau)dz. \tag{8.42}$$

Then Equation (8.36) can be integrated over the amplifier length L to provide the following equation for $h(\tau)$:

$$\frac{dh}{d\tau} = -\frac{I_{in}(\tau)}{I_s}(e^h - 1). \tag{8.43}$$

Equation (8.43) can be integrated to obtain $h(\tau)$ for different input pulse shapes $P_{in}(\tau)$. The parameter of interest is the evolution of total gain $G(t)$ as a function of time. From Equation (8.43), it is given by [45, 46]

$$G(\tau) = \exp(h(\tau)) = \frac{G_0}{G_0 - (G_0 - 1)\exp\left[-\frac{I_0(\tau)}{I_s}\right]}, \tag{8.44}$$

where G_0 is the unsaturated gain of the amplifier and the quantity $I_0(\tau)$ is given by

$$I_0(\tau) = \int_{-\infty}^{\tau} I_{in}(\tau)d\tau. \tag{8.45}$$

Equation (8.44) shows the gain is time dependent i.e., different portions of the pulse experience different amounts of gain. This leads to a modification of the shape of the pulse after transmission through the amplifier.

8.6 Chirped Pulse Amplification

The short pulse laser is important for exploring the interaction between ultrahigh intensity pulse laser and matter. Such high-intensity short pulses also have various applications [51–62]. The chirped pulse amplification (CPA) system [61, 62] plays an important role in the generation of the ultrashort, ultrahigh-intensity pulse laser [51–62]. Pulse compression grating is the most critical element in the CPA system. Strickland and Mourou were the first to demonstrate the CPA technique [61]. The pulse compression is carried out using a grating which determines the performance of the whole laser system. The gratings need to have high diffraction efficiency, broad bandwidth, and high laser-induced damage threshold, particularly for high-power applications [52].

8.6.1 *Grating pulse compressor (GPC)*

An optical pulse has a broad spectrum i.e., it consists of a range of wavelengths. When the pulse is incident on the grating, these wavelengths diffract at different angles. Consider a "chirped" pulse, where the wavelength of the rising edge of the pulse is higher (frequency lower) and the wavelength of the trailing edge of the pulse is lower (frequency higher). If the wavelength is high, the diffraction angle is higher and the opposite is true if the wavelength is low. Thus if a set of diffraction gratings are arranged in the form shown in Figure 8.10, a pulse shorter than the incident pulse will exit the grating system. The shorter exit pulse will also have a higher peak intensity than the incident pulse.

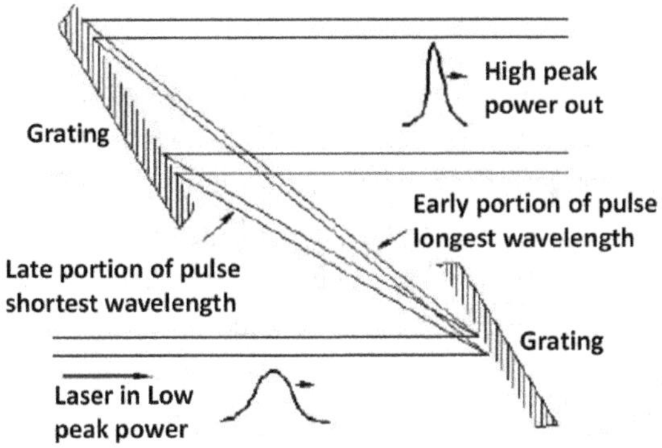

Figure 8.10. Schematic of the principle of a pair of gratings used for pulse compression.

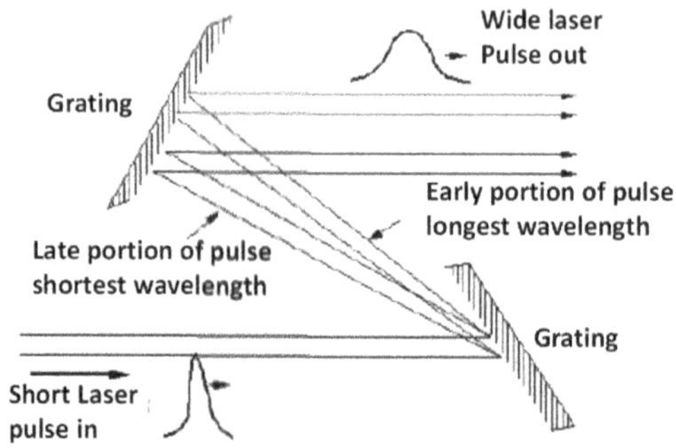

Figure 8.11. Schematic of the principle of a pair of gratings used for pulse stretching.

Alternatively, an incident pulse can be stretched in time (transformed into a longer pulse) using the same principle as shown in Figure 8.11. The peak power of the exit pulse is lower than that of the incident pulse. For a chirped pulse amplification (CPA) system, the incident pulse goes through a pulse stretcher which reduces the peak intensity followed by an amplifier which increases the total intensity (time-integrated intensity) of the pulse. This is followed by a pulse compressor which increases the peak intensity of the exit pulse [61]. A CPA system can result in 10–100 times peak intensity increase. This is the principle of a chirped pulse compression (CPA) system.

Three types of grating pulse compressors (GPCs) have been used [63, 64]. Metal grating, usually an Au-coated grating, is the common type of GPC used in the CPA system. The metal-coated gratings have high diffraction efficiency over a broad bandwidth. However, due to the high absorption of metal, the use is limited for high-power operation. Multilayer dielectric gratings have low absorption, but their bandwidth is low; hence, the use is limited to ps pulses. The hybrid grating viz. metal multilayer dielectric grating is a promising scheme for both broad bandwidth and high damage threshold. Fiber-based pulse compressors and other nonlinear pulse compressors can also be used, but they are limited to low power.

The principle of the CPA system is shown in Figure 8.12 [51]. It has three main elements as stated earlier: (a) Pulse stretching: The low-energy, ultrashort pulse from a laser is transformed to a long pulse by a pulse stretcher. The long pulse duration is generally between 100 ps and 3 ns,

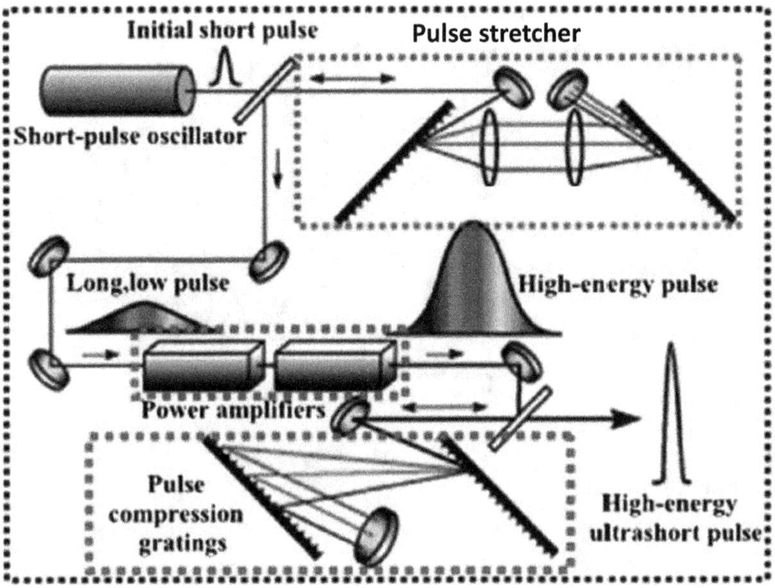

Figure 8.12. Schematic of operating principle for CPA. Its operation consists of (a) pulse stretching, (b) pulse amplification, and (c) pulse compression [51].

and the peak intensity is reduced by several orders of magnitude after stretching. (b) Pulse amplification: The stretched long pulse is amplified 10–100 times by a laser amplifier, and it carries sufficient energy. (c) Pulse compression: The amplified long pulse with high total energy is recompressed to an ultrashort pulse with the same or shorter duration than the original pulse, consequently it shows ultrahigh peak intensity and short pulse. Using the CPA system, the laser with petawatt level energy in picosecond pulses can be achieved. The process produces an ultrahigh-intensity laser source.

A short pulse source with high beam quality which can be focused to a small spot is important for many applications e.g., high harmonic generation (HHG, Chapter 11). For a fiber laser, the beam quality is determined by the guided mode and is nearly power-independent. Diffraction-limited CW powers of ~10 kW under CW operation have been reported [65]. A rod-type double-clad fiber design can produce very large powers. Rare-earth-doped photonic crystal fibers (also known as microstructure fibers)

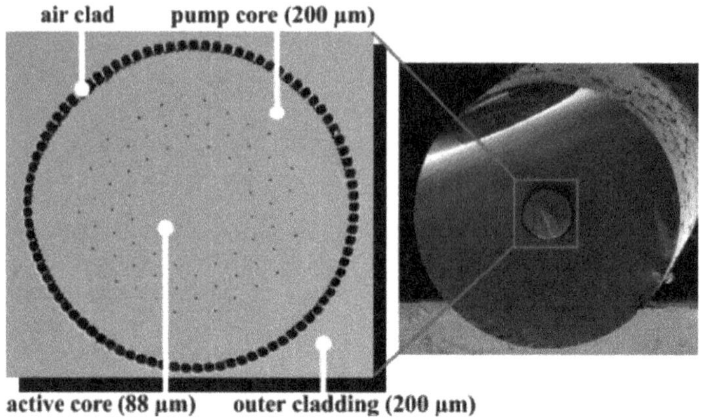

Figure 8.13. Cross-section of the large-mode-area rod-type PCF with a zoom into the embedded microstructure [66].

with large core and cladding sizes can produce high powers due to very good pump coupling. An example of a core design is shown in Figure 8.13.

The fiber has a 200 μm inner cladding which serves as a pump waveguide surrounded by air cladding which consists of 400 nm thick, 10 mm long silica bridges. Nineteen missing holes in the center of the fiber form the Yb/Al-doped core active region with a corner-to-corner distance of 88 μm. The structure has a numerical aperture of 0.6 for efficient coupling of multimode 976 nm pump lasers. The mode area of the fundamental mode is 4000 μm^2 and the small signal pump absorption is ~30 dB/m [66]. The whole inner structure is surrounded by a stiff 1.5-mm diameter fused silica outer cladding. This large outer cladding is introduced to keep the fiber straight, hence preventing bend-induced losses.

The schematic setup of a high-energy high average-power fiber chirped pulse amplification (CPA) system is shown in Figure 8.14 [66, 67]. It consists of a passively mode-locked Yb-doped laser, a dielectric grating stretcher/compressor unit, an acousto-optical modulator used as a pulse selector, and two ytterbium-doped photonic crystal fibers used in single-pass configuration as amplification stages, providing an overall gain factor of approximately 25000.

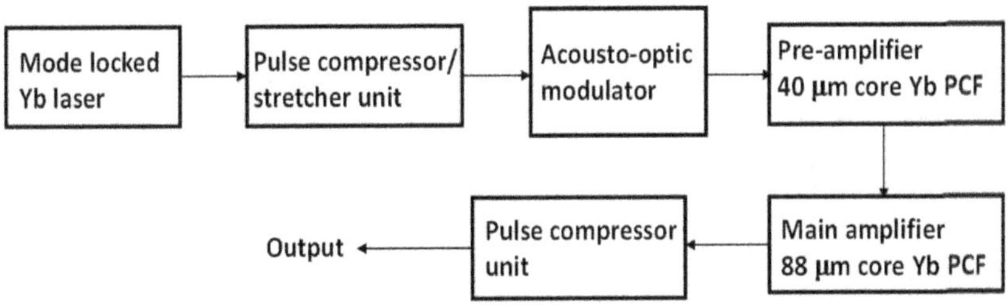

Figure 8.14. Schematic of a fiber-chirped pulse amplification (CPA) system. *Note*: OI: optical isolator. AOM: acosto-optic modulator. PCF: photonic crystal fiber [66].

The laser produces 400-fs pulses at a repetition rate of 9.7 MHz and with an average power of 1.6 W at 1030 nm. The stretcher-compressor unit employs two 1740-lines/mm dielectric diffraction gratings and stretches the 3.3 nm bandwidth pulses to 2 ns. An acousto-optical modulator is used to reduce the pulse repetition rate. The preamplifier consists of a 1.2-m-long 40-μm core single polarization air-clad photonic crystal fiber. The main amplifier is built using a 1.2-m-long piece of the low-nonlinearity air-cladding photonic crystal rod-type fiber as described above (Figure 8.13). The main amplifier was seeded with 0.5 W of average power, corresponding to 2.5 μJ pulse energy. At a launched pump power of 230 W, an average output power of 145 W is obtained with a slope efficiency as high as 66%. Compressed average power of 100 W implies a pulse energy of 500 μJ. The compressed pulses are about 1.2 ps wide.

High-field physics experiments require ultrashort pulses (~100 fs) and a high peak power of several gigawatts. In particular, the requirement on the pulse duration cannot be met with fiber CPA (FCPA). Nonlinear pulse compression techniques are needed to shorten the pulses further. The schematic of such a system is shown in Figure 8.15. The method uses self-phase modulation (for pulse broadening) in noble-gas-filled hollow fibers followed by pulse compression with a chirped mirror compressor [68, 69]. Using a waveguide for self-phase modulation reduces any spatial chirp.

For one case, 600-μJ pulses from the FCPA output at a repetition rate of 30.3 kHz were coupled to the fiber. The pressure of the noble gas was adjusted for optimal spectral broadening (~3 bar). For 18 W of input

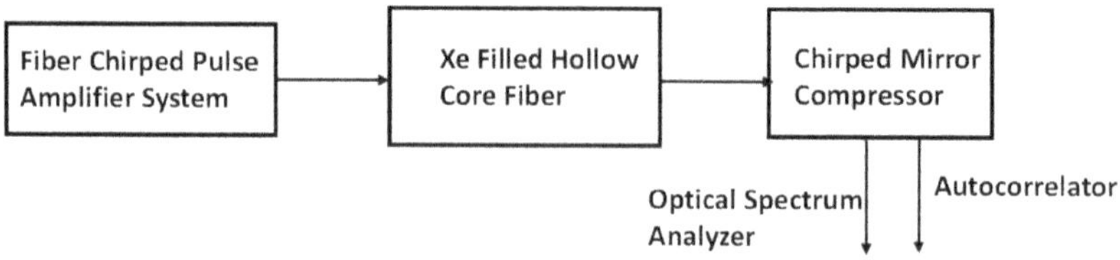

Figure 8.15. The setup for nonlinear compression in a noble gas-filled hollow fiber. *Note*: OSA: optical spectrum analyzer. AC: autocorrelator [66].

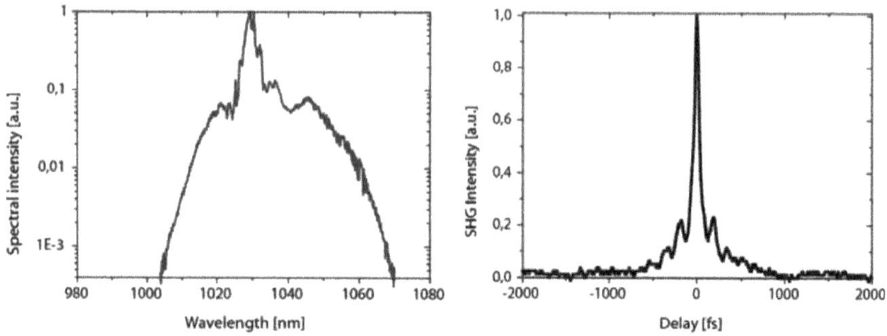

Figure 8.16. Experimental results on nonlinear compression of a fiber chirped-pulse amplification system. (a) Broadened spectrum. (b) Autocorrelation trace of the compressed pulses [66].

average power, the measured output power is 3.4 W, which equals 112 μJ (at 30.3 kHz). The spectral broadening of the input signal is shown in Figure 8.16(a). The chirp induced by the self-phase modulation in the hollow fiber is removed by a chirped mirror compressor. The pulses are compressed to an autocorrelation width of 100 fs (Figure 8.16(b)) corresponding to a pulse duration of 68 fs. This resulted in 105-μJ compressed pulse energy and a peak power of more than 1GW. The compression efficiency is ~94%.

8.7 Supercontinuum Generation using Fibers

Broadband supercontinuum generation (SCG) schemes have gained significant interest due to their applications in optical coherence tomography, spectroscopy, and dense wavelength division multiplexing (DWDM) sources for high-capacity optical communication systems. In SC generation, extreme broadening of pulse spectrum can be realized for sufficiently intense narrow-band incident pulses. The primary mechanism of spectral broadening is considered as a combined action of self-phase modulation (SPM), cross-phase modulation (XPM), stimulated Raman scattering (SRS), parametric four-wave mixing (FWM), and the dispersive properties of the fiber. The processes of the SCG are sensitive to the dispersive and nonlinear properties of the optical fiber.

Development of a relatively high-power mode-locked fiber laser is important for supercontinuum generation. Pumping by a CNT mode-locked Tm-doped fiber laser around 2.0 μm, Thapa *et al.* reported ultrawide mid-IR supercontinuum generation spanning from 0.9 μm to 4.2 μm in a low-loss W-type tellurite fiber [70]. Tellurite glass has a transmission window of about 5 μm. Chalcogenide materials such as As_2Se_3, As_2S_3, and GaSe not only have a nonlinear index but also have a much broader transmission window up to ~15 μm. This makes them important for broad supercontinuum generation. Using a mode-locked Ho/Pr-doped ZBLAN fiber laser with a dispersion-engineered As_2Se_3/As_2S_3-tapered fiber device, Hudson *et al.* achieved a supercontinuum with a bandwidth of over 10000 nm (125 THz) in the mid-IR region [71]. The schematic and the results are shown in Figure 8.17. The pump laser emitted 230 fs pulses around 2.9 μm at a repetition rate of 42 MHz with an average power of 140 mW. It generated a supercontinuum covering from ~2 μm to 12 μm range.

The mechanism of SCG has been analyzed numerically [72, 73]. The results showed that a dispersion-decreasing fiber with a convex dispersion profile can generate a flat broadened spectrum. The progress in the design of photonic crystal fiber (PCF) enables the fabrication of fibers with two zero dispersion wavelengths (ZDWs) and SCG in these fibers has been studied extensively [74–79]. In [78–80], the authors investigated the possibility of obtaining the efficient mid-IR supercontinuum sources based on nonsilica microstructured optical fiber (MOF) technology. The results show that small core MOFs with two ZDWs can have large efficiency for mid-IR continuum generation using suitable pump wavelengths by phase-matching across the upper (longer wavelength) ZDW. They found that a seed wavelength approximately midway between the ZDWs can produce

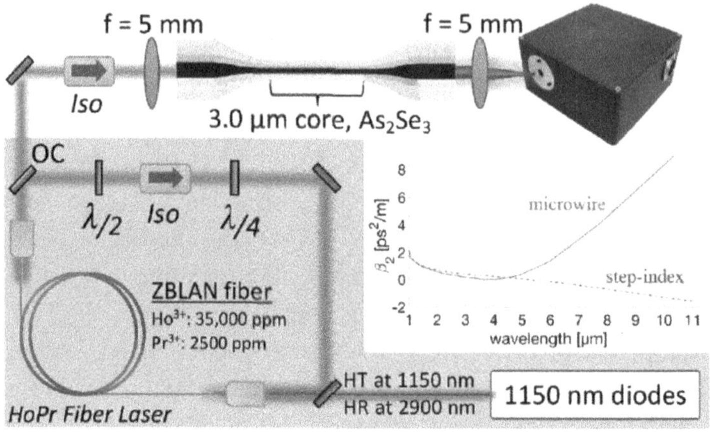

(a)

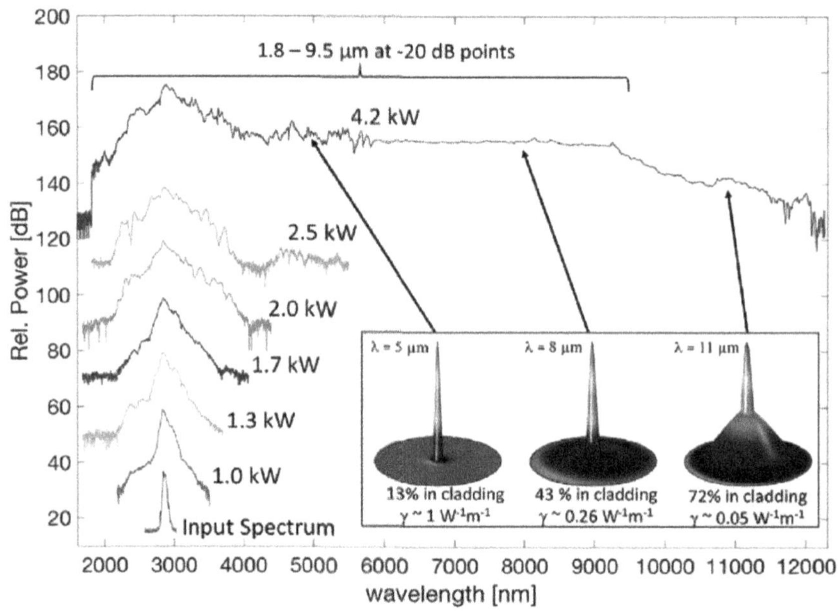

(b)

Figure 8.17. (a) Schematic of the mode-locked Ho/Pr-doped ZBLAN fiber laser cavity and the subsequent supercontinuum generation. Inset: Dispersion of both the untampered step-index fiber and the 3.0 μm core section. (b) Spectrum of generated supercontinuum at different pump peak powers. Note the wide spectrum produced at high power [71].

the broadest spectrum. In [81], a ZBLAN fiber taper with exponentially decreasing core size was numerically modeled and the coherent continuum could be generated. The authors showed that the spectrum generated across the upper ZDW can be progressively shifted with this nonuniform fiber taper design.

In Ref. [82], a broad bandwidth, mid-IR SC generation using an 8 mm length of highly nonlinear telluride microstructured PCF has been demonstrated. It is pumped with 100 fs pulses of energy 1.9 nJ at 1550 nm. The resulting bandwidth is from 789 nm to 4870 nm which covers over 4000 nm spectrum range. The experimental data are shown in Figure 8.18. For the tellurite PCF and pump pulses (β_2 = 80 ps^{2}km^{-1}, P_0 = 7954 W, γ = 596 km^{-1}W^{-1}, T_{FWHM} = 110 fs), the nonlinear length is calculated to be 0.21 mm, and the dispersion length is 38 mm [82]. The soliton number found in the tellurite fiber is N = 14 indicating that higher-order soliton dynamics affect the spectral characteristics of the supercontinuum

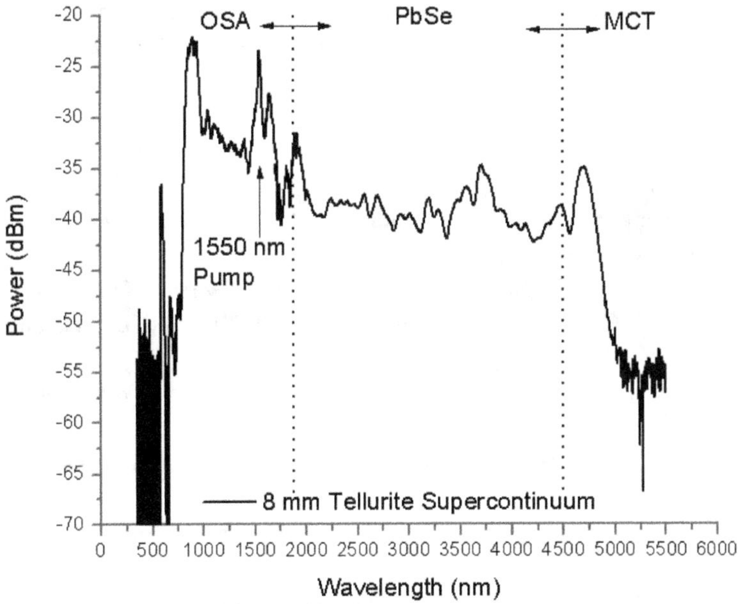

Figure 8.18. Spectrum of SC generated in an 8 mm-long tellurite microstructured fiber. The different detector types are noted on top.

Note: OSA: optical spectrum analyzer. PbSe detector. MCT: mercury cadmium telluride [82].

radiation generated in this fiber [82]. For a tapered fiber, a spectrum of width ~4000 nm is predicted with ~1/3 input pulse energy (~0.6 nJ) than that for tellurite fiber [15]. For some applications, it is important to understand the coherence as studied by Dudley and Coen [83].

Nonlinear processes in single-mode fibers using mode-locked high-power lasers are important for coherent supercontinuum generation. These types of sources provide new laser wavelengths for applications in spectroscopy, tomography, and sensors. Very high peak powers can be obtained using the chirped pulse amplification technique described in Section 8.6.

References

1. S. G. Anikitchev, K. E. Lindsay, and A. Starodoumov, Method for manufacturing a multimode fiber pump power combiner. U.S. Patent 7 272 956, (2007).
2. D. J. Digiovanni and A. J. Stentz, Tapered fiber bundles for coupling light and out of cladding-pumped fiber devices. U.S. Patent 5 864 644, (1999).
3. A. B. Grudinin, *et al.*, Multi-fibre arrangements for high power fibre lasers and amplifiers. U.S. Patent 6 826 335, (2004).
4. P. Valentin, V. F. Gapontsev, and N. Platonov, Powerful fiber laser system. U.S. Patent 7 593 435, (2009).
5. D. J. Ripin and L. Goldberg, *Electron. Lett.*, **31**, (1995) 2204.
6. W. Dazheng, W. Yonggang, L. Suping, and M. Xiaoyu, *Acta Opt. Sci.*, **29**, (2009) 974.
7. W. M. Sean, J. P. Koplow, and A. Hansen, in Proc. 21st LEOS, (2008) 870.
8. Z. Fan, W. Chun-can, N. Tigang, L. Chu, G. Rui, and L. Yuchun, *Opt. Commun.*, **282**, (2009) 3325.
9. C. Jauregui, S. Böhme, G. Wenetiadis, J. Limpert, and A. Tünnermann, *J. Opt. Soc. Am B*, **27**, (2010) 1011.
10. Y. Jaouen, S. Bordais, E. Olmeda, and J. Y. Allain, *Ann. Telecommun.*, **58**, (2003) 11.
11. N. K. Dutta, *Fiber Amplifiers and Fiber Lasers* (World Scientific, Singapore, 2014).
12. A. Kosterin, V. Temyanko, M. Fallahi, and M. Mansuripur, *Appl. Opt.*, **43,** (2004) 3893.
13. A. Wetter, M. Faucher, M. Lovelady, and F. Séguin, *Proc. SPIE Fiber Lasers IV: Technol. Syst. Appl.*, **6453**, (2007) 645301.

14. F. Séguin, A. Wetter, L. Martineau, M. Faucher, C. Delisle, and S. Caplette, *Proc. SPIE Fiber Lasers III: Technol. Syst. Appl.*, **6102**, (2006) 61021N.
15. H. Yu, D. A. V. Kliner, and K.-H. Liao, *Proc. SPIE*, **8237**, (2012) 82370G.
16. Q. Xiao, P. Yan, J. He, Y. Wang, X. Zhang, and M. Gong, *Laser Phys.*, **21**, (2011) 1415.
17. A. Salokatve, Optical fiber combiner and method of manufacturing there of U.S. Patent 0 154 881, (2009).
18. A. Tunnermann and H. Zellmar, Chapter 4.2 In C. E. Webb and J. D. C. Jones (Eds.), *Handbook of Laser Technology and Applications* (Institure of Physics publishing, London, 2004).
19. Qirong Xiao, Haichui Ren, Xiao Chen, Pin Yan, and Mali Gong, *IEEE Photon. Technol. Lett.*, **25**, (2013) 2442.
20. P. Peterka, I. Kašík, V. Matějec, P. Honzátko, and J. Sláníčka, *Proc. Optical Fiber Commun.*, **ME4**, (San Jose, CA, USA, 2004).
21. L. Goldberg, Method and apparatus for side pumping an optical fiber. U.S. Patent 5 854 865, (1998).
22. T. Weber, W. Luthy, H. P. Weber, V. Neuman, H. Berthou, and G. Kotrotsios, *Opt. Commun.*, **115**, (1995) 99.
23. J. P. Koplow, S. W. Moore, and D. A. Kliner, *IEEE J. Quantum Electron.*, **39**, (2003) 529.
24. J. J. Larsen and G. Vienne, *Opt. Lett.*, **29**, (2004) 436.
25. L. Goldberg, B. Cole, and E. Snitzer, *Electron. Lett.*, **33,** (1997) 2127.
26. Cheng Li, Deyuan Shen, Jie Song, Nam Seong Kim, and Ken-ichi Ueda, *CLEO, Pacific Rim*, **ThV4**, (1999) 805.
27. C. W. Huang, C. L. Chang, D. Y. Jheng, K. Y. Hsu, S. L. Huang, and D. W. Huang, *IEEE Photon. J.*, **4**, (2012) 411.
28. Yushi Kaneda, Jie Zong, Oscar Romero, Ziping Deng, Bryson Case, Arturo Chavez-Pirson, Gary Paysnoe, Jim Whitwham, Patrick Moran, Pete Rohan, Zuolan Wang, Shibin Jiang, William Eaton, Martin Brutsch, Ping Li, and Ike Song, *Proc. Opt. Fiber Commun.*, **THD**, (2000) 798.
29. J. P. Koplow, Sean W. Moore, and Dahv A. V. Kliner, *IEEE JQE*, **39**, (2003) 529.
30. T. Y. Fan, A. Sanchez-Rubio, J. N. Walpole, R. C. Williamson, I. Melngailis, J. R. Leger, and W. C. Goltsos, Multiple-laser pump optical system. U.S. Patent 5 185 758, 1993.
31. T. Y. Fan, *Appl. Opt.*, **30**, (1991) 630.
32. F. Hakimi and H. Hakimi, In Conf. Lasers and Electro-Optics, (Baltimore, MD, USA, 2001) p. 116.

33. T. Aoyagi and K. Shigihara, U.S. Patent 5 170 458, 1992.
34. J. P. Koplow, L. Goldberg, and D. A. V. Kliner, *IEEE Photon. Technol. Lett.*, **10**, (1998) 793.
35. L. Goldberg and J. P. Koplow, *Electron. Lett.*, **34**, (1998) 2027.
36. J. P. Koplow, D. A. V. Kliner, and L. Goldberg, *Opt. Lett.*, **25**, (2000) 442.
37. H. M. Pask, J. L. Archambault, D. C. Hanna, L. Reekie, P. St. J. Russell, J. E. Townsend, and A. C. Tropper, *Electron. Lett.*, **30**, (1994) 863.
38. M. Muendel, B. Engstrom, D. Kea, B. Laliberte, R. Minns, R. Robinson, B. Rockney, Y. Zhang, R. Collins, P. Gavrilovic, and A. Rowley, In Conf. Lasers and Electro-Optics, paper CPD30-2 (Baltimore, MD, USA, 1997).
39. S. V. Chernikov, J. R. Taylor, N. S. Platonov, V. P. Gapontsev, P. J. Nacher, G. Tastevin, M. Leduc, and M. J. Barlow, *Electron. Lett.*, **33**, (1997) 787.
40. R. Paschotta, D. C. Hanna, P. De Natale, G. Modugno, M. Inguscio, and P. Laporta, *Opt. Commun.*, **136**, (1997) 243.
41. J. P. Koplow, L. Goldberg, and Dahv A. V. Kliner, *IEEE Photon. Technol. Lett.*, **10**, (1998) 793.
42. C.-H. Liu, B. Ehlers, F. Doerfel, S. Heinemann, A. Carter, K. Tankala, J. Farroni, and A. Galvanauskas, *Electron. Lett.*, **40**, (2004) 1000.
43. G. P. Agrawal and N. A. Olsson, *Opt. Lett.*, **14**, (1989) 500.
44. N. A. Olsson and G. P. Agrawal, *Appl. Phys. Lett.*, **55**, (1989) 13.
45. G. P. Agrawal and N. A. Olsson, *IEEE J. Quantum Electron.*, **QE-25**, (1989) 2297.
46. G. P. Agrawal, *Nonlinear Fiber Optics* (Academic Press, NY, 1989).
47. N. K. Dutta, *Fiber Amplifies and Fiber Lasers* (World Scientific, Singapore, 2014).
48. T. Saitoh, H. Itoh, Y. Noguchi, S. Sudo, and T. Mukai. *IEEE Photon. Technol. Lett.*, **1**, (1989) 297.
49. G. P. Agrawal, *Electron. Lett.*, **27**, (1991) 620; *IEEE J. Quantum Electron.*, **QE-27**, (1991) 1843.
50. G. P. Agrawal and N. K. Dutta, *Semiconductor Lasers*, Chapter 11 (van Nostrand Reinhold, NY, USA, 1992).
51. W. Zhang, W. Kong, G. Wang, F. Xing, F. Zhang, H. Zhang, and S. Fu, *Opt. Eng.*, **60**, (2021) 02902-1.
52. J. Zhang, J. Wen, Y. Wang, W. Wang, W. Lin, X. Wei, and Z. Yang, *Opt. Lett.*, **47**, (2022) 4941.
53. P. Maine, *et al.*, *IEEE J. Quantum Electron.*, **24**, (1988) 398.
54. J. Kmetec, J. J. Macklin, and J. F. Young, *Opt. Lett.*, **16**, (1991) 1001.
55. P. Beaud, *et al.*, *Opt. Lett.*, **18**, (1993) 1550.

56. T. Ditmire and M. D. Perry, *Opt. Lett.*, **18**, (1993) 426.
57. M. D. Perry, *et al.*, *Opt. Lett.*, **24**, (1999) 160.
58. M. Aoyama, *et al.*, *Opt. Lett.*, **28**, (2003) 1594.
59. I. V. Yakovlev, Quantum Electron. **44**, (2014) 393.
60. M. Pessot, *et al.*, *Opt. Lett.*, **14**, (1989) 797.
61. D. Strickland and G. Mourou, *Opt. Commun.*, **56**, (1985) 219.
62. M. D. Perry and G. Mourou, *Science*, **264**, (1994) 917.
63. Y. X. Jin, K. Yi, and J. D. Shao, In OSA Adv. Solid State Lasers Conf., p. ATu2A.30 (2015).
64. N. Bonod and J. Néauport, *Adv. Opt. Photon.*, **8**, (2016) 156.
65. D. Gapontsev, In Proc. Conf. Solid State and Diode Laser Techn. Rev. Directed Energy Prof. Soc., Albuquerque, NewMexico, (2008) 95–99.
66. J. Limpert, S. Hadrich, J. Rothhardt, M. Krebs, T. Eidam, T. Schreiber, and A. Tunnermann, *Laser Photon. Rev.*, **5**, (2011) 634.
67. F. Roser, T. Eidam, J. Rothhardt, O. Schmidt, D. N. Schimpf, J. Limpert, and A. Tunnermann, *Opt. Lett.*, **32**, (2007) 3495.
68. M. Nisoli, S. De Silvestri, and O. Svelto, *Appl. Phys. Lett.*, **68**, (1996) 2793.
69. M. Nisoli, S. Stagira, S. De Silvestri, O. Svelto, S. Sartania, Z. Cheng, M. Lenzner, Ch. Spielmann, and F. Krausz, *Appl. Phys.*, **B 65**, (1997) 189.
70. R. Thapa, D. Rhonehouse, D. Nguyen, Z. Yao, J. Zong, and A. Chavez-Pirson, In Frontiers in Optics, OSA Technical Digest (Optical Society of America, 2012), Paper No. FW4D.2 (2012).
71. D. D. Hudson, S. Antipov, L. Li, I. Alamgir, T. Hu, M. E. Amraoui, Y. Messaddeq, M. Rochette, S. D. Jackson, and A. Fuerbach, *Optica*, **4**, (2017) 1163.
72. K. Mori, H. Takara, S. Kawanishi, M. Saruwatari, and T. Morioka, *Electron. Lett.*, **33,** (1997) 1806.
73. K. Mori, H. Takara, and S. Kawanishi, *J. Opt. Soc. Am. B*, **18**, (2001) 1780.
74. T. Schreiber, T.V. Andersen, D. Schimpf, J. Limpert, and A. Tunnermann, *Opt. Express*, **13**, (2005) 9556.
75. M. H. Frosz, P. Falk, O. Bang, *Opt. Express*, **13**, 6181 (2005) 6181.
76. A. Mussot, M. Beaugeois, M. Bouazaoui, and T. Sylvestre, *Opt. Express*, **15**, (2007) 11553.
77. J. H. V. Price, T. M. Monro, H. Ebendorff-Heidepriem, F. Poletti, P. Horak, V. Finazzi, J. Y. Y. Leong, P. Petropoulos, J. C. Flanagan, G. Brambilla, X. Feng, and D. J. Richardson, *IEEE J. Quantum Electron.*, **13**, 738 (2007) 738.
78. Z. Chen, A. J. Taylor, and A. Efimov, *Opt. Express*, **17**, (2009) 5852.

79. V. Skryabin, F. Luan, J. C. Knight, and P. St. J. Russell, *Science*, **301**, (2003) 1705.
80. P. Petropoulos, T. M. Monro, H. Ebendorff-Heidepriem, K. Frampton, R. C. Moore, and D. J. Richardson, *Opt. Express*, **11**, (2003) 3568.
81. Z. Chen, S. Ma, and N. K. Dutta, *Opt. Commun.*, **283**, (2010) 3076.
82. P. Domachuk, N. A. Wolchover, M. Cronin-Golomb, A. Wang, A. K. George, C. M. B. Cordeiro, J. C. Knight, and F. G. Omenetto, *Opt. Express*, **16**, (2008) 7161.
83. J. M. Dudley and S. Coen, *IEEE J. Sel. Top. Quantum Electron.*, **8**, (2002) 651.

Chapter 9

Optical Transmission and Optical Logic Studies

9.1 Introduction

The semiconductor injection laser and photodetectors were originally developed in the 1970s for optical fiber transmission system application. Early systems used GaAs-based lasers emitting near 850 nm and later systems used InGaAsP-based lasers emitting near 1300 and 1550 nm. The 1500 nm systems use distributed feedback single wavelength lasers. These systems evolved to wavelength division multiplexed (WDM) transmission which uses multiple distributed feedback lasers for transmission. Current systems use WDM-based transmission where the output of 8 or 16 or more single frequency (DFB) lasers carrying data is multiplexed (combined) into a single fiber. Polarization multiplexing can be used to double the data rate per wavelength channel. The typical data rate per channel is in the 10–40 Gb/s range which aggregates to a transmission capacity of $(2 \times 16 \times 40)$ of ~1 Tb/s. An alternative to the WDM system is an optical time division multiplexed (OTDM) system [1–14].

9.2 OTDM Transmission

In an optical time division multiplexing (OTDM) system, the data rates are high enough that the process of multiplexing (the generation of high data rate signals from several low data rate signals) and demultiplexing (the process of conversion of high data rate signals to several low data rate signals) is carried out optically.

For the demultiplexing process, an optical clock (a timing signal) needs to be extracted from the incoming high speed (>100 Gb/s optical data). This clock recovery process has been carried out using cross-gain modulation, cross-phase modulation, and four-wave mixing [14].

9.2.1 *Signal generation*

The optical time division multiplexing process involves the generation of high data rate signals. The process involves the generation of short pulses (~1 ps wide) at relatively high repetition rate (e.g., 20 GHz) from a mode-locked laser (possibly a mode-locked fiber laser). Then it is transmitted through multiple delay lines of suitable values and encoded with a synchronized 20 Gb/s optical modulator to create a high data rate signal (e.g., 100 Gb/s to 1 Tb/s) [12]. A typical optical transmission system is shown in Figure 9.1

The initial pulses in Ref. [12] are generated using a mode-locked Er-doped fiber laser with an intracavity $LiNbO_3$ modulator as described in previous chapters for active mode locking. The laser schematic is shown in Figure 9.2. A tunable filter is used in the ring cavity to generate pulses at different wavelengths. The pump laser emits near 1480 nm (1.48 μm).

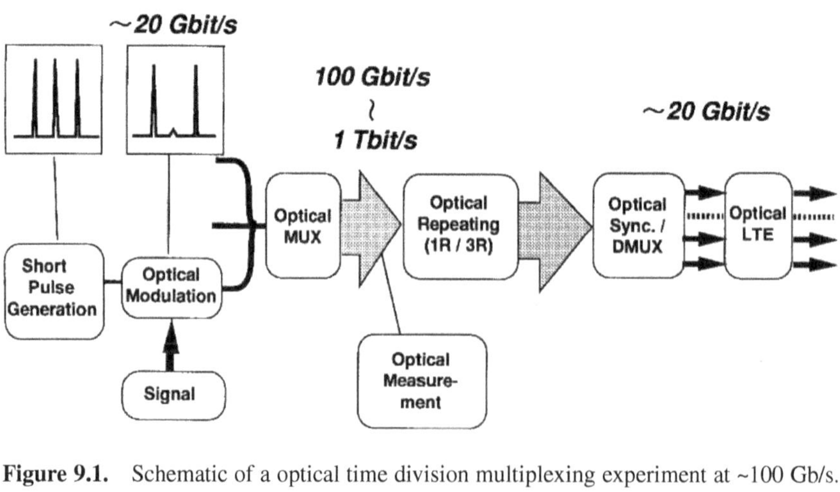

Figure 9.1. Schematic of a optical time division multiplexing experiment at ~100 Gb/s. The front part up to the first arrow is the signal generation unit [12].

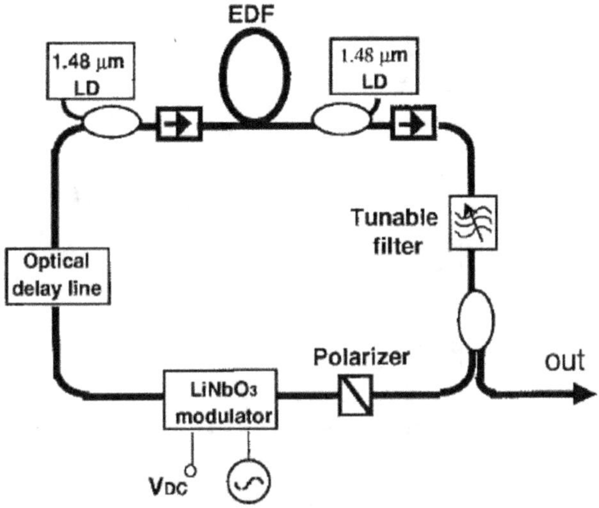

Figure 9.2. Schematic of an actively mode-locked Er-doped fiber laser [12].

There are two main methods for optical multiplexing: parallel type [15] and serial type [15–17], as shown in Figure 9.3. The parallel-type multiplexer uses all passive components. However, some stabilization scheme is required to control the relative time delay between each channel which complicates the configuration. The serial-type multiplexer has a simple configuration. In this case, a high-speed optical clock pulse train and modulated signal pulses are combined and introduced to an all-optical switch to generate a modulated channel in the high-bit-rate signal. Cascading this process allows all the channels to be independently modulated; all that is needed then is to change the relative time delay of each channel. Although, all-optical modulation/time-division multiplexing is achieved using either method, the serial method is often preferred.

9.2.2 *Clock recovery*

The schematic of an experimental setup used for clock recovery from high-speed data signals is shown in Figure 9.4. The input optical signal and the local oscillator optical signal produces a four-wave mixing (FWM) signal at the output of the SOA.

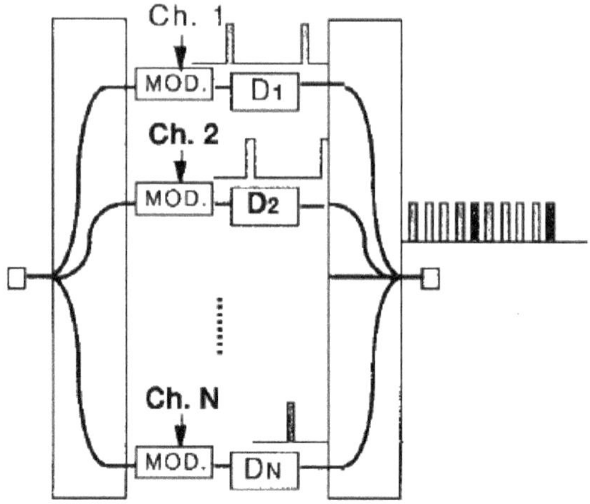

(a)

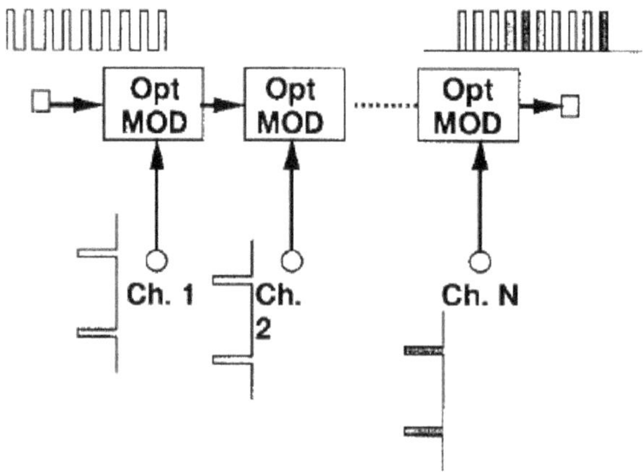

(b)

Figure 9.3. Schematic of an all-optical time division multiplexer: (a) parallel type, (b) serial type [12].

In a transmission system, the timing signal (clock) is extracted from the input optical data and is used to demultiplex the data to lower data rate channels, the error rate of which is then measured using traditional electronic circuits. The clock recovery process could be carried out using an optical phase-locked loop in conjunction with an electrical low-frequency phase-locked loop. This is shown schematically in Figure 9.4. In this

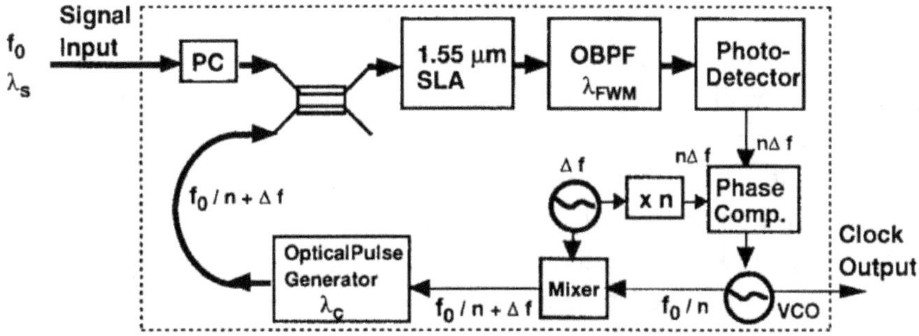

Figure 9.4. Schematic of a clock recovery diagram used for recovering the clock from a high speed optical signal using FWM in SOA (SLA in the figure). The high-speed optical input (at >100 Gb/s) is down converted to a low-frequency (at ~1 MHz) electrical signal using optical to electrical nonlinear down converter. This low-frequency signal is phase locked using a conventional electrical phase-locked loop [12].

process, the high data rate optical signals are converted to a low data rate electrical signal using a four-wave mixing process followed by a photodetector. The phase and frequency of this electrical signal are locked to that of a voltage-controlled local oscillator using conventional electrical phase-locked loop techniques.

The key to the operation of this optical phase-locked loop is the four-wave mixing optical to the electrical converter. It is a semiconductor optical amplifier which has two optical signal inputs at wavelengths λ_1 and λ_2. The wavelength λ_1 is the input wavelength carrying data and the wavelength λ_2 is the optical clock generated by a laser at the demultiplexer. The wavelength λ_1 carries the data at B Gb/s and the wavelength λ_2 is the optical clock at a frequency F ~ B/N GHz + f kHz. The four-wave mixing process produces an optical signal at a wavelength of $2\lambda_1-\lambda_2$. This optical signal has several electrical components including a low-frequency component close to Nf kHz produced by mixing of the electrical *N*th harmonic of the clock (NF ~ B + Nf) and the incoming signal at B Gb/s. This low-frequency component in the optical signal is converted to an electrical signal using an avalanche photodiode, filtered, and amplified. This low-frequency electrical signal and the Nf kHz signal from a signal generator are used in a conventional phase-locked loop configuration to lock the frequency F to B/N. The conventional phase-locked loop has a phase detector (electrical mixer) and a voltage-controlled oscillator. A stable optical clock is generated using a low-frequency electrical phase-locked loop which would provide the needed stability and an optical front end (semiconductor optical amplifier) which would convert the high data rate signal to a low-frequency optical signal. Although this conversion from a high data rate optical signal to a low-frequency optical signal is described above using a four-wave mixing process, it can also be carried out using the gain saturation process in a semiconductor amplifier.

In the latter process, the clock signal will be an input and the data signal will be the second input to the amplifier. The gain of the data signal is influenced by the high-power clock signal and this gain modulation has a component at the *N*th harmonic (Nf) of the difference frequency. The phase variance of a phase-locked loop (PLL) fabricated using the four-wave mixing (FWM) and the gain modulation schemes have been analyzed [11–14]. The performance of the gain/phase modulation scheme is limited by the gain recovery time of the amplifier to ~100 Gb/s using a differential scheme. The FWM scheme can in principle operate at very high data rates [14, 18].

9.2.3 *Optical demultiplexing*

Optical demultiplexing i.e., the process of generating several lower data rate signals from a high data rate signal (e.g., four 20 Gb/s signals from an 80 Gb/s signal), can be carried out using four-wave mixing in a semiconductor optical amplifier (SOA).

Four-wave mixing occurs both in semiconductor optical amplifiers [19] and in optical fibers. Four-wave mixing in fibers needs long fiber lengths. It does not lead to compact devices and it is not discussed here.

A schematic of optical demultiplexing using four-wave mixing (FWM) is shown in Figure 9.5. This technique allows any regularly spaced set of bits to be simultaneously extracted from the incoming data stream using a semiconductor optical amplifier (SOA) [14, 17]. Consider the high-speed data (at a data rate B) at wavelength λ_1 and an optical clock (at a clock rate B/N) at wavelength λ_2 injected into an optical amplifier. The output of the amplifier will include four-wave mixing signals at

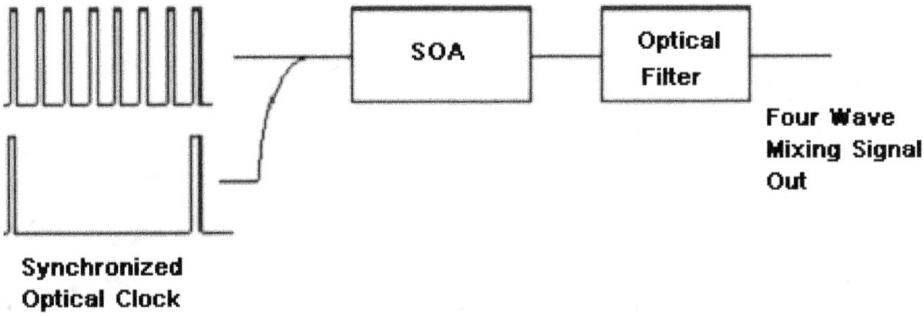

Figure 9.5. Schematic of a four-wave mixing-based demultiplexer. The four-wave mixing takes place in the semiconductor amplifier (SOA) only when there is a temporal overlap between the signal and the clock pulse.

wavelengths $2\lambda_1-\lambda_2$ and $2\lambda_2-\lambda_1$. Typically, one of the FWM signals is filtered out using an optical filter at the exit port of the semiconductor amplifier (any nonlinear element can be used) and it represents the data signal before further processing. Thus, a signal representing the data every *N*th bit time slot (where the mixing occurs with the clock signal) is generated by the four-wave mixing process (Figure 9.5). The clock signal can then be delayed by 1 additional bit each time to retrieve the original data in successive time slots as described in the following or a bit-delayed parallel system (similar to Figure 9.4) could be set up for simultaneous extraction of *N* data streams each at B/N data rate.

A complete system would need *N* SOA elements in order to extract *N* data bits simultaneously. The FWM process can have modulation bandwidths as high as 1.5 Tb/s [18]. Thus, in principle, demultiplexing from Tb/s optical data is feasible using the FWM process.

9.2.4 *OTDM transmission experiment*

The schematic of an optical time division multiplexed (OTDM) transmission setup is shown in Figure 9.6.

A stabilized mode-locked fiber laser generates short pulses at 10 GHz repetition rate. The 3.5 ps wide pulses are compressed to 0.98 ps pulses and are coded with data by transmission through an $LiNbO_3$ modulator.

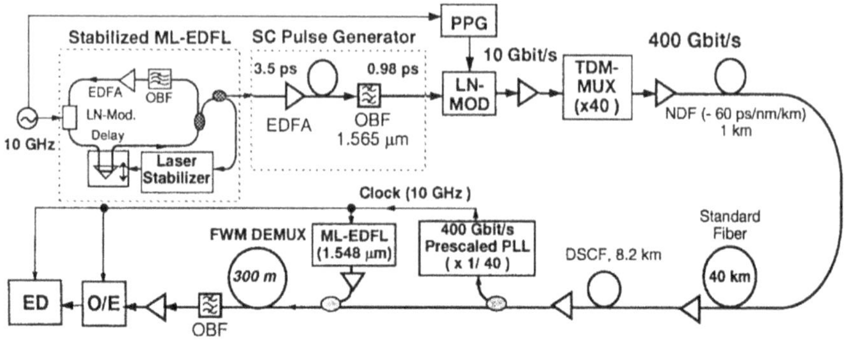

Figure 9.6. Schematic of a 400 Gb/s optical transmission experiment setup [12].

It then goes through a time division multiplexer (TDM) which converts the data stream to 400 Gb/s data. The TDM consists of a suitably designed sequence of optical delay lines.

The 400 Gb/s data are transmitted through various lengths of the fiber. The clock is extracted from the transmitted data using a four-wave mixing (FWM) in an SOA-based clock recovery process. A mode-locked fiber laser operating at 10 GHz, the exact frequency of which is recovered by phase locking to the received 400 Gb/s data, provides the optical clock. The demultiplexing in this experiment is carried out using also an FWM process in an optical fiber which serves as the nonlinear element. The error rate performance of the demultiplexed lower data rate channels is shown in Figure 9.7. The bit error rate of $<10^{-9}$ is obtained suggesting good performance.

Very high-speed OTDM transmission systems are now being developed in several research laboratories for future installment in commercial traffic. Semiconductor optical amplifiers play a key role in the development of these systems. They can be used for both optical demultiplexing and optical clock recovery.

9.2.5 *Soliton transmission*

The demand for ultra-large capacity transmission systems i.e., with capacities approaching Tb/s, continues to increase. For such systems, high-speed transmission with the bit rate of multiples of 10 Gb/s is quite

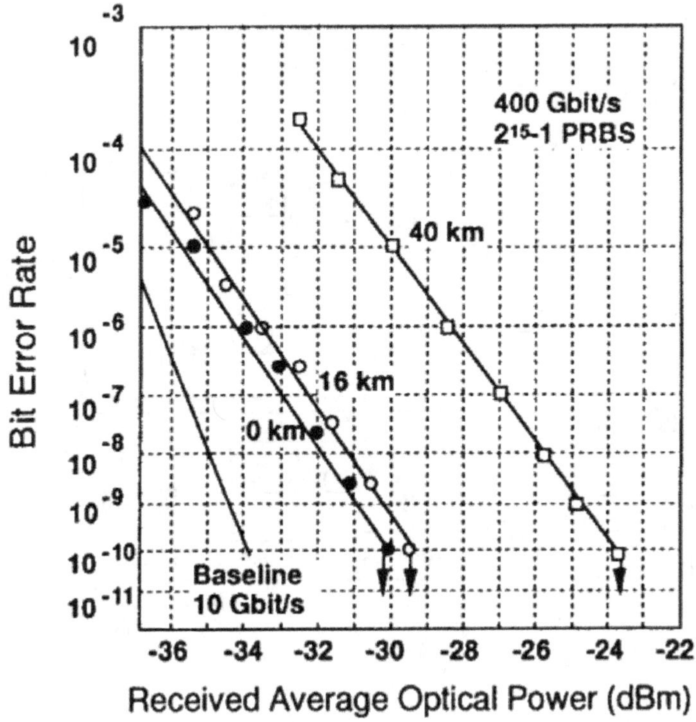

Figure 9.7. Measured bit-error-rate (BER) results at 10 Gb/s after demultiplexing of the 400 Gb/s data into forty 10 Gb/s data channels. The baseline is the BER data after the modulator. The three other lines are the data after transmission through ~10 m (labeled 0 km in the figure), 16 km, and 40 km of fiber [12].

attractive since it can reduce the number of WDM channels. Time-division multiplexing and transmission using optical solitons are important for these systems. Solitons allow error-free ultra-long-distance transmission in dispersion-managed systems [20–26].

A schematic of an optical soliton transmission system experiment at 40 Gb/s is shown in Figure 9.8. In this scheme, the cumulative chromatic dispersion is offset periodically by dispersion compensation fiber (DCF) with negative (normal) dispersion so that the total system dispersion is close to zero. The Gordon–Haus timing jitter is also reduced in this system with close to zero dispersion [27]. The experimental system uses a recirculating loop (described in the following) for ultra-long-distance transmission.

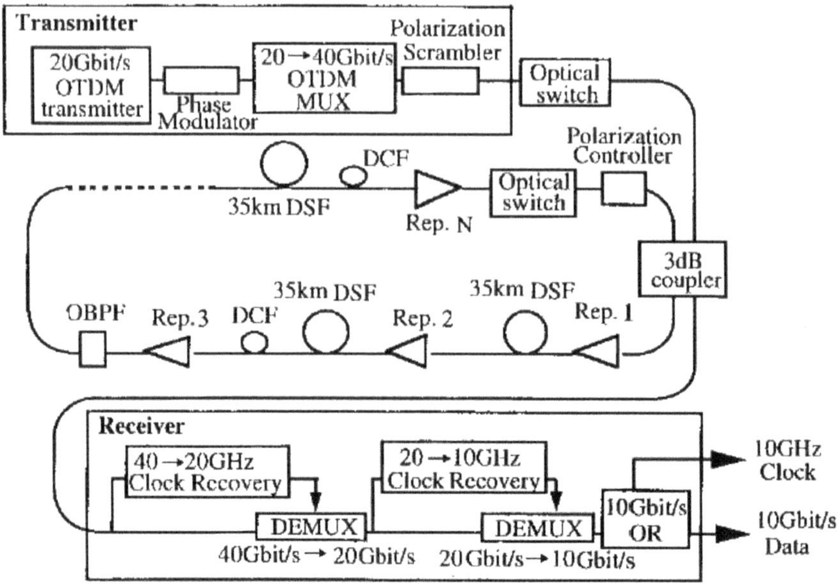

Figure 9.8. Soliton transmission experiment at 40 Gb/s [26].

To simulate very long-distance transmission (many thousands of km), a recirculating loop experiment is done [19]. In a loop experiment, the optical data circulates in a loop many times. The loop typically has all the elements of a transmission experiment, such as fibers and chain of amplifiers. Typical loop lengths are a few tens of km to a few hundreds of km. The transmitter, the receiver, and the error rate test equipment are outside the loop. The experiment starts with the transmitter switch on to load the loop with data. The data are allowed to circulate in the loop for some specified number of revolutions. The number of revolutions corresponds to the total transmission distance. The delay per loop circulation is known from loop length (~4.9 ps/km of fiber). After a specific total transmission distance, the data are let out of the loop by a switch and the error rate is measured.

The recirculating loop in the soliton experiment consists of several units. Each unit has a dispersion-shifted fiber (DSF), a 980 nm laser-pumped erbium-doped amplifier (EDFA), and a dispersion-compensating fiber (DCF). In addition, it has a polarization controller and an optical band-pass filter (OBPF). The latter reduces the spontaneous emission noise.

For the transmitter, a 20 Gb/s optical soliton data stream is produced by optical time division multiplexing (OTDM) of 10 Gb/s RZ (return to zero) data pulses. These data pulses are generated by a distributed feedback (DFB) laser, sinusoidally driven electroabsorption (EA) modulators, and two $LiNbO_3$-intensity modulators operated at 10 Gb/s with a pseudorandom binary sequence (PRBS) input. The 20 Gb/s data are generated by multiplexing the two 10 Gb/s PRBS signals. Alternately, a 20-GHz pulse from a stabilized mode-locked fiber laser followed by a 20 GHz $LiNbO_3$ modulator (or electroabsorption modulator) driven by a 20GHz electrical PRBS data will produce the optical 20GHz PRBS signal. The generated pulse width is about 9 ps. These pulses pass through an $LiNbO_3$ phase modulator driven by a 20 GHz clock. The phase condition is optimized by adjusting the delay between the optical data bits and the phase modulation signal using an electric phase shifter. This initial phase modulation is effective in improving the transmission performance. Then 40 Gb/s signal is generated by optical time division multiplexing (OTDM) of 20 Gb/s signals.

In the receiver, the clock signal is recovered from the transmitted signals. The transmitted 40 Gb/s signals are demultiplexed in the optical domain using EA modulators in two stages: 40–20 Gb/s followed by 20–10 Gb/s. The bit error rate (BER) for the demultiplexed 10 Gb/s signals is measured. The transmission measurements (i.e., BER data) are carried out for different dispersion maps. The dispersion map is varied by changing the signal wavelength which varied the average dispersion of DSF (D_{local}). The system average dispersion (D_{ave}) changes as D_{local} changes and as the compensating dispersion, by DCFs, are changed. BERs of less than 10^{-9} for all four 10 Gb/s demultiplexed channels are observed for ~9000 km of transmission for D_{local} values of less than 0.3 ps/km-nm. The experiment showed that 40 Gb/s transmission over long distances is feasible without active dispersion compensation.

System performance impairment due to soliton–soliton interaction was reduced by the polarization division multiplexing. By using the polarization division multiplexing and optimum dispersion map, 40 Gb/s single-channel transmission over 10200 km was successfully demonstrated without any active inline transmission control.

9.3 Optical Logic Studies

All-optical signal processing is expected to become increasingly important in future ultrahigh-capacity telecommunication networks.

The development of all-optical logic technology is important for a wide range of applications in all-optical networks, including high-speed all-optical packet routing and optical encryption. An important step in the development of this technology is a demonstration of optical logic elements and circuits, which can also operate at high speeds.

These logic elements include the traditional Boolean logic functions, such as XOR, OR, AND, and INVERT, and circuits, such as parity checker, all-optical adder, and shift register. Several articles on principles of all-optical logic have been published. Many of these are described in Ref. [14]. All-optical logic schemes were recognized as a possibility in early nonlinear optics studies. Semiconductor optical amplifiers (SOAs) exhibit nonlinearity in their gain and phase response to input optical signals. This property is used to construct optical logic gates using SOAs. The optical logic studies at high speeds often involve mode-locked fiber lasers as sources as they can be used to produce a series or patterned (using an $LiNbO_3$ modulator in series) optical pulses.

9.3.1 *Optical logic XOR*

All-optical exclusive OR (XOR) logic gate is an important component in OTDM communication networks. Its truth table is shown in Table 9.1. It can be used in all-optical signal processing, such as bit pattern matching [28], pseudo random number generation [29], and label swapping [30]. All-optical logic gates fundamentally rely on nonlinearities [31].

Table 9.1. The truth table of XOR.

Data B	Data A	Output XOR
0	0	0
1	0	1
0	1	1
1	1	0

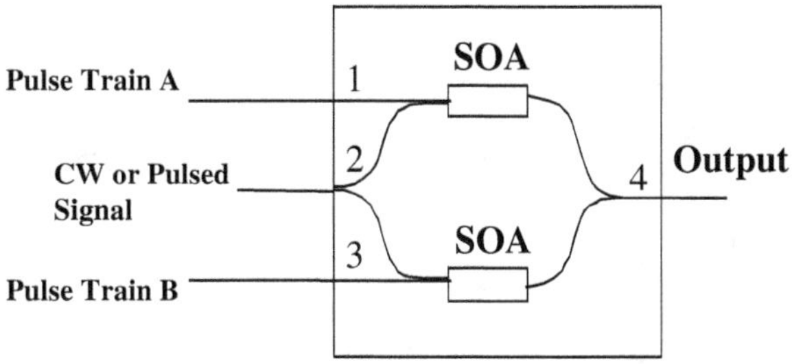

Figure 9.9. Schematic setup of the XOR gate.

Semiconductor optical amplifier has the advantages of high nonlinearity and ease of integration. The SOA-based Mach–Zehnder interferometer is the most promising candidate due to its attractive features of low-energy requirement, low latency, high stability, and compactness. This makes SOA-MZI suitable for producing complex photonic logic circuits, such as shift registers.

The optical XOR gate with SOA-MZI consists of a symmetrical MZI with two SOAs placed in the upper and lower arms of the interferometer, as shown in Figure 9.9. To perform the XOR function as shown in the truth table, two optical control beams A and B are injected into the two arms separately. The signal, a clock stream of continuous series of "1" s, or a CW beam is split into two equal parts and injected into the two SOAs. Initially, the MZI is unbalanced i.e., when A = 0 and B = 0, the signal at port 2 traveling through the two arms of the SOA acquires a phase difference of π when it recombines at the output port, thus the output is "0".

When A = 1, B = 0, the signal traveling through the arm with signal A acquires a phase change due to the cross-phase modulation (XPM) between the pulse train A and signal, and the signal traveling through the lower arm does not have this additional phase change. This results in an output "1". The same phenomenon happens if A = 0 and B = 1. However, when A = 1 and B = 1, the phase change for the signal traveling both arms is equal, hence the output is "0". When the phase shift is optimum, the best contrast ratio is achieved at the output port. Figure 9.10 shows the schematic of an SOA-based MZI chip as fabricated.

Figure 9.11 shows the results of the all-optical XOR logical gate operation. The top figure shows the results of a 20 Gb/s pattern, and the

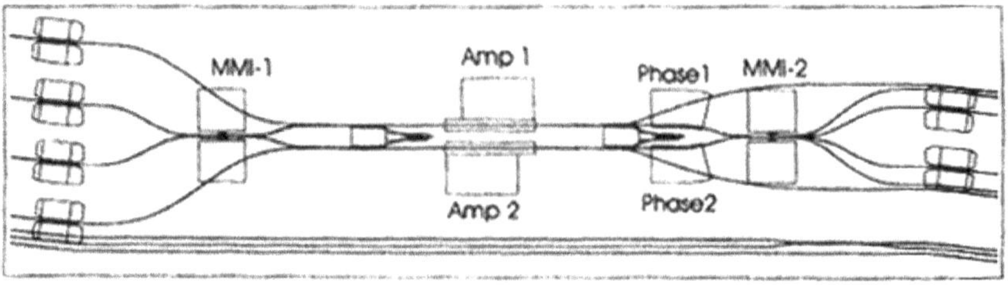

Figure 9.10. Schematic of a Mach–Zehnder interferometer (MZI) with SOA in each arm. The MMI (multimode interference) couplers are used to couple the light into two waveguides. The phase shifters are needed to balance the phase difference between the two arms. The waveguides are angled at the facet to reduce the effect of reflections from the facet.

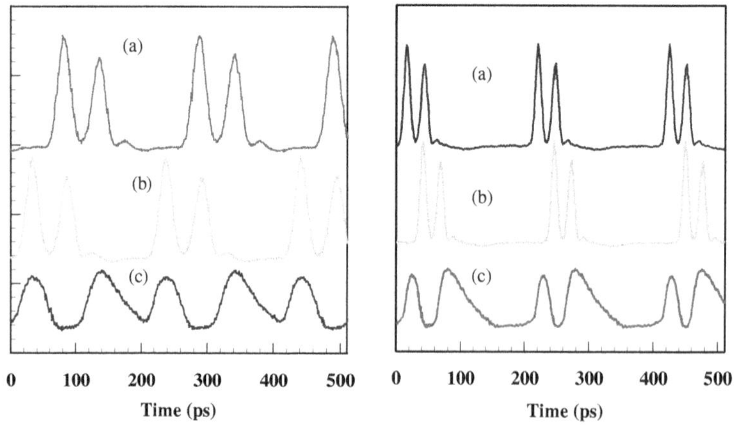

Figure 9.11. Left figure is for 20 Gb/s pattern and the right figure is for 40 Gb/s pattern. (a) Input data A. (b) Input data B. (c) XOR output [14, 32].

bottom figure shows the results of a 40 Gb/s pattern. Figure 9.11(a) and Figure 9.11(b) show the control signal A, "11001100...", and its 1-bit period delayed control signal B, "10011001". The final XOR result is clearly seen in Figure 9.11(c). However, the XOR output pulse is broad and pattern dependent, which is due to the long carrier recovery time of SOA in the SOA-MZI, and it is measured to be ~150 ps in this experiment for the 40 Gb/s case when the SOAs are operated at high current.

For laboratory experiments, the pulse trains are created using typical pulsed sources, such as gain-switched distributed feedback lasers or mode-locked semiconductors, or fiber lasers followed by optical multiplexing. An efficient method to overcome the limitation imposed by a long carrier lifetime is using a differential phase modulation scheme [32, 33]. With the differential phase modulation scheme, ultrahigh-speed (>100 Gb/s) all-optical processing, such as wavelength conversion, and all-optical demultiplexing have been demonstrated with an SOA-based interferometer [14].

Several all-optical logic functions such as XOR, AND, OR, NAND, and INVERT have been demonstrated [14, Chapter 9]. These demonstrations use mode-locked fiber lasers or semiconductor lasers as optical sources. All-optical Boolean logic operations such as XOR, OR, AND, and INVERT are important for various high-speed all-optical signal processing applications. All-optical XOR and all-optical AND gates are fundamental in the implementation of various all-optical systems, such as packet switches, binary adders, binary counters, decision circuits, optical processors, data encoders, and bit-pattern recognition circuits. These circuits are important for optical communication networks. The demonstration and higher-speed implementation of several of these operations are described [14, Chapter 10].

Optical logic circuits are being investigated by several research groups. Many optical logic circuits require several devices to be integrated on the same chip. Thus much of the advances in the optical logic circuit development would not have been possible without the advances in materials and processing technology. An avenue that is being pursued by several researchers involves hybrid integration where the optical waveguides are fabricated using silica-based technologies and the semiconductor amplifier uses InGaAsP/InP materials technology. Both sets of chips are fabricated in the form of arrays and then they are optically aligned [14]. The challenges of much of the current optical logic circuit research are intimately linked with the challenges in materials growth which include not only the investigation of new material systems but also the improvements in existing technologies, such as quantum dots, to make them more reproducible.

References

1. S. Kawanishi and M. Saruwatari, *IEEE J. Lightwave Technol.*, **11**, (1993) 2123.

2. R. S. Tucker, G. Eisenstein, and S. K. Korotky, *J. Lightwave Technol.*, **6**, (1988) 1737.
3. S. Nakamura, Y. Ueno, and K. Tajima, *IEEE Photon. Technol. Lett.*, **10**, (1998) 1575.
4. K. L. Hall and K. A. Rauschenbach, *Opt. Lett.*, **23**, (1996) 1271.
5. S. Kawanishi, In 2nd Optoelectronics & Commun. Conf. (OECC '97) Technical Digest, 12, Seoul, Korea, 1997.
6. K. E. Stubkjaer, *et al.* In 2nd Optoelectronics & Commun. Conf. (OECC '97) Tech. Dig. 178, Seoul, Korea, 1997.
7. K. L. Hall and K. A. Rauschenbach, *Electron. Lett.*, **32**, (1996) 1214.
8. N. S. Patel, K. L. Hall, and K. A. Rauschenbach, *Appl. Opt.*, **37**, (1998) 2831.
9. O. Kamatani and S. Kawanishi, *J. Lightwave Technol.*, **14**, (1996) 1757.
10. S. Kawanishi, T. Morioka, O. Kamatani, H. Takara, J. M. Jacob, and M. Saruwatari, *Electron. Lett.*, **30**, (1994) 981.
11. O. Kamatani, S. Kawanishi, and M. Saruwatari, *Electron. Lett.*, **30**, (1994) 807.
12. S. Kawanishi, *IEEE J. Quantum Electron.*, **QE-34**, (1998) 2064.
13. A. D. Ellis, T. Widdowson, X. Shan, G. E. Wickens, and D. M. Spirit, *Electron. Lett.*, **29**, (1993) 990.
14. N. K. Dutta, *Semiconductor Optical Amplifiers*, (2nd edn.) (World Scientific, Singapore, 2014).
15. M. Eiselt, W. Pieper, and H. G. Weber, *Electron. Lett.*, **29**, (1993) 1167.
16. S. Kawanishi and O. Kamatani, *Electron. Lett.*, **30**, (1994) 1697.
17. S. Kawanishi, K. Okamoto, M. Ishii, O. Kamatani, H. Takara, and K. Uchiyama, *Electron. Lett.*, **33**, (1987) 976.
18. C. Wu and N. K. Dutta, *J. Appl. Phys.*, **87** (2000) 2076.
19. N. K. Dutta, *Optoelectronic Devices* (World Scientific, Singapore, 2018), Chapter 11.
20. M. Suzuki, N. Edagawa, H. Taga, H. Tanaka, S. Yamamoto, and S. Akiba, *Electron. Lett.*, **30**, (1994) 1083.
21. F. Favre and D. LeGuen, *Electron. Lett.*, **31**, (1995) 991.
22. G. Aubin, E. Jeanney, T. Montalant, J. Moulu, F. Pirio, J.-B. Thomine, and F. Devaux, *Electron. Lett.*, **31**, (1995) 1079.
23. J. King, I. Hardcastle, J. Harvey, P. Greene, B. Shaw, M. Jones, D. Forbes, and M. Wright, *Electron. Lett.*, **31**, (1995) 1090.
24. M. Suzuki, I. Morita, N. Edagawa, S. Yamamoto, H. Taga, and S. Akiba, *Electron. Lett.*, **31**, (1995) 2027.

25. G. Aubin, T. Montalant, J. Moulu, F. Pirio, J.-B. Thomine, and F. Devaux, *Electron. Lett.*, **32**, (1996) 2188.
26. I. Morita, K. Tanaka, M. Suzuki, and N. Edagawa, *IEEE J. Lightwave Technol.*, **17**, (1999) 2506.
27. J. Gordon and H. Haus, *Opt. Lett.*, **11**, (1986) 665.
28. A. J. Poustie, K. J. Blow, R. J. Manning, and A. E. Kelly, *Opt. Commun.*, **159**, (1999) 208.
29. T. Fjelde, A. Kloch, D. Wolfson, B. Dagens, A. Coquelin, I. Guillemot, F. Gaborit, F. Poingt, and M. Renaud, *IEEE Photon. Technol. Lett.*, **13**, (2001) 750.
30. N. S. Patel, K. L. Hall, and K. A. Rauschenbach, *Appl. Opt.*, **37**, (1998) 2831.
31. M. Jinno and T. Matsumoto, *Opt. Lett.*, **16**, (1991) 220.
32. H. Chen, G. Zhu, Q. Wang, J. Jaques, J. Leuthold, A. B. Piccirilli, and N. K. Dutta, *Electron. Lett.*, **38,** (2002) 1271.
33. Q. Wang, G. Zhu, H. Dong, J. Jaques, J. Leuthold, A. B. Piccirilli, and N. K. Dutta, *IEEE J. Quantum Electron.*, **40**, (2004) 703.

Chapter 10

Planar Waveguide-Based Mode-Locked Lasers

10.1 Introduction

Rare-earth-doped lasers are important devices for both terrestrial and submarine optical communication systems, for sensor systems, and as laser sources at various wavelengths for various applications including high-power applications. Optical fiber gain mediums are typically 10 m–30 m in length. The core and cladding of the fiber form a waveguide through which the light travels and undergoes amplification. The core is typically doped with the rare-earth dopant. The length of a planar waveguide amplifier is typically limited to tens of centimeters both due to fabrication limitation and higher optical attenuation in planar waveguides. In order to achieve high gain in ~10 cm-long planar waveguides, the rare-earth concentration is typically two orders of magnitude higher than that for fibers. This may introduce nonradiative mechanisms in some rare earths such as Er due to up-conversion.

Planar waveguide amplifiers have been fabricated in many glass materials by several methods [1–8]. These include modifications of refractive index at the surface by ion exchange [9], ion implantation [11], or ion diffusion [12]. Also, a waveguide can be formed by the growth or deposition of one material on top of another with a different refractive index [13, 14]. Usually, the waveguides formed by these techniques are in the form of a planar thin film which may be patterned into an array of channel waveguides using photolithographic techniques.

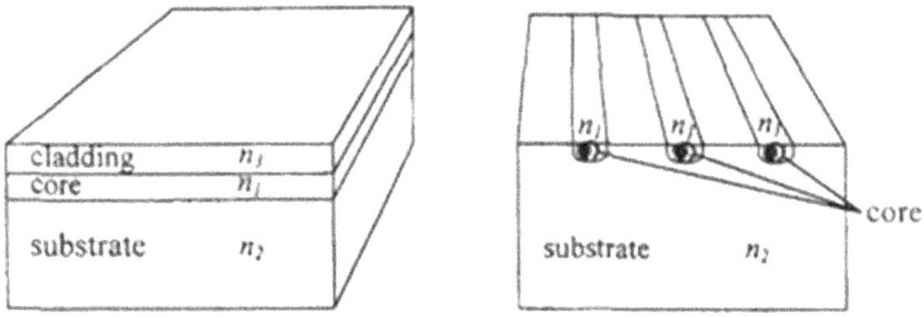

Figure 10.1. Schematic of typical planar and channel waveguide structures [8]. The left figure shows a planar waveguide and to the right is channel waveguide. The core index $n_1 > n_2$: cladding index.

The optical confinement of the pump light in the waveguide results in a low threshold for net gain or laser action for rare-earth-doped planar waveguide devices.

Planar waveguide-based mode-locked lasers are fabricated using the rare-earth-doped planar waveguide amplifier as the gain medium. These devices can be made quite compact if the reflectors needed for laser action are fabricated by material deposition on the facet of the waveguide. These waveguides are optically pumped by the other lasers to produce gain. However, the key to achieving low threshold operation is low optical loss and high gain in the material. The schematics of typical planar and channel waveguide structures are shown in Figure 10.1.

Efficient coupling of high-power diode pump sources is important for low pump threshold. Methods such as optical fiber-based couplers for single-mode waveguides, specialized lens optics, and diode beam shapers have been used for efficient coupling [15–18]. In general, the pump light can be coupled from the end faces of the waveguide, from the side, or from the surface. This is shown schematically in Figure 10.2.

Typically, a planar waveguide has a symmetric core and cladding region with the high index core region being rare-earth-doped bounded by the low index cladding regions. Similar to the case for fibers, a double-cladded waveguide can be fabricated for the planar waveguide geometry (Figure 10.3). In the latter case, the inner core is doped with rare earth, and the outer layers (Figure 10.3) are not doped with rare earth and have nearly the same index as the inner core. The entire region is bounded by

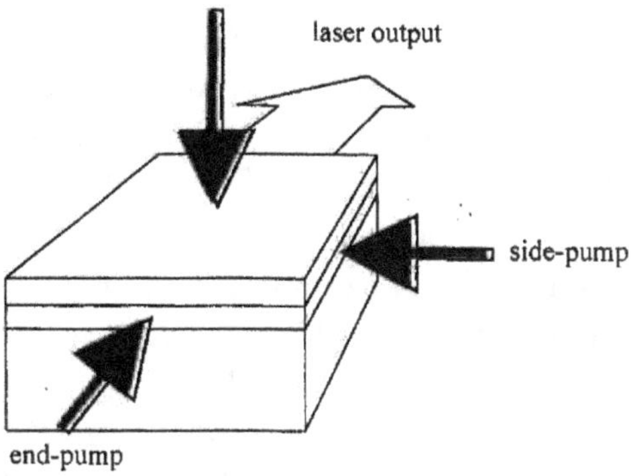

Figure 10.2. Schematics of pump laser coupling into the waveguide [8].

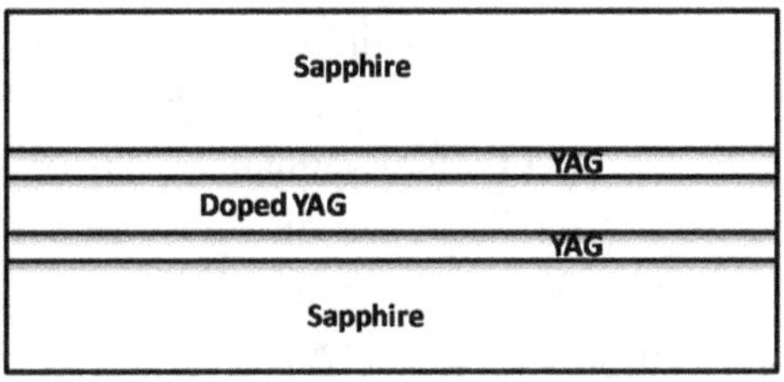

Figure 10.3. Schematic of a double-clad waveguide. Both of these planar structures can be processed into channel waveguides. The undoped YAG regions have higher index than the sapphire.

lower index cladding layers. Figure 10.3 shows an Nd-doped YAG (Yttrium-Aluminum-Garnet) double-clad structure on a sapphire substrate.

Diode lasers are suitable for end pumping because the laser mode is laterally spread like the waveguide allowing good overlap between the gain region in the waveguide and the diode laser mode.

10.2 Er-Doped Planar Waveguide

As mentioned earlier, in order to achieve significant gains in ~10 cm-long waveguides, the concentration of rare-earth dopant (e.g., Er) in planar waveguides needs to be two orders of magnitude greater than that for optical fibers. At higher concentration, the average spacing between Er ions decrease. At close spacing, the Er ions lose the exited energy by a nonradiative process known as up-conversion. Consider two Er ions in their upper energy levels; in the up-conversion process, one electron (in one Er ion) decays nonradiatively to the ground state, transferring its energy to the other Er ion, which gets excited to a higher-energy level. The clustering of Er ions in silica glass doped to high Er concentrations accentuates this nonradiative decay mechanism. Up-conversion is less pronounced in other glass types (e.g., soda–lime glass due to its cage-like structure) where the atoms in the glass surround each Er ion and prevent clustering. These types of glasses have been used in the study of planar waveguide amplifiers.

Er-doped silica-based nearly single-mode waveguide has been reported by several authors. One of the early reports is by Shuto *et al.* [19]. They reported a waveguide which has an Er-doped phosphosilicate glass core formed by PECVD (plasma-enhanced chemical vapor deposition). The doped core has a silica glass cladding layer formed by flame hydrolysis deposition (FHD) on a Si substrate. The core layer is deposited on the FHD-grown bottom cladding layer at a substrate temperature of 400°C using the PECVD growth technique using SiH and N_2O gases. P-alkoxide and Er-chelate are used as sources for P and Er doping. The doping densities (of P and Er) are controlled (using mass flow controllers) by transporting the vapor of the heated sources into the reaction region using N_2 carrier gas. The core is annealed after deposition. The lateral size of the core regions is etched using reactive ion etching and then the core layer is embedded in the cladding region grown by the FHD process. Thus the Er forms a buried waveguide region. The refractive index of the core is ~1.6% higher than that of the surrounding cladding regions. The Er concentration in the core is 0.48%.

Prior to amplification studies, the propagation loss of the waveguide is measured. The absorption spectrum is shown in Figure 10.4. The optical loss at 1530 nm is ~1.1 dB/cm. The waveguide is 7.5 cm long and has a spot size of 8×11 μm^2 at 1550 nm. Pump laser light at a wavelength of 980 nm is coupled into the waveguide. The output signal at 1530 nm is measured as a function of pump power. The optical gain is determined

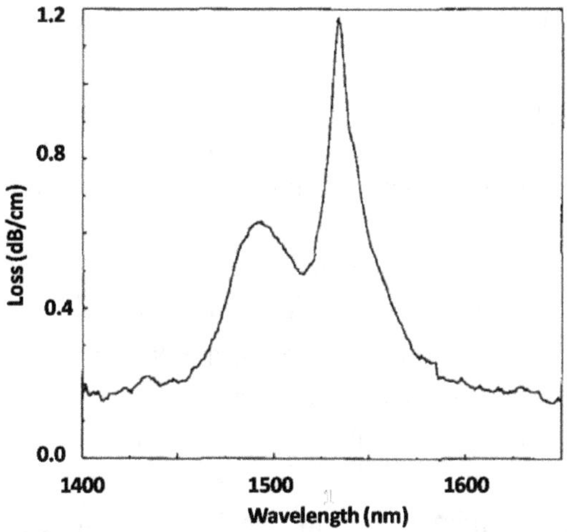

Figure 10.4. Absorption loss as a function of wavelength [19].

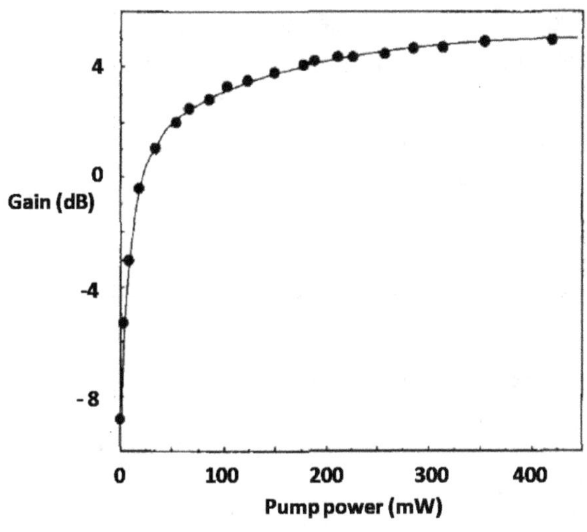

Figure 10.5. Net gain at 1530 nm as a function of input pump power at 980 nm [19].

from the ratio of the output to input signal power. The data are shown in Figure 10.5.

Using an S-shaped waveguide and a similar fabrication process as described earlier, a gain of ~10 dB has been reported at a pump power of

100 mW and a gain as high as 16 dB has been reported at a pump power of ~210 mW [20].

Soda–lime glass has been used as a core for Er doping [21]. For high gain at low pump power optimizing Er concentration is important in addition to waveguide dimensions and index difference between the core and the cladding region. Although high Er concentration provides high gain, it also increases the up-conversion rate [22–28]. For the experiment of Ref. [21], an Er concentration of $0.7 \times 10^{20} cm^{-3}$ is used. Er-doped films are deposited by RF sputtering from a soda–lime glass target (doped with Er) on a silica lower cladding layer. This layer is 15 μm thick and is deposited on a silicon substrate by thermal oxidation. Ion milling is used to produce 3–9 μm-wide waveguides on the deposited Er-doped film. The etching-induced side wall roughness is removed by reflowing the surface of the waveguides at a high temperature. The refractive index difference between the core and lower cladding layer is 3%. The end faces of the Er-doped waveguides (6 cm long) are polished to optimize the fiber coupling.

The optical gain of the amplifier is measured as a function of pump power. The experimental data on small signal gain with an input power of –42 dBm at 1530 nm is shown in Figure 10.6. The low threshold (~10 mW) for net gain is due to low scattering loss, tightly confined mode, and low glass absorption loss [21].

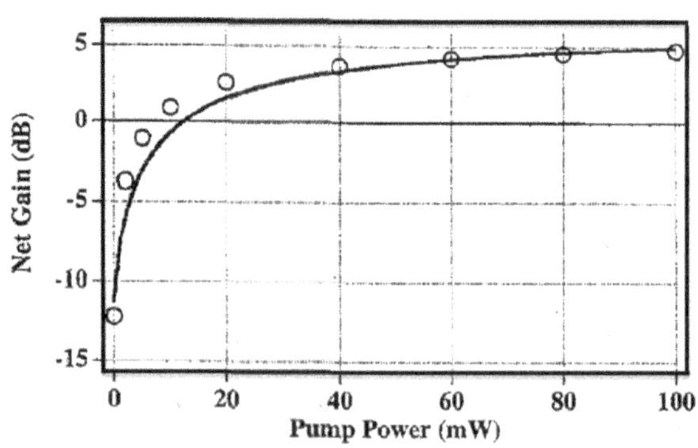

Figure 10.6. Measured optical gain as a function of pump power. Pump wavelength is 980 nm [21].

10.3 Nd-Doped Planar Waveguide

Single-mode Nd-doped waveguide amplifiers and lasers on glass have been investigated [29–36]. In an early experimental study, Nd-doped buried channel waveguides were used [37]. The waveguides are formed by a field-assisted ion migration process in the phosphate glass substrate (HOYA glass doped with Nd, 3.3% by weight). The waveguides are buried with sodium–silver ion exchange in a molten salt (20% NaNO+ 80% CsNO). The waveguides are ~1 cm long and the ends of the waveguides are polished. The multimode waveguides have elliptical cross-sections with long and short axes of 33 μm and 17 μm, respectively.

For amplification measurement, pump light from an Ar laser (514 nm) or an 808 nm diode laser is coupled to the waveguide using a single-mode optical fiber. The signal light (~10 μW) is coupled using a dichroic mirror coupler to the same waveguide. The measured gain as a function of optical power is shown in Figure 10.7.

Single-mode Nd-doped waveguide lasers fabricated on a Si substrate have been reported [38–41]. To fabricate the doped waveguide, $SiCl_4$ and

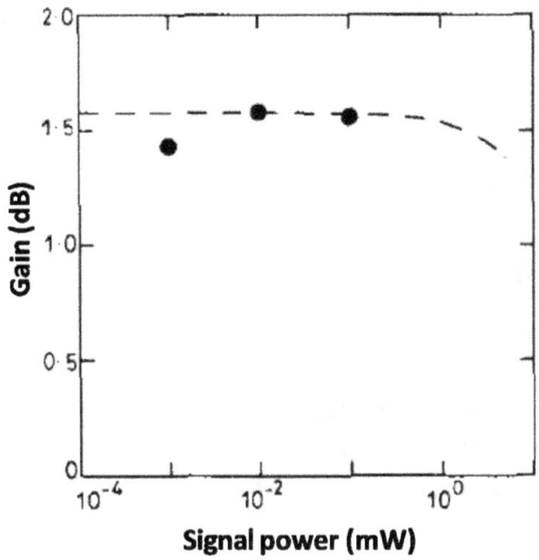

Figure 10.7. Optical gain as a function of pump power for an Nd glass waveguide amplifier using an 808 nm pump [37].

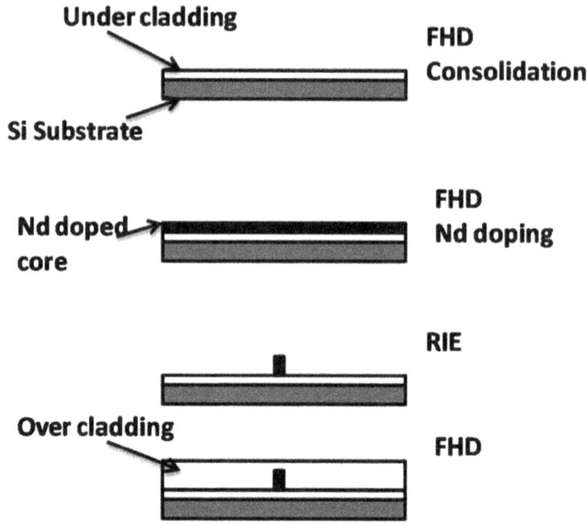

Figure 10.8. Schematic of the fabrication process [41]. The process starts at the top.

PCl_3 are hydrolyzed (FHD) using an oxy-hydrogen burner to form a low-density soot on a Si substrate with a 10 μm thick SiO_2 layer (thermally grown). The soot is fused and immersed in an alcohol solution of NdCl. After drying, the porous structure is fused to form the Nd-doped SiO_2-P_2O_5 glass. Ridge waveguides are formed on the glass using standard photolithography and reactive ion etching (RIE). Flame hydrolysis technique is then used to form a cladding layer of SiO_2-P_2O_5-B_2O_3. The cladding layer is index matched to the buffer layer and is ~20 μm thick. The schematic of the fabrication process is shown in Figure 10.8.

In one experiment, the waveguide dimensions are 12 μm × 8 μm and the Nd doping density is 0.38 wt%. The index difference between core and cladding is 0.87% and the waveguide device length is 5.7 cm. The laser cavity is formed by using dielectric-coated mirror in close proximity to the waveguide ends using index-matching liquids. The input mirror has a reflectivity of 99.9% at the laser wavelength (~1060 nm) and transmission of 90% at the pump wavelength (804 nm). The output mirror has a reflectivity of 95% at the laser wavelength and transmission of 95% at the pump wavelength. The laser output as a function of pump power is shown in Figure 10.9. The threshold power is ~20 mW.

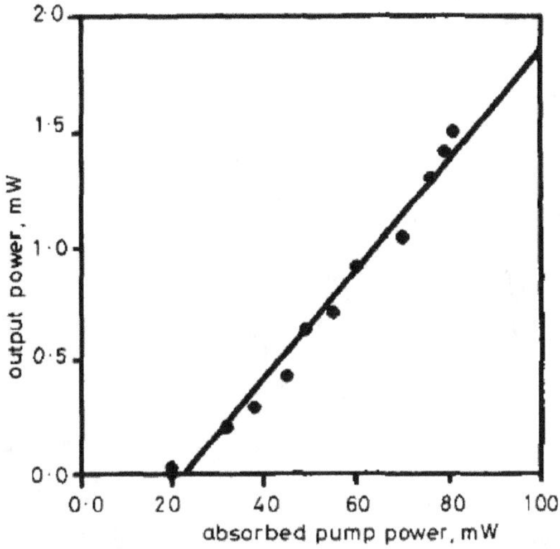

Figure 10.9. Laser output power as a function of absorbed pump power [42].

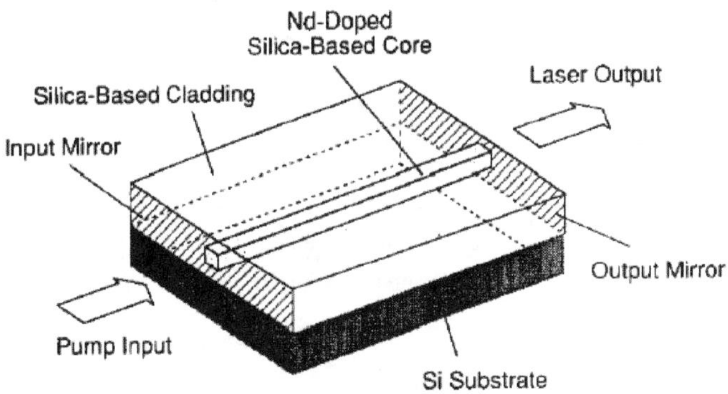

Figure 10.10. Schematic of the Nd-doped waveguide laser [41].

The schematic of another waveguide laser design is shown in Figure 10.10. For this laser, the Nd-doped core has an Nd doping level of 0.2 wt% and a thickness of 6 μm. The index difference between the core and cladding is 0.5%. The waveguide length is 4 cm and the widths varied in the 6–30 μm range. Dielectric mirrors are deposited on the ends of the waveguide cavity. The input mirror had high reflectivity (R = 99.9%) at the lasing wavelength and high transmittance (>95%) in the pump wavelength range of 730–820 nm.

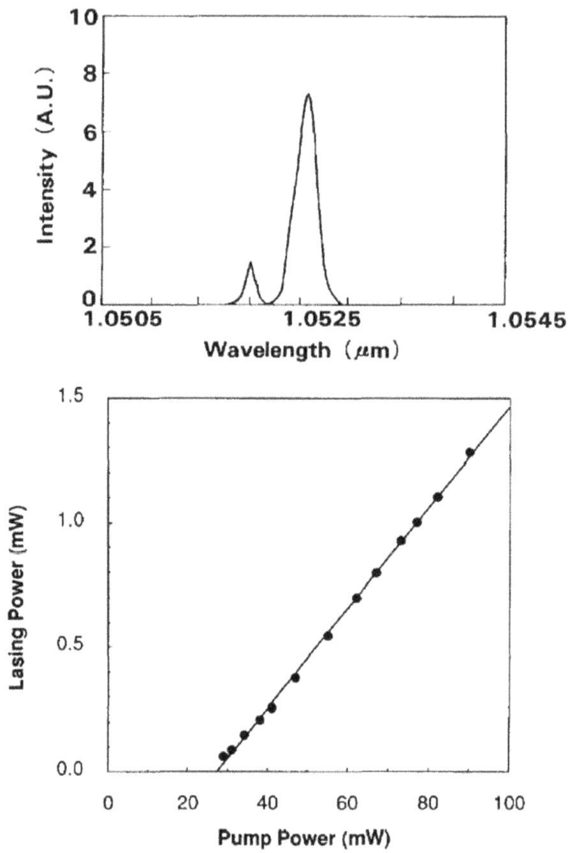

Figure 10.11. Lasing spectra (top) and laser output power as a function of pump power (bottom). The waveguide width is 8 μm [41].

The reflectivity of the output mirror is 95% at the lasing wavelength with a transmittance of >95% at the pump wavelengths. The lasing spectrum and the output power as a function of pump power for an Nd-doped waveguide laser in Figure 10.10 are shown in Figure 10.11. The threshold power is ~26 mW. The threshold pump power is found to vary with the pump wavelength.

10.4 Er–Yb Co-Doped Planar Waveguide

Similar to the case of Er-doped fiber amplifiers, Er–Yb co-doped waveguide amplifiers have been investigated [43–50]. This has the potential of reducing the effect of up-conversion which is a nonradiative process at high Er densities. In Er–Yb co-doped amplifiers, energy transfer from excited Yb to upper Er level results in gain for the Er transition.

The Er–Yb co-doped glasses containing ZnO are prepared by standard melting technique. The oxides used in the melt are SiO_2, Na_2O (14.2 mol%), ZnO, Al_2O_3, Er_2O_3, and Yb_2O_3. The samples prepared had 0.6 at% of Er but differing amounts of Yb in the 0–1.2 at% range [47]. Although typically Al_2O_3 is used to enhance the solubility of Er, ZnO is used in this case. The glass components are melted in a crucible at 1470°C. Mechanical stirring and quenching at 520°C is used to produce the bulk glass which is cut into 40 × 20 × 2 mm pieces for waveguide fabrication. Waveguide boundaries are formed by standard contact lithography. The widths of the waveguides vary from 2 μm to 6 μm. Waveguides are formed by the ion exchange process using K to Na or Ag to Na ion exchange. The process of ion exchange with K ions was performed in a melt of potassium nitrate at 400°C for various times, depending on the exact Yb concentration in the glass substrate. The Ag ion exchange is performed in eutectic $NaNO_3$–KNO_3 melt with an additional 3.6 wt% of silver nitrate at 280°C. The experimental setup used for measurements is shown in Figure 10.12.

The pump laser (980 nm) and the signal (used for gain spectrum measurement) are coupled into the waveguide using a WDM multiplexer.

The intensity profiles of the guided modes and the waveguide scattering losses are experimentally determined. The mode field profiles at the signal and pump wavelengths are obtained by near-field IR imaging. In potassium and silver ion-exchanged waveguides, the mode dimensions are found to be 11 × 14 μm and 3.5 × 5 μm at 1550 nm, respectively. In the Er emission band, the K and Ag ion-exchanged channel waveguide losses are estimated to be 0.18 dB/cm and 0.8 dB/cm, respectively.

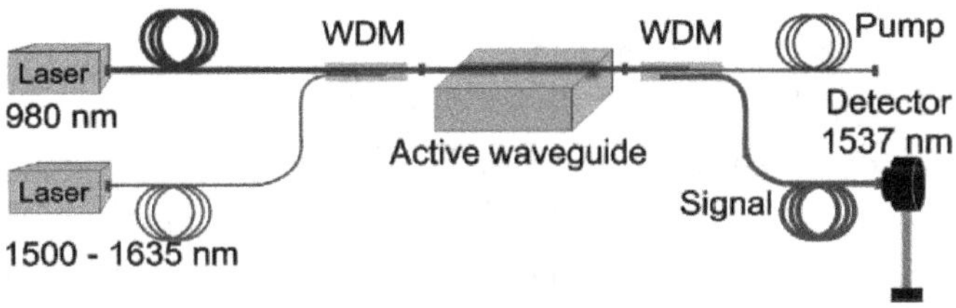

Figure 10.12. Schematic of the experimental setup used for Er–Yb co-doped ion-exchanged planar waveguide measurements [47].

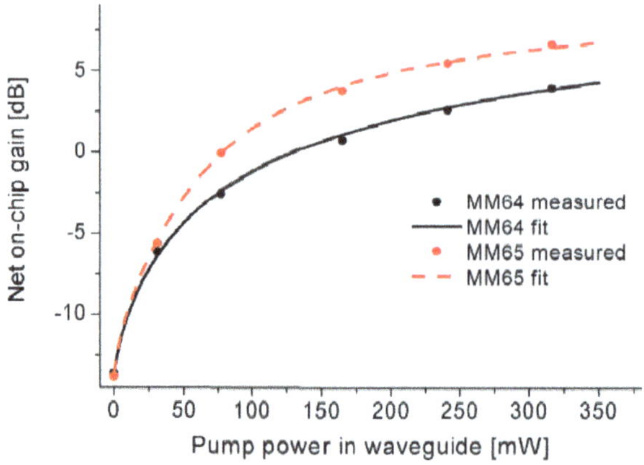

Figure 10.13. Net gain as a function of pump power for two cases: (1) Er-doped glass (solid line), (2) Er–Yb co-doped glass with equal amounts of Er and Yb (dashed line) [47].

The net on-chip gain as a function of pump power for K-ion-exchanged waveguide is shown in Figure 10.13. The data are shown for two cases: (1) Er-doped glass only and (2) Er–Yb co-doped glass with equal amounts (both 0.6 at%) of Er and Yb. Optical gain at various input wavelengths is measured by using a tunable input laser. The spectral dependence of gain for several pump powers for K-ion-exchanged waveguide is shown in Figure 10.14. An up-conversion coefficient for Er of 1×10^{-24} m^{-3} s^{-1} is determined from the data on different Yb concentrations and from known model parameters.

10.5 Yb-Doped Planar Waveguide

Yb-doped waveguide amplifiers have been investigated [51–55]. They generally have much higher gain due to high solubility and lack of nonradiative up-conversion process such as that present in Er. High optical gain has been reported in Yb-doped micro structured $KGd_xLu_yYb_{1-x-y}(WO_4)_2$ channel waveguides [51]. Yb-doped $KY(WO_4)_2$ layers were grown by liquid-phase epitaxy onto an undoped, (010)-oriented, polished, 1-cm^2-sized $KY(WO_4)_2$ substrate in a $K_2W_2O_7$ solvent at temperatures of 920–923°C [52]. Co-doping the layers with appropriate percentages of optically inert Gd^{3+} and Lu^{3+} ions enables the growth of lattice-matched layers [53]. The replacement of a fraction of Lu^{3+} by Yb^{3+} ions of similar

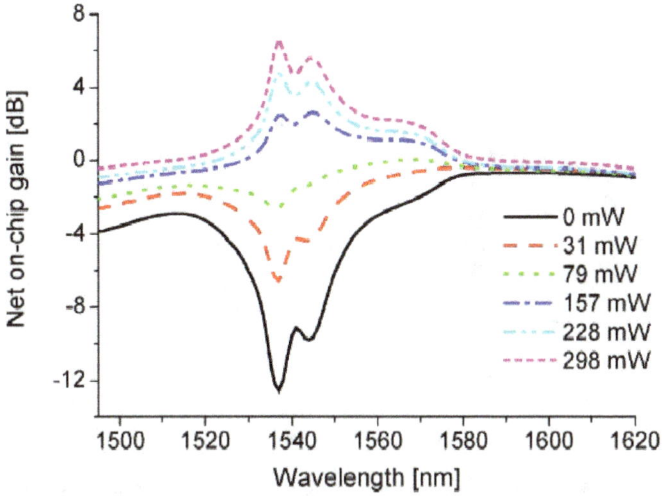

Figure 10.14. Spectral dependence of gain/absorption at various pump powers for the Er–Yb co-doped glass. Er and Yb concentrations at 0.6 at% and 1.2 at%, respectively [47].

ionic radius resulted in up to 47.5 at% Yb^{3+}-doped $KGd_xLu_yYb_{1-x-y}(WO_4)_2$ layers. The index difference between the Yb-doped layer and substrate is ~2×10^{-2}. The samples were between 180 and 320 μm long and the Yb-doped layer surface was polished down to 2.2–3.5 μm thickness. The ridge waveguides are fabricated using Ar beam etching. This produced ~1.4 μm deep ridges about 6 μm wide. Undoped $KY(WO_4)_2$ is then grown over the sample burying the ridges.

The gain of the sample is measured using pump light from a Ti-sapphire laser emitting near 932 nm. The signal light is at 980 nm. The measured gain for different at% of Yb is shown in Figure 10.15. The measured maximum gain is ~900 dB/cm at 60 mW of pump power.

The spectral dependence of gain is measured using a tunable laser for signal input. The measured data and results of a model are shown in Figure 10.16. The line with arrows indicates a 55 nm band for which the optical gain exceeds 150 dB/cm.

The above result illustrates that high gain can be obtained in Yb-doped planar waveguides. This suggests the possibility of high-power laser emission in Yb doped with appropriate thermal management and pump input geometry.

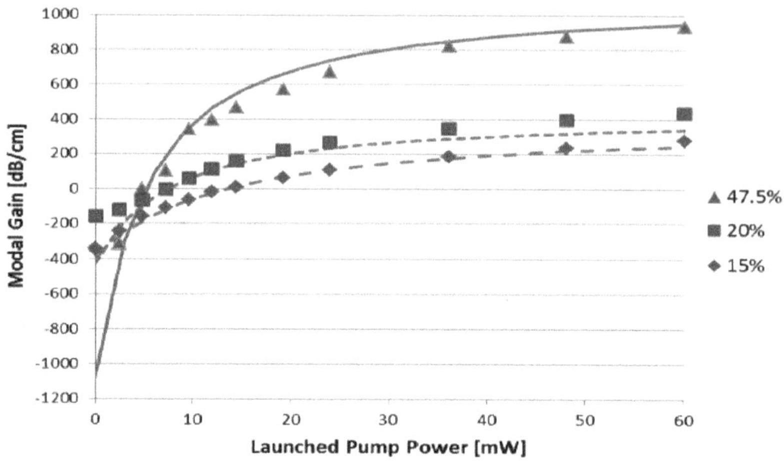

Figure 10.15. Measured small signal modal gain as a function of pump power for different at% of Yb. Signal wavelength is at 980 nm [51].

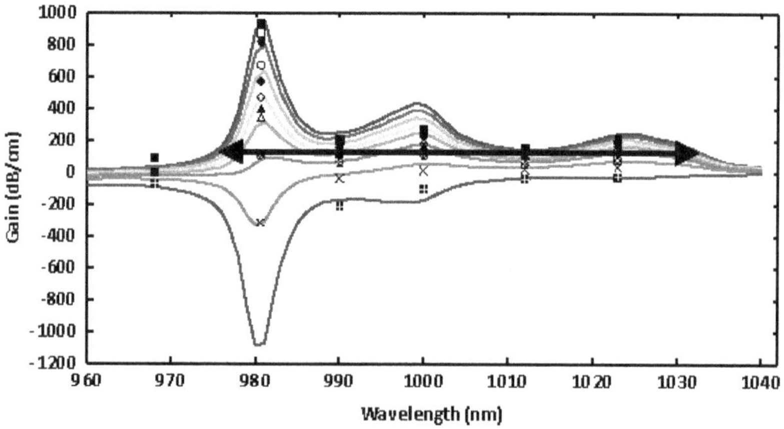

Figure 10.16. The modal gain for different launched powers: solid lines are calculated results, and the points indicate experimental data. The line with arrows indicates a 55 nm band for which the optical gain exceeds 150 dB/cm [51].

10.6 Planar Waveguide Mode-Locked Laser

Both passive and active mode locking for planar waveguide lasers have been demonstrated. For active mode locking, the modulator is integrated with a gain waveguide. For passive mode locking, a fiber-coupled saturable absorber mirror is generally used.

10.6.1 *Passive mode locking*

Passively mode-locked Er/Yb mode-locked laser has been produced using Er/Yb co-doping in phosphate glass [52, 53]. It is shown in Figure 10.17. The waveguide is 2 cm long. A laser diode emitting near a wavelength of 977 nm is used as a pump laser. A partially reflecting mirror served as an output coupler (OC). The other facet was cut and polished at Brewster's angle to minimize reflections. A microscope objective collimated the light from the waveguide, and this beam was focused onto the semiconductor saturable absorber mirror (SESAM) with a second lens. The SESAM had a multiple quantum-well (MQW) layer made of three low-temperature-grown (LT) InGaAs–GaAs quantum wells in an antiresonant configuration on top of a 22.5-period distributed Bragg reflector (DBR) consisting of alternating AlAs–GaAs quarter-wave layers [54, 55]. It is mounted on a piezoelectric transducer (PZT) for active cavity-length control.

The mode-locked laser produced pulses with a pulse width of 6-ps full-width at half-maximum (FWHM) and a spectral bandwidth of 0.65-nm FWHM centered at 1533.5 nm. The time–bandwidth product was 0.5, indicating the presence of some pulse chirp. The average output power for this laser was 4.5 mW with an estimated launched pump power

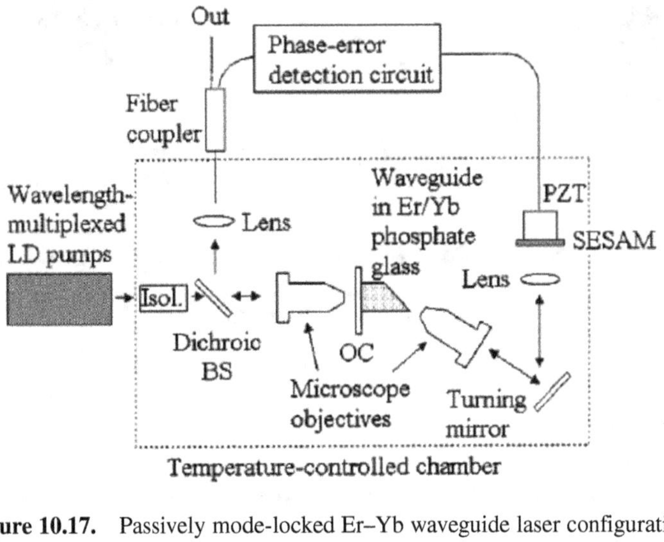

Figure 10.17. Passively mode-locked Er–Yb waveguide laser configuration.

Note: LD: laser diode. Isol.: optical isolator. BS: optical beam splitter. OC: output coupler. PZT: piezo-electric transducer. SESAM: semiconductor saturable absorber mirror [52].

of 180 mW at 977 nm. Shorter pulse durations are possible with SESAMs that have shorter recovery times.

Er and Yb co-doped (0.5 mol% Er3+, 2 mol% Yb3+, and 5 mol% Ce) Fluorozirconate glasses (ZBLAN) are also used as a waveguide gain medium [56]. ZBLAN has a high rare-earth solubility and low phonon energies which make it an excellent candidate for compact and efficient waveguide lasers. ZBLAN also have high transparency from the UV to ~4 μm wavelength [56]. Depressed cladding waveguides are directly imprinted in the glass substrate (length = 13 mm) by ultrafast laser pulse (~250 fs) inscription (5 MHz rate, λ = 520 nm), produced from a frequency-doubled laser. The schematic of the setup is shown in Figure 10.18.

A fiber-coupled diode laser is used as a pump. The repetition rate of pulses can be varied by varying the laser cavity length. The mode-locked output is reflected out of the beam path by a dichroic mirror (DM) and

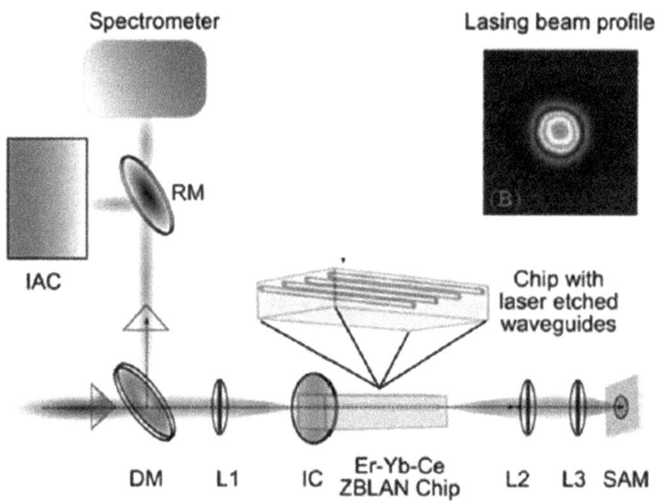

Figure 10.18. Schematic of the passively mode-locked ultrafast Er–Yb–Ce ZBLAN waveguide laser.

Note: DM: dichroic mirror. L1: F = 30 mm. IC: input coupler (reflectivity at 1550 nm = 95%). L2: F = 25 mm. L3: (F = 4.5 mm). SAM: semiconductor-based saturable absorber. RM: retractable mirror to get intensity autocorrelation traces; optical spectrum and rf spectrum. Inset: near-field beam profile of the mode-locked laser output. The two mirrors of the ML laser are IC and SAM [56].

passes through a combination of half-wave plate and polarization beam-splitter to get the autocorrelation trace and spectrum. For a cavity length of ~96 cm, the pulse repetition rate is 156 MHz and the pulse width is 180 fs. For a shorter cavity which gives a pulse repetition rate of 1.3 GHz, the pulse width is 410 fs.

Mode locking has also been demonstrated using direct laser writing of Er-doped planar waveguide [57]. The material is a commercial phosphate glass doped with 2 wt% of Er_2O_3 and 4 wt% of Yb_2O_3. Direct waveguide writing is carried out by a diode-pumped cavity-dumped femtosecond Yb:KYW laser, operating at 1040 nm wavelength [58] with 885 kHz repetition rate, 250 nJ pulse energy, and 350 fs pulse duration. An objective lens is used to focus the femtosecond pulses 170 mm inside the glass. The sample is translated in a direction perpendicular to the laser beam. Both end facets of the waveguides are polished after laser writing. The parameters are optimized to obtain the best waveguide performance. The waveguide is connected with fiber pigtails and the usual WDM element is used for pump coupling. The schematic of the laser setup is shown in Figure 10.19. The setup uses carbon nanotubes (CNT) as a saturable absorber for mode locking.

As expected, the laser has a narrow spectrum under CW operation and has a broad spectrum under mode-locked operation. The autocorrelation trace of the mode-locked pulse is shown in Figure 10.20. The laser

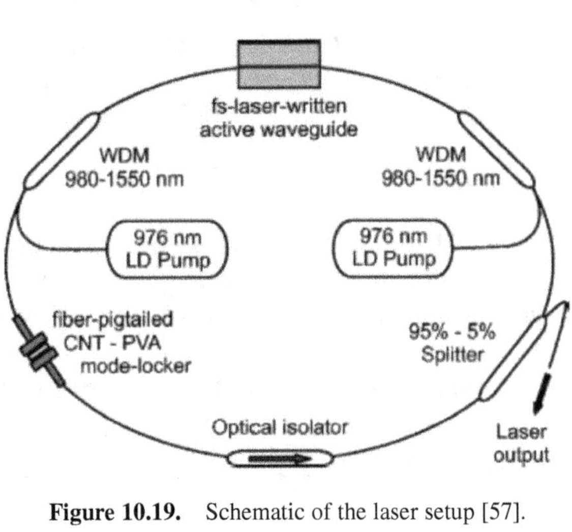

Figure 10.19. Schematic of the laser setup [57].

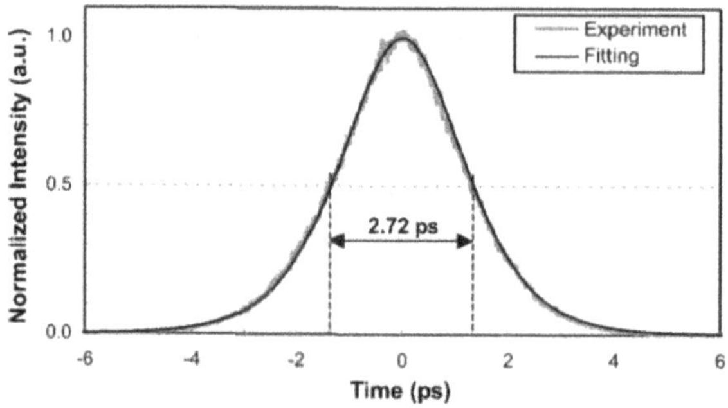

Figure 10.20. Autocorrelation trace of the mode-locked pulse [57].

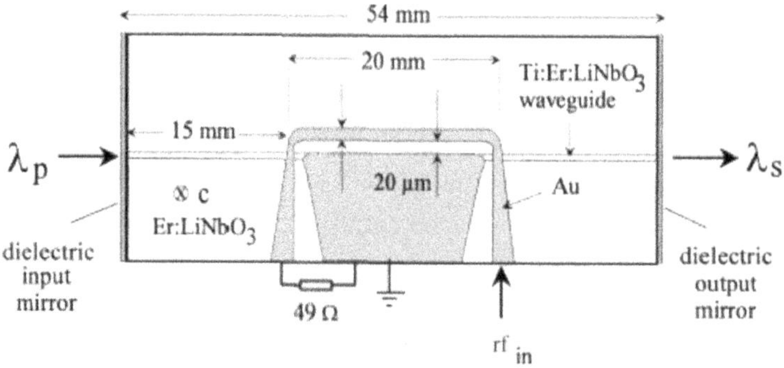

Figure 10.21. Schematic of an Er:Ti:LiNbO$_3$ mode-locked laser [59].

operates near 1500 nm with a 16.7 MHz repetition rate and the derived FWHM pulse width from the autocorrelation trace is 2.7 ps.

10.6.2 *Active mode locking*

Active mode locking of an Er-doped LiNbO$_3$ waveguide pumped by a diode laser has been demonstrated. Ti-LiNbO$_3$ allows phase modulation of the generated signal. Thus, combining Ti doping and Er doping active mode locking has been demonstrated [59, 60]. Laser design with an integrated phase modulator is shown in Figure 10.21. Z-LiNbO3 (Z-cut) is the substrate material. Er is diffused at 1100°C by sputtering for 100 h to a depth of 5 μm. Following this, Ti is diffused in 7 μm stripes (and 95 nm

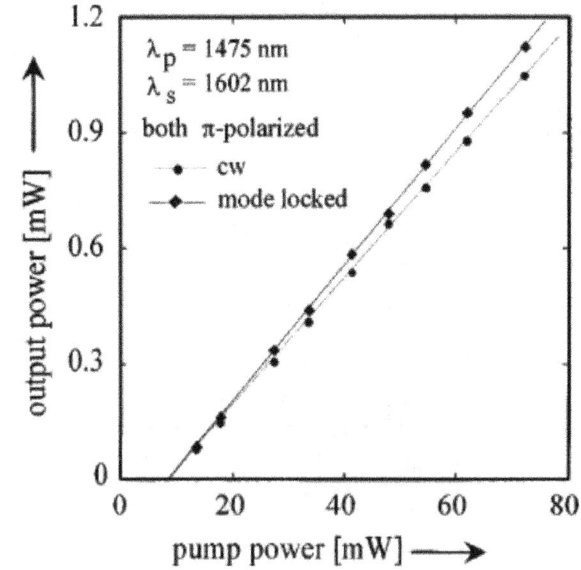

Figure 10.22. Laser output power as a function of pump power for Er:Ti:LiNbO3 laser for both CW and mode-locked operation [59].

thick) by using standard photolithographic techniques. The phase modulator has Au electrodes deposited using standard photolithographic techniques. To create the dielectric mirrors for the laser, SiO_2/TiO_2 pairs are deposited at the polished end faces of the laser waveguide as shown in Figure 10.21.

A diode laser emitting near 1475 nm is used as a pump laser (λ_p) to create gain in the Er:Ti:$LiNbO_3$ waveguide. Laser operation was observed at a pump power of ~10 mW. The laser output as a function of pump power is shown in Figure 10.22. To achieve mode-locked operation, the phase modulator is driven at a fundamental frequency of 1.282 GHz. The most stable operation is achieved at the third harmonic frequency of 3.823 GHz because the phase modulator is optimized for this frequency. The autocorrelation trace of the third harmonic mode-locked pulse is shown in Figure 10.23. The autocorrelation trace is plotted as the second harmonic (SH) power.

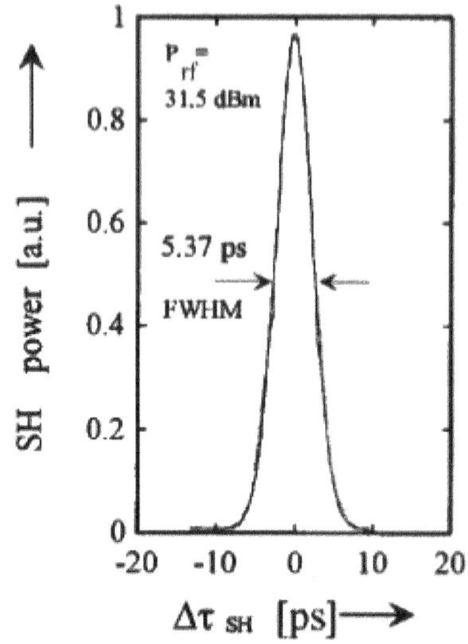

Figure 10.23. Autocorrelation trace of the third harmonic mode-locked laser pulse [59].

10.7 Supercontinuum Generation

As mentioned earlier, supercontinuum generation (SCG) is the spectral broadening of narrow-band incident pulses by propagation through nonlinear media. The physical mechanism of supercontinuum (SC) is a combination of linear and nonlinear processes including self-phase modulation (SPM), cross-phase modulation (XPM), stimulated Raman scattering (SRS), four-wave mixing (FWM), and soliton fission and dispersive wave radiation [61–74]. Various types of glasses such as lead silicate, fluoride [69], and chalcogenide glass [70–74] have been introduced as the core material, which benefit from higher nonlinearity and transparency than conventional silica fiber [72–83]. For example, photonic crystal fibers (PCFs) using lead silicate SF57 glass have been fabricated and a nonlinear parameter of $\gamma = 1860\ W^{-1}km^{-1}$ was obtained, approximately 2000 times that of a standard single-mode fiber [66, 67].

Supercontinuum generation in planar waveguides has gained much interest recently. Fedotova *et al.* [77] simulated SCG in Air-TaFD5-SiO_2 and Air-SiO_2-Ag planar rib waveguides and predicted output spectra spanning up to two octaves. In [78], SC generated in silicon-on-insulator waveguides was numerically studied. It turned out that the spectrum only

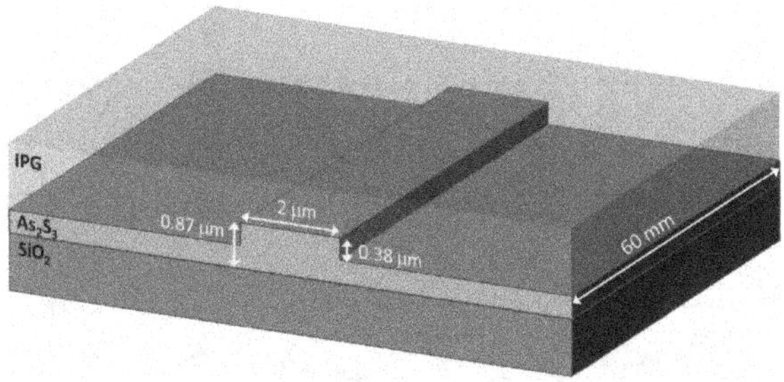

Figure 10.24. Schematic of the planar waveguide used for supercontinuum generation. The waveguide cross-section is 2 μm × 0.87 μm [79].

extended over 400 nm, limited by two-photon absorption and free-carrier effects. In [79], dispersion-engineered highly nonlinear chalcogenide planar waveguides were experimentally demonstrated to have SC with a bandwidth of 750 nm. The schematic of the waveguide is shown in Figure 10.24. A 0.87 μm layer of As_2S_3 is deposited by thermal evaporation onto a thermally oxidized silicon substrate. The 2 μm-wide waveguides are defined using photolithography and created using reactive ion etching with CHF_3 gas. The waveguide chip is then coated in a protective coating layer of inorganic polymer glass (IPG). The waveguides are 6 cm long and the optical loss (α) of the waveguide is ~0.6 dB/cm.

The nonlinear parameter of the material is 10 $W^{-1}m^{-1}$, and the dispersion is 29 ps/km-nm. A mode-locked fiber laser operating at a repetition rate of 10 MHz with ~610 fs-wide pulses at 1550 nm is used as the input pulsed source. The maximum peak power is ~136 kW. The input pulses could be coupled into TE and TM modes of the waveguide using a polarization controller. The measured SC spectrum for various peak powers for TM coupled power is shown in Figure 10.25. TE-coupled powers produced considerably less broadened spectrum.

A study of tapered planar rib waveguide which allows dispersion management for SCG is now discussed. The Air-SF57-SiO_2 rib waveguide continuously increases its etch depth along the propagation distance, which shifts the second zero dispersion wavelength. By launching hyperbolic secant pulses at 1.55 μm into this waveguide, supercontinuum with 30 dB bandwidth of ~1–~4.6 μm is obtained, limited by the multiphonon absorption edge of the material.

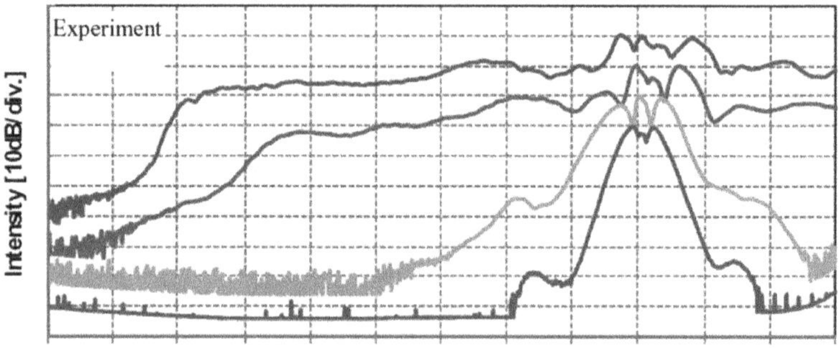

Figure 10.25. SC spectrum for different coupled powers. The peak powers are 68, 46, 29, and 18 W from top to bottom curves, respectively. High peak produces broader spectrum. The wavelength range in the figure is from 1100 nm to 1700 nm (50 nm/div.) [79].

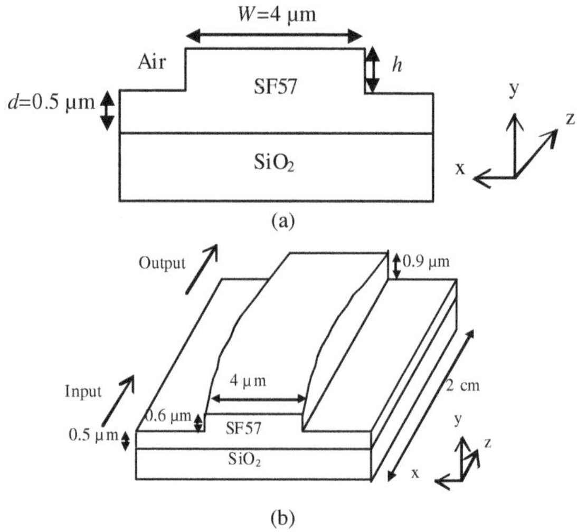

Figure 10.26. (a) Cross-section of the Air-SF57-SiO_2 planar rib waveguide with the rib width *W*, outer slab thickness *d*, and etch depth *h*, respectively. (b) Schematic of a tapered waveguide.

A numerical study of SCG in an Air-SF57-SiO_2 planar rib waveguide has been carried out. The waveguide schematic (with a rib width *W*, an outer slab thickness *d*, and an etch depth *h*) is shown in Figure 10.26. Scott SF57 is a commercially available lead-silicate glass mainly composed of SiO_2 and PbO, with a Pb cationic content of 40–50% [80]. The linear refractive index of SF57 at 1.55 μm is n = 1.8 [81], which results in an index contrast of 17.6% to the SiO_2 substrate, allowing for tight modal confinement. The nonlinear refractive index of SF57 measured at 1.06 μm is $n_2 = 4.1 \times 10^{-19}$ m^2/W, much higher than that of pure silica glass (2.7×10^{-20} m^2/W at 1.06 μm) [82]. The material loss of SF57 glass is 1.6 dB/m at 1.55 μm [82]. The transmission window of SF57 glass is limited by the multiphonon absorption edge to ~5 μm.

The wavelength-dependent dispersion profile of the rib waveguide for various h values (etch depth) is shown in Figure 10.25. W and d are set to be 4 μm and 0.5 μm, respectively. The computation is done by a full-vector finite difference mode solver for the fundamental TM mode. Since the waveguide dispersion depends strongly on the geometrical design, the total dispersion characteristic, which is the combined effect of material and waveguide dispersion, can be controlled by varying the etch depth. The model follows the pulse propagation described in Ref. [83].

For the model calculation, the etch depth h continuously increases from 0.6 μm to 0.9 μm along the propagation distance z by the following relationship:

$$h = h_0 \left[1 + \left(\frac{z}{z_0} \right)^{\frac{1}{3}} \right]. \tag{10.1}$$

For small h (<0.83 μm), a second zero dispersion wavelength (ZDW) exists, and it shifts to shorter wavelengths while h is decreasing. The results of the model are now described [84, 85]. Figure 10.27 (top) shows the spectrum evolution of the incident pulse (P = 5 kW, FWHM = 100 fs) traveling in the tapered waveguide (with etch depth given by Equation (10.1)) if no optical loss is included in the simulation. The effective mode area A_{eff} at the input is equal to 3.2 μm^2, yielding a nonlinear parameter γ of 0.52/W/m at 1.55 μm, about 500 times that of a conventional single-mode fiber. The generated spectrum has a 30 dB bandwidth from ~1–~6 μm. Since the continuum extends over the multiphonon edge, the loss should be considered. The spectrum including loss is shown in Figure 10.27 (bottom). The output spectrum reaches ~4.6 μm, which is considerably larger than that of uniform waveguides.

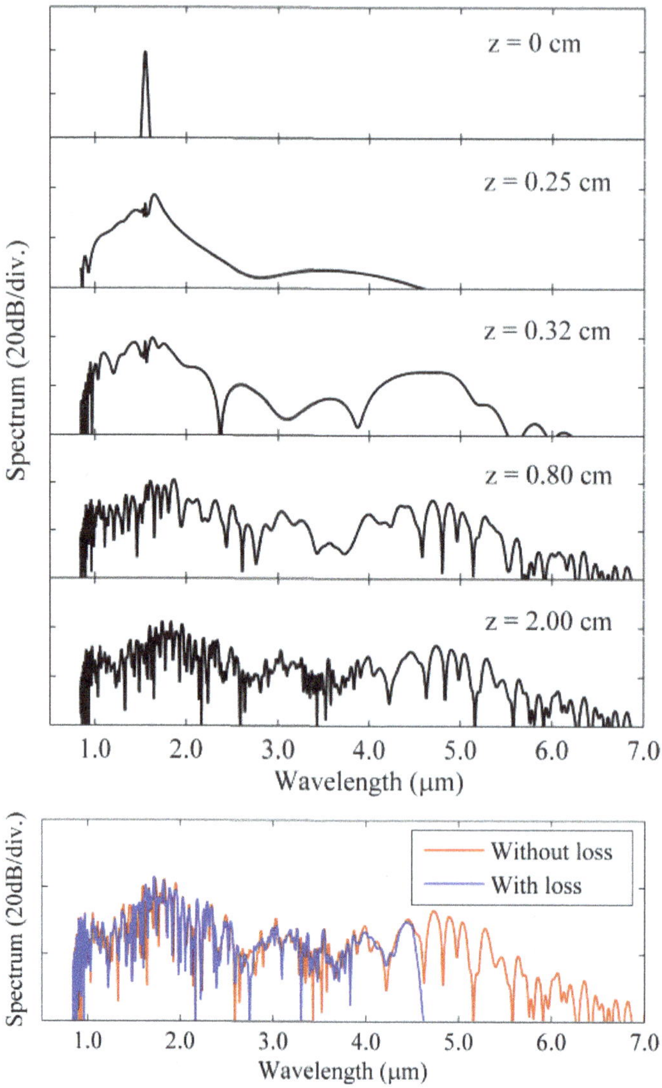

Figure 10.27. (Top) Spectrum evolution along the propagation distance for the tapered SF57 rib waveguide without loss. (Bottom) Effect of loss on the output continuum.

The waveguide is carefully designed, so the center wavelength of the incident pulse is within the anomalous dispersion regime. As a result, the input pulses undergo soliton fission and dispersive wave (DW) emission across the upper ZDW, as described before. For the tapered waveguide with an increasing value of *h*, the second ZDW shifts along the propagation distance, leading to a continuous modification of the phase-matched wavelength for DW emission. This allows the DW to be produced toward longer wavelengths and thus significantly broadens the supercontinuum at the output. Clearly, different tapered structures other than that described here may be used, but the tapering rate should be optimized to ensure efficient energy transfer from solitons to dispersive waves (DWs).

Planar waveguides, such as CW or pulsed lasers, amplifiers, or a broadband source, promise scalable geometries, low-cost fabrication, and the potential for integrated optical chip solutions.

References

1. J. Shmulovich, Y. H. Wong, G. Nykolak, P. C. Becker, R. Adar, A. J. Bruce, D. J. Muehlner, G. Adams, and M. Fishteyn, OSA Optical Amplifier Meet. Opt. Soc. Am., paper PD18-1, (Toronto, CA, 1993).
2. T. Kitagawa, K. Hattori, K. Shuto, M. Yasu, M. Kobayashi, and M. Horiguchi, In Opt. Amplifiers and Their Appl. Tech. Dig. Ser. Opt. Soc. Am., paper PD-1, **17**, (1992).
3. T. Kitagawa, K. Hattori, M. Shimizu, Y. Ohmori, M. Kobayashi, and M. Horiguchi, *Electron Lett.*, **28**, (1992) 1818.
4. K. Shuto, K. Hattori, T. Kitagawa, and M. Horiguchiin, In *Proc. Nineteenth Eur. Conf. Opt. Commun.*, **53**, (Montreux, Switzerland, 12–16 September 1993).
5. T. Feuchter, E. K. Mwarania, J. Wang, L. Reekie, and J. S. Wilkinson, *IEEE Photon. Technol. Lett.*, **4**, (1992) 542.
6. P. Becker, R. Brinkmann, M. Dinand, W. Sohler, and H. Suche, *Appl. Phys. Lett.*, **61**, (1992) 1257.
7. G. N. van den Hoven, E. Snoeks, A. Polman, C. van Dam, J. W. M. van Uffelen, and M. K. Smit, In Proc. Seventh Eur. Conf. Integrat. Opt., April 3–6, 1995, (Delft, Netherlands) p. 229.
8. D. P. Shepherd, Chapter B4.6, In C. Webb, and J. Jones (Eds.), *Handbook of Laser Technology and Applications* (Institute of Physics, London, UK, 2004).
9. R. V. Ramaswamy and R. Srivastava, *IEEE J. Lightwave Tech.*, **6**, (1988) 984.
10. P. J. Chandler, L. Zhang, and P. D. Townsend, *Nucl. Intrument. Methods*, **B59**, (1991) 1223.

11. R. V. Schmidt and I. P. Kaminow, *Appl. Phys. Lett.*, **25**, (1974) 458.
12. M. Svalgaard, C. V. Poulsen, A. Bjaerklev, and O. Paulsen, *Electron. Lett.*, **30**, (1994) 1401.
13. C. T. A. Brown, C. L. Bonner, T. J. Warburton, D. P. Shepherd, A. C. Tropper, and D. C. Hanna, *Appl. Phys. Lett.*, **71**, (1997) 1139.
14. M. Kawachi, *Opt. Quantum Electron.*, **22**, (1990) 391.
15. J. I. Mackenzie and D. P. Shepherd, *Opt. Lett.*, **27**, (2002) 2161.
16. T. Kitagawa, K. Hattori, K. Shuto, M. Yasu, M. Kobayashi, and M. Horiguchi, *Electron. Lett.*, **28**, (1992) 1818.
17. J. Shmulovich, Y. H. Wong, G. Nykolak, P. C. Becker, R. Adar, and A. J. Bruce, *Electron. Lett.*, **28**, (1992) 1181.
18. T. Feuchter, E. K. Mwarania, I. Wang, L. Reekie, and J. S. Wilkinson, *IEEE Photon. Technol. Lett.*, **4**, (1992) 542.
19. K. Shuto, K. Hattori, T. Kitagawa, Y. Ohmori, and M. Horiguchi, *Electron. Lett.*, **29**, (1993) 139.
20. K. Hattori, T. Kitagawa, M. Oguma, M. Wada, J. Temmyo, and M. Horiguchi, *Electron. Lett.*, **29**, (1993) 357.
21. R. N. Ghosh, J. Shmulovich, C. F. Kane, M. R. X. de Barros, G. Nykolak, A. J. Bruce, and P. C. Becker, *IEEE Photon. Tech. Lett.*, **8**, (1996) 518.
22. G. Nykolak, P. C. Becker, J. Shmulovich, Y. H. Wong, D. J. DiGiovanni, and A. J. Bruce, *IEEE Photon. Technol. Lett.*, **5**, (1993) 1014.
23. O. Lumholt, A. Bjarklev, T. Rasmussen, and C. Lester, *J. Lightwave Technol.*, **13**, (1995) 275.
24. E. Snoeks, P. G. Kik, and A. Polman, *Opt. Muter.*, **5**, (1996) 159.
25. J.-M. P. Delavaux, S. Granlund, O. Mizuhara, L. D. Tzeng, D. Barbier, M. Rattay, F. Saint Andre, and A. Kevorkian, *IEEE Photon. Tech. Lett.*, **9**, (1997) 2.
26. J. Shmulovich, Technical Digest, OFC99, paper PD42, 1999.
27. E. Desurvire, *Erbium-Doped Fiber Amplifiers, Principles and Applications* (Wiley, New York, 1994).
28. J.-M. P. Delavaux, C. McIntosh, J. Shmulovich, A. J. Bruce, R. K. Pattnaik*, and B. R. Wirstiuk, Technical Digest, OFC 2000, paper WA2-1, 2000 (Optical Fiber Conference (OFC) San Jose, CA, UK).
29. M. Saruwatari and T. Izawa, *Appl. Phys. Lett.*, **24**, (1974) 603.
30. Y. Hibino, T. Kitagawa, M. Shimizu, F. Hanawa, and A. Sugita, *IEEE Photon. Technol. Lett.*, **1**, (1989) 349.
31. H. Aoki, O. Maruyama, and Y. Asahara, *IEEE Photon. Technol. Lett.*, **2**, (1990) 459.

32. N. A. Sanford, K. J. Malone, and D. R. Larson, *Opt. Lett.*, **15**, (1990) 366.
33. E. K. Mwarania, L. Reekie, J. Wang, and J. S. Wilkinson, *Electron. Lett.*, **26**, (1990) 317.
34. N. Takato, K. Jinguji, M. Yasu, H. Toba, and M. Kawachi, *IEEE J. Lightwave Technol.*, **6**, (1988) 1003.
35. M. J. F. Digonnet and C. J. Gaeta, *Appl. Opt.*, **24**, (1985) 333.
36. Y. Hibino, T. Kitagawa, M. Shimizu, A. Sugita, and Y. Ohmori, In Proc. Integrated Photon. Res., Hilton Head, SC, paper TuJ2, (1990).
37. H. Ioki, E. Ishikawa, and Y. Asahara, *Electron. Lett.*, **27**, (1991) 2352.
38. G. R. J. Robertson and P. E. Jessop, *Appl. Opt.*, **30**, (1991) 276.
39. R. Tumnunelli, F. Halumi, and J. Haavisto, *Opt. Lett.*, **16**, (1991) 1098.
40. K. Hattori and T. Kitagawa, *IEEE Photon. Technol. Lett.*, **4**, (1992) 973.
41. T. Kitagawa, K. Hattori, Y. Hibino, and Y. Ohmori, *IEEE J. Lightwave Technol.*, **12**, (1994) 436.
42. J. R. Bonar, J. A. Bebbington, J. S. Aitchison, G. D. Maxwell, and B. J. Ainslie, *Electron. Lett.*, **30**, (1994) 229.
43. F. Ondracek, *Opt. Mater.*, (2007). doi:10.1016/j.optmat.2006.12.004.
44. J. Jágerská, F. Ondráček, L. Salavcová, M. Míka, J. Špirková, and J. Ctyroký, In 13th Eur. Conf. Integrated Optics ECIO (2007), Paper ThG14, 2007 (Copenhagen, Denmark).
45. F. Ondráček, M. Skalský, and J. Čtyroký, *Proc. SPIE*, **6180**, (2006) 414.
46. M. P. Hehlen, N. J. Cockroft, and T. R. Gosnell, *Phys. Rev. B*, **56**, (1997) 9302.
47. F. Ondráček, J. Jágerská, L. Salavcová, M. Míka, J. Špirková, and J. Čtyroký, *IEEE JQE*, **QE-44**, (2008) 536.
48. F. D. Patel, S. DiCarolis, P. L. Lum, S. Venkatesh, and J. N. Miller, *IEEE Photon. Technol. Lett.*, **16**, (2004) 2607.
49. D. L. Veasey, D. S. Funk, P. M. Peters, N. A. Sanford, G. E. Obarski, N. Fontaine, M. Young, A. P. Peskin, W. C. Liu, S. N. Houde-Walter, and J. S. Hayden, *J. Non-Cryst. Solids*, **263 & 264**, (2000) 369.
50. F. Di Pasquale and M. Federighi, *IEEE J. Quantum Electron.*, **30**, (1994) 2127.
51. D. Geskus, S. Aravazhi, E. H. Bernhardi, L. Agazzi, S. M. García-Blanco, and M. Pollnau, CLEO Technical Digest, paper CMA1.6, 2012 (Conference on Lasers and Electro-Optics (CLEO), Baltimore, MD, USA).
52. J. B. Schlager, B. E. Callicoatt, R. P. Mirin, and N. A. Sanford, *IEEE Photon. Tech. Lett.*, **14**, (2002) 1351.

53. J. B. Schlager, B. E. Callicoatt, K. L. Silverman, R. P. Mirin, N. A. Sanford, and D. L. Veasey, In Proc. Conf. Lasers and Electro-Optics, May 6–11, (Baltimore, MD, USA, 2001), pp. 87–88.
54. D. L. Veasey, D. S. Funk, P. M. Peters, N. A. Sanford, G. E. Obarski, N. Fontaine, M. Young, A. P. Peskin, W. Liu, S. N. Houde-Walter, and J. S. Hayden, *J. Non-Crystal. Solids*, **263 & 264**, (2000) 369.
55. B. E. Callicoatt, J. B. Schlager, K. L. Silverman, R. K. Hickernell, R. P. Mirin, N. A. Sanford, J. S. Hayden, S. Conzone, and R. Simpson, in Proc. IEEE Lasers and Electro-Optics Society 2001 Annu. Meeting (LEOS 2001), San Diego, CA, November 11–15, 2001, paper WW1.
56. D. Lancaster, C. Khurmi, N. B.-Hebert, J. Genest, G. Chen, W. Zhang, S. Afshar, and T. M. Monro, *SPIE Lase*, (2017) 10082.
57. G. Valle, *et al.*, *Proc. SPIE*, **301**, (2007) 301–310.
58. R. Osellame, N. Chiodo, G. Della Valle, G. Cerullo, R. Ramponi, P. Laporta, A. Killi, U. Morgner, and O. Svelto, *IEEE J. Select. Top. Quantum Electron.*, **12**, (1992) 277.
59. H. Suche, R. Wessel, S. Westenhofer, W. Sohler, S. Bosso, C. Carmannini, and R. Corsini, *Opt. Lett.*, **20**, (1995) 596.
60. Y. Katano, S. Shinada, S. Nakajima, T. Kawanishi, and H. Nakajima. *IEICE*, **9**, (2012) 245.
61. J. M. Dudley, G. Genty, and S. Coen, *Rev. Mod. Phys.*, **78**, (2006) 1135.
62. J. M. Dudley and J. R. Talor, *Supercontinuum Generation in Optical Fibers*. (Cambridge University Press, New York, 2010).
63. S. Coen, A. H. L. Chau, R. Leonhardt, J. D. Harvey, J. C. Knight, W. J. Wadsworth, and P. S. J. Russell, *J. Opt. Soc. Am. B*, **19**, (2002) 753.
64. W. J. Wadsworth, A. Ortigosa-Blanch, J. C. Knight, T. A. Birks, T. P. M. Man, and P. S. J. Russell, *J. Opt. Soc. Am. B*, **19**, (2002) 2148.
65. T. Yamamoto, H. Kubota, S. Kawanishi, M. Tanaka, and S. Yamaguchi, *Opt. Express*, **11**, (2003) 1537.
66. J. Y. Y. Leong, P. Petropoulos, J. H. V. Price, H. Ebendorff-Heidepriem, S. Asimakis, R. C. Moore, K. E. Frampton, V. Finazzi, X. Feng, T. M. Monro, D. J. Richardson, and J. Lightwave, *Technology*, **24**, 183 (2006) 183.
67. Z. Chen, S. Ma, and N. K. Dutta, *Opt. Commun.*, **283**, (2010) 3076.
68. J. H. V. Price, X. Feng, A. M. Heidt, G. Brambilla, P. Horak, F. Poletti, G. Ponzo, P. Petropoulos, M. Petrovich, J. Shi, M. Ibsen, W. H. Loh, H. N. Rutt, and D. J. Richardson, *Opt. Fiber Technol.*, **18**, (2012) 387.
69. Z. Chen, A. J. Taylor, and A. Efimov, *Opt. Express*, **17**, (2009) 5852.

70. D. I. Yeom, E. C. Mägi, M. R. Lamont, M. A. Roelens, L. Fu, and B. J. Eggleton, *Opt. Lett.*, **33**, (2008) 660.
71. J. Hu, C. R. Menyuk, L. B. Shaw, J. S. Sanghera, and I. D. Aggarwal, *Opt. Express*, **18**, (2010) 6722.
72. N. Granzow, S. P. Stark, M. A. Schmidt, A. S. Tverjanovich, L. Wondraczek, and P. S. J. Russell, *Opt. Express*, **19**, (2011) 21003.
73. R. R. Gattass, L. Brandon Shaw, V. Q. Nguyen, P. C. Pureza, I. D. Aggarwal, and J. S. Sanghera, *Opt. Fiber Technol.*, **18**, (2012) 345.
74. A. Marandi, C. W. Rudy, V. G. Plotnichenko, E. M. Dianov, K. L. Vodopyanov, and R. L. Byer, *Opt. Express*, **20**, (2012) 24218.
75. A. V. Husakou and J. Herrmann, *Appl. Phys. B*, **77**, (2003) 227.
76. K. M. Hilligsøe, T. Andersen, H. Paulsen, C. Nielsen, K. Mølmer, S. Keiding, R. Kristiansen, K. Hansen, and J. Larsen, *Opt. Express*, **12**, (2004) 1045.
77. O. Fedotova, A. Husakou, J. Herrmann, *Opt. Express*, **14**, (2006) 1512.
78. L. Yin, Q. Lin, G. P. Agrawal, *Opt. Lett.*, **32**, (2007) 391.
79. M. R. Lamont, B. Luther-Davies, D. Y. Choi, S. Madden, and B. J. Eggleton, *Opt. Express*, **16**, (2008) 14938.
80. G. Genty, M. Surakka, J. Turunen, and A. T. Friberg, *Opt. Lett.*, **35**, (2010) 3057.
81. V. Diez-Blanco and J. Siegel, *J. Solis Appl. Surf. Sci.*, **252**, (2006) 4523.
82. Schott Glass Catalogue, 2003.
83. G. P. Agrawal, *Nonlinear Fiber Optics*, (4th edn.) (Academic Press, Burlington, MA, London, 2007).
84. H. Hu, W. Li, and N. K. Dutta, *Optics Communications*, **324**, (2014) 252.
85. H. Hu, X. Zhang, W. Li, and N. K. Dutta, *Appl. Opt.*, **54**, (2015) 3448.

Chapter 11
Applications

11.1 Introduction

Application of mode-locked and -pulsed lasers in various domains such as harmonic generation, surgery, microscopy, and materials processing are described in this chapter. Pump–probe experiments using pulsed laser systems allow starting and tracing of chemical reactions and studies of ultrafast relaxation processes in materials. High-power fiber lasers are candidates for metal processing applications. Very high-power pulses are being utilized for nuclear fusion studies. Although significant developments have occurred in high-power optical fiber lasers, the need for higher power for commercial (metal cutting and shaping) applications will continue to drive the research on higher-power fiber lasers using Nd- or Yb-based systems and possibly others.

11.2 Harmonic Generation

Ultrashort laser pulses opened the way for many interesting and important applications and fiber lasers play an important role here. For example, few-cycle laser pulses can be used for the study of high harmonic generation. At pulse energies on the order of several hundred microjoules, these ultrafast pulses can easily be focused to intensities that allow the study of exciting light–matter interaction at high electric fields. High-harmonic generation (HHG) is one of the key processes of high-field physics.

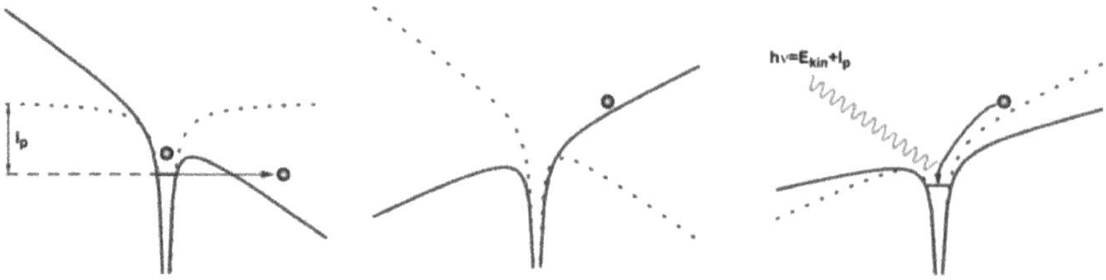

Figure 11.1. A model for high-harmonic generation. At high intensities, the electric field deforms the Coulomb barrier of the atom such that an electron can tunnel through. After being released, the electron is accelerated by the external field before it recombines with the parent ion leading to the emission of a high-energy photon [1, 2].

When ultrashort light pulses are focused onto noble gases to intensities on the order of 10^{13}–10^{15} W/cm^2, the resulting high electrical field strength can depress the Coulomb barrier of the atom, as shown in Figure 11.1. The figure illustrates the three-step model for high-harmonic generation: (i) tunnel ionization of the electron, (ii) acceleration in the laser electric field, and (iii) recombination and emission of a high-energy photon. In the three-step model, as soon as the electric field reverses, electrons are first decelerated and then accelerated back toward their parent ions. The electron can recombine with its parent ion with a certain probability, leading to the emission of a broadband extreme ultraviolet photon. One photon per electron is emitted. The energy of the emitted photon ($h\nu = I_p + K$) depends on the ionization potential (I_p) of the atom and on the kinetic energy (K) of the electron upon its return to its parent ion [1].

The kinetic energy depends on the phase of the electric field at the moment of ionization. The maximum photon energy i.e., cutoff energy, that can be achieved in this process can be calculated using classical [3, 4] or quantum mechanics [5] and is given by $E_{\text{cutoff}} = h\nu_{\text{max}} = I_p + 3.17\ U_p$. The quantity U_p is the average "quiver" energy of an electron in the laser field is called the ponderomotive energy U_p. It is proportional to the intensity I of the driving laser and is given by [1]

$$U_p = e^2E^2/(4\ m\omega^2) \sim I\lambda^2, \tag{11.1}$$

where *e* and *m* are the charge and the mass of the electron, *E* is the electric field of the laser, and ω is its angular frequency. Thus, U_p is proportional to the intensity I and square of the wavelength of incident laser light.

With high peak powers in short pulses (which can be focused to a small spot), fiber lasers can be used for high-field experiments. For HHG generation, high field is needed. The schematic of the setup is shown in Figure 11.2. The schematic shows the HHG in a noble gas (AR or Kr). The fundamental light is separated from the harmonics by a 200-nm aluminum filter and the radiation is analyzed by a grazing incidence monochromator. The result of the experiment with Kr gas is shown in Figure 11.3. A 50-μm gas nozzle enables high particle densities of either argon or krypton gas in the focal region while keeping the overall gas load minimal.

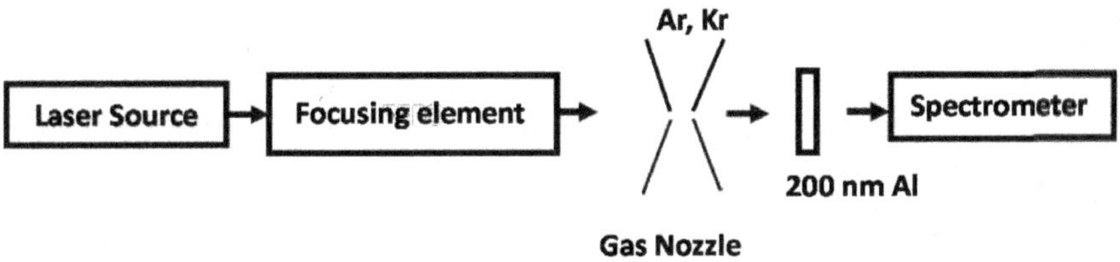

Figure 11.2. The schematic for high repetition-rate high-harmonic generation with fiber laser systems. The driving laser source (high power short, pulsed laser) is focused by a lens onto a gas target [2].

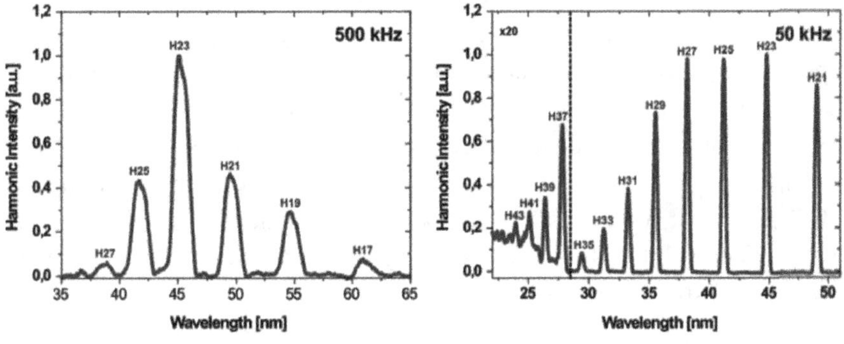

Figure 11.3. High-harmonic generation with a fiber chirped-pulse amplification system at a repetition rate of 500 kHz (100 μJ) and 50 kHz (400 μJ) in krypton [2].

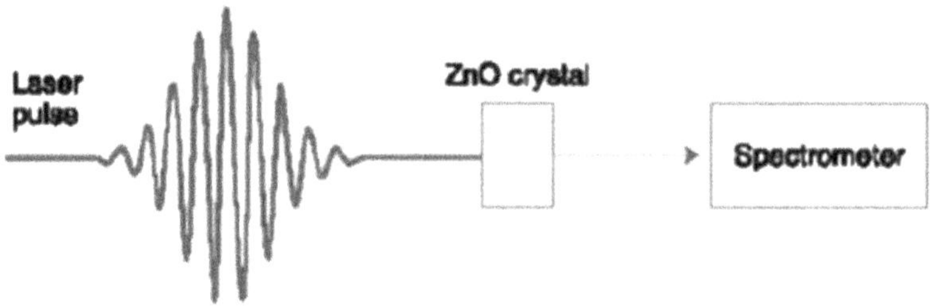

Figure 11.4. Schematic of harmonic generation from a ZnO crystal [7].

HHG has also been observed in bulk crystals using intense mid-infrared laser excitation [6–8]. A schematic of the experimental setup is shown in Figure 11.4 [7]. High harmonics up to the twenty-fifth order are produced in the transmission through a 500 μm thick ZnO crystal at peak intensity 5 TW cm^{-2} (field strength ~0.6V A-1, pulse energy ~3 μJ), without damage for a repetitive excitation at 1 kHz. The observed spectrum extended beyond the bandgap and exhibited a high energy cutoff, even for the sub-μJ level of laser pulse energy.

The intensity of harmonics also deviates strongly from the power law (harmonic order) expected from small signal nonlinear optics. Compared to this, the typical gas-phase HHG sources use mJ-class excitation lasers. The high-energy cutoff from solids is found to scale linearly with the peak field strength in contrast to the quadratic dependence expected from the three-step re-collision model stated earlier (Figure 11.1). Figure 11.5 shows for a wide range of materials and pump laser wavelengths the reported approximate high-harmonic cutoff energies. The figure shows that narrow-bandgap materials are suitable for mid-infrared laser excitation, and for wide-bandgap dielectrics, HHG can be excited by strong near-infrared lasers. Both experimental and theoretical investigations of HHG in several types of solids are being investigated by many researchers. HHG in solids offers the possibility of optimizing HHG using two-dimensional structures or engineered structures.

11.3 Surgical Applications

Ultrafast mid-IR laser sources have been demonstrated to have great advantages for the applications for surgery and tissue diagnostics [9–14].

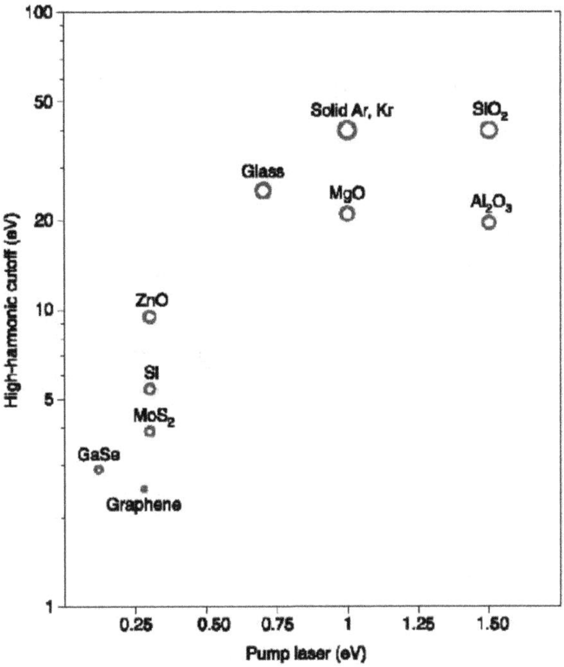

Figure 11.5. HHG cutoff wavelength for various excitation laser wavelengths for several types of solids [7].

The mid-IR wavelength (~2–3.0 μm) is within the vibrational absorption band of water molecules within tissue. Thus, laser radiation at these wavelengths can be substantially absorbed within a few microns of tissue. In addition, the picosecond-level pulse duration is short enough to drive tissue ablation faster than the thermal and acoustic energy transfer processes that can cause damage to the surrounding tissue but long enough to avoid tissue ionization and the risks of creating toxic-free radicals and fragmenting constituent proteins [9]. Therefore, high-energy mid-IR picosecond laser pulses around 3 μm are of great interest for minimally invasive surgery and may be better than conventional medical lasers with pulse durations longer than nanoseconds.

The mechanical surgical scalpel causes a damage zone of more than 400 μm from the borders of the incision and the conventional laser (~10 ns pulses) damages skin up to 800 μm away from the ablated edge. In contrast, cuts using the mid-IR picosecond laser system have sharp edges and a damage zone of less than 10 μm [15]. Also, an investigation of wound healing has demonstrated that wounds produced by a mid-IR picosecond laser healed faster than those produced via either a conventional surgical laser or a mechanical surgical scalpel. The scars formed were much smaller which decreases the risk of infection after surgery.

Ultrafast mid-IR laser sources have significant potential for real-time bio-diagnostics [16–18]. Mouras *et al.* [18] recently used a high-repetition-rate (80MHz), tunable ultrafast mid-IR laser to develop a high-speed, label-free bench-top mid-IR microscope for chemical imaging with sub-wavelength resolution [17]. This microscope could be applied to living cells and tissue, where it offered real-time tracking of drugs and drug carriers within cells and tissues.

11.3.1 *Eye surgery*

Femto-second (fs)-pulsed laser systems have been found to be suitable for eye surgery. In ophthalmic fs-laser systems, a large range of pulse energies is available so that optimized amounts of pulse energy can be used for each tissue type and geometry. By using just the required amount of pulse energy, not more, the mechanical side effects (such as bubble formation and shock waves [12, 19]) are minimized. In a study of *n*J and μJ pulse energy fs-laser cutting of cornea [20], it was observed that the lower pulse energy regime avoided cell damage and reduced inflammation reactions in the adjacent tissue.

For fs-laser systems that are for clinical use, some critical supporting technologies are also needed. They are (i) laser beam scanning technologies, (ii) imaging of tissue structures required to place cuts at a desired location, and (iii) the patient interface system that connects the eye via suction to the laser beam delivery system and ensures a stable relative position during treatment.

The laser system generally has an integrated camera system that provides a frontal view of the eye, while some laser systems provide depth ranging using optical coherence tomography (OCT). OCT allows imaging of structures inside tissues to a depth of 5–20 μm [21, 22].

To achieve the desired cut pattern inside the eye, the individual laser spots need to be arranged in suitable patterns. The calculation of such patterns is achieved by software, which also controls the scanning systems to position the laser foci in lines and planes. Optical scanning of the laser focus in two dimensions can be achieved by galvanometer-driven scanning mirrors, combined with a mechanism to change the focus position along the beam axis.

For patient interface, the laser can be focused on the eye using a long working distance objective lens or with an articulated arm going directly to the eye. The contact can be made directly to the cornea using a liquid gel interface.

The types of eye surgeries for which lasers are used are corneal surgery and cataract surgery. Both of these surgeries have several steps each of which involves an optimized procedure for a particular patient.

11.4 Microscopy

Due to the development of commercial turnkey mode-locked laser (ML) systems, multiphoton microscopy is now available without extreme complexity. The important concepts underlying the most commonly used type of multiphoton microscopy (two-photon excitation) are described here [23]. The mode-locked lasers used here generate short light pulses (~100 fs) with a repetition rate of ~100 MHz. This provides high pulse peak power with low average power. Also, high numerical aperture objectives are used in a two-photon excitation setup. An objective with a high numerical aperture concentrates the light in a diffraction-limited focal volume with a size of ~1 μm^3 [24]. The probability of an absorption is proportional to the square of the instantaneous laser intensity, $P \sim I^2$, which is highest at the focal spot in the microscope. The intensity-squared dependence of two-photon excitation sharply decreases outside the focal volume.

Thus, in a two-photon setup, it is possible to produce nonlinear events only in a precise region of the sample. If the ML source is combined with a scanning system, nonlinear excitation (and hence imaging) in a thin optical section of the sample is feasible, without influencing regions

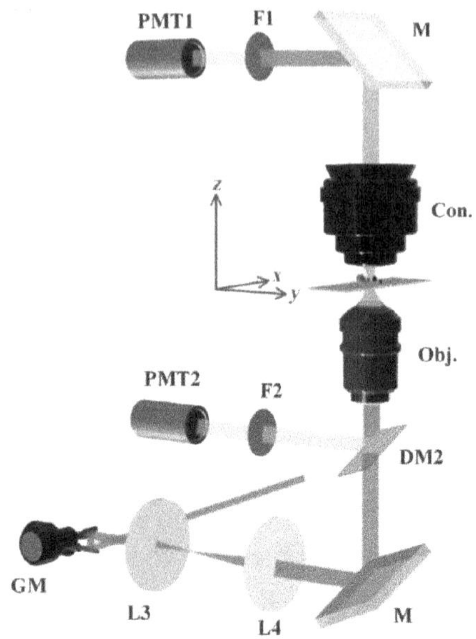

Figure 11.6. Schematic of the two-photon microscopy experimental setup with both illuminated and transmitted detections.

Note: GM: galvanometric mirror. L: lens. M: mirror. Obj.: objective. Con.: condenser. DM: dichroic mirror. F: filters. PMT: photomultiplier tube [27].

outside the focal plane [25, 26]. Schematic of a two-photon scanning system is shown in Figure 11.6. A two-dimensional galvanometric mirror (GM) is used for scanning the incident laser in the x–y plane. A telescope consisting of two lenses, L3 and L4, serves to focus the scanned laser to the back focal plane of the objective lens. The two-photon fluorescence signals generated are collected by the condenser in the transmitted path and the objective in the epi-path and detected by the photomultipliers PMT1 and PMT2.

The laser (not shown in Figure 11.6) is a thulium-doped fiber laser mode locked using a semiconductor absorber mirror (SESAM). The laser is followed by an amplifier and a frequency doubler (Figure 11.7). The laser emits near 1800 nm.

Similar mode-locked lasers are described in earlier chapters. Both the laser and amplifier are pumped by 1570 nm diode lasers. A 10-mm-long periodically poled lithium niobate (PPLN) crystal was used for frequency doubling to generate the signal at ~937 nm. Based on the average power,

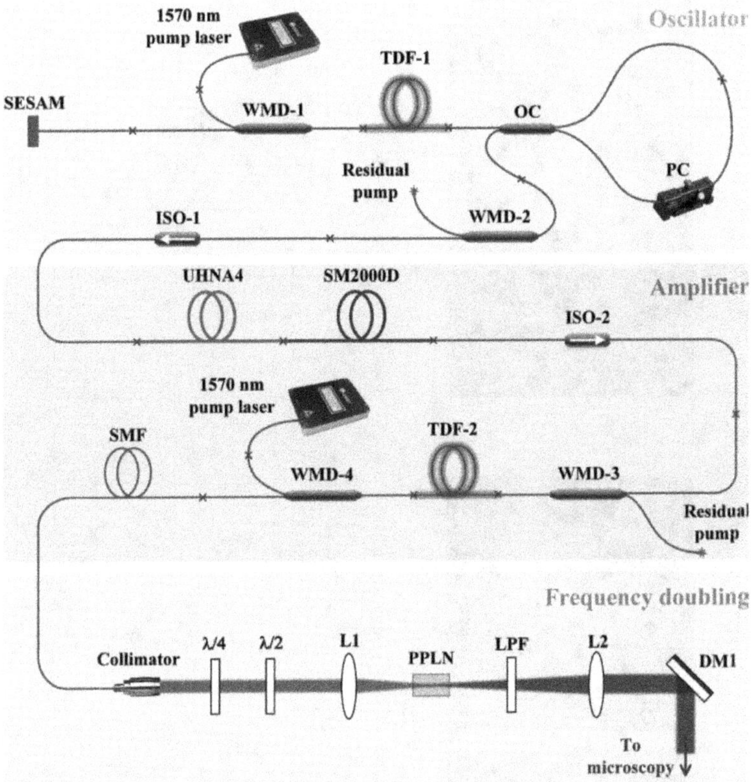

Figure 11.7. The schematic of the mode-locked laser with three elements laser, amplifier, and frequency doubler is shown [27].

repetition rate, and pulse width of the output signal, the peak power is estimated to be ~285 kW after the amplifier. The lateral and axial resolutions for the reflected and transmitted directions are shown in Figure 11.8. The system has been used to image the brain of rats.

The third harmonic generation has also been used to generate high-resolution deep tissue images using an fs mode-locked fiber laser source [28]. The laser produced ~100 fs pulses. The setup is similar to that in Figure 11.6. The average transmitted power through the objective is measured to be 9 mW, or 80 pJ per pulse, and the intensity at the specimen (focus of the objective) is estimated to be 92 GW/cm^2. The full width at half-maximum of the focal spot is ~7.5 μm. The third harmonic generation intensity is proportional to the cube of the intensity.

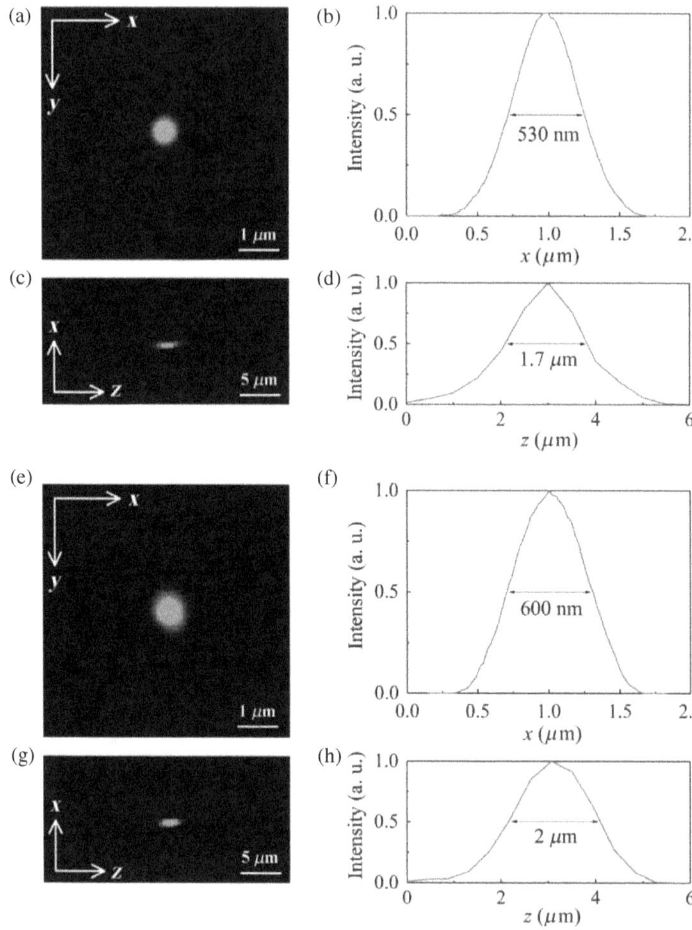

Figure 11.8. (a), (c): Lateral and axial resolution in reflected direction. (b), (d): Trace plots for images (a) and (c). (e), (g): Lateral and axial resolution in the transmitted direction. (f), (g): Trace plots for images (e) and (g) [27].

11.5 Materials Processing

Fiber lasers are becoming important in materials processing because they offer low price per Watt, have good beam quality and good efficiency, and are low maintenance compared to other types of lasers. There are three types of fiber lasers in use: (1) very high peak power lasers with ps pulse width, (2) high peak power ~20 kW and average power of ~2000 W, and

(3) essentially quasi CW lasers of 50 W–>10 kW. All of these lasers use telecom-grade highly reliable pump lasers. This has resulted in numerous applications in materials processing. The automotive industry uses 2–10 kW fiber lasers for cutting and welding. Pulsed fiber lasers are widely used in the marking of products.

Carbon fiber-reinforced plastic (CFRP) composites have attracted significant interest from many industrial sectors including the aerospace sector. These fiber composites can be easily patterned or drilled using fiber lasers with about 0.5 millisecond pulses with a peak power of ~10 kW, and an average power of ~100 W is sufficient for this purpose.

Ultrafast lasers are important tools for advanced micro-machining and materials processing. The shortness of the pulses facilitates material removal without imposing thermal effects on the remaining structure. The pulse width requirement (for nonthermal ablation) depends on the properties of the material. The pulse width regime for nonthermal ablation is given by [29]

$$\tau_p << \delta^2/D, \tag{11.2}$$

where τ_p is the laser pulse duration full width at half-maximum (FWHM) power, δ is the optical skin depth, and D is the thermal diffusivity of the material [30]. There is some variety in the pulse width requirements for different materials since the electron–phonon coupling time can range from a few picoseconds to several tens of picoseconds [31]. It is believed that an ultrafast laser source that can be widely used for nonthermal processing of most materials should have sub-picosecond FWHM pulses and a tightly confined temporal energy distribution.

The following is an example of a typical fiber laser used in materials processing applications [32]. A diode-pumped, Q-switched pulsed Yb-doped fiber laser. The laser emits near 1064 nm. The average beam power is ~20 W and the energy of a 200 ns pulse is 0.95 mJ with a peak power of 10 kW at a repetition rate of 20 kHz. The beam diameter at the output is 5.0 ± 1.0 micrometer, and the divergence is ±0.5 mrad. A visible laser of <1 mW emitting near 600–700 nm is often included with the high-power laser for alignment purposes. A marker head is included that allows working on a 120 × 120 mm^2 area. It contains two mirrors that move the beam emitted by the laser (using "galvo" system of mirrors) in the x and y axes before it is focused on the working surface. The system is computer controlled. Pulse length and pulse energies can be varied for the desired application. The main use of this laser is the high-speed and high-precision marking of metallic or polymer materials.

The high brightness ytterbium fiber laser operating at 1060–1080 nm spectral range has a unique combination of high-power output and high beam quality, Consequently, Yb-doped fiber laser built using double-clad fibers (Chapter 8) has now become important for many materials processing which was difficult using traditional Nd-doped YAG rods [33–39] due to low beam quality at high power. Yb-doped fiber lasers to output powers beyond 1 kW in CW operation with near diffraction-limited beam quality are feasible due to the availability of high-power diode pumps. Laser beam combination techniques are also used for power scaling of the Yb fiber laser by bundling multiple single-mode fiber laser elements to provide up to 10 kW multimode power output.

Thus, Yb fiber lasers delivering over 10 kW power output are now available for materials processing applications. Absorption of the laser beam by metals increases for wavelengths toward the visible and ultraviolet regions and decreases toward the longer infrared wavelengths. The near-infrared wavelength (~1060 nm) of the high-brightness ytterbium fiber laser has a higher absorption by metals than that of the longer wavelength (~10600 nm) CO_2 laser. Also, the shorter wavelength offers lower sensitivity to laser-induced plasmas during materials processing when compared to the CO_2 laser. The schematic of a metal cutting process is shown in Figure 11.9.

Laser cutting of metals requires very high power intensities (~10^{10} W/m^2) to melt the metallic material to the required penetration depth at a high cutting speed. Therefore, the material's surface reflectivity, thermal conductivity, and thickness are the material parameters that affect the efficiency of the laser-cutting process.

The ease with which a metallic material can be cut depends on the absorptivity of the material to the incident laser beam and the melting temperature of the material or that of the oxide formed [39]. Metallic materials like stainless steel and aluminum are often laser cut using an

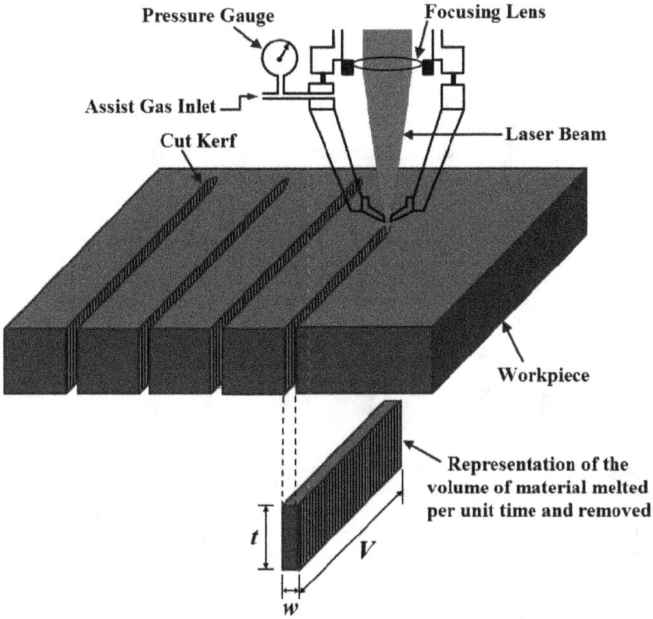

Figure 11.9. Schematic of a metal cutting process using inert gas for removal of molten metal [33].

inert assist gas jet (usually nitrogen) to give clean unoxidized cut edges. This process uses a focused high-intensity laser beam to melt the kerf volume and a high-pressure inert gas jet to blow out the molten metal to form a cut kerf. In this cutting process, the melt temperature is not raised to boiling point, and the melt removal process is solely dependent on the drag force supplied by the high-pressure inert gas jet.

Using 5 kW fiber laser power, stainless steel of 10 mm thickness can be cut at a maximum cutting speed of 1.5 m/min with a nitrogen assist gas jet and aluminum of 4 mm thickness can be cut at 10.2 m/min using a nitrogen assist gas jet. The cutting speed increases with laser power. The data for Yb fiber laser and CO_2 laser are shown in Figure 11.10. The cutting rate using fiber laser is about 30–50% higher than that using CO_2 laser. Optimization of the cutting process parameters is needed for good cut-edge quality in thick-section metal cutting at high cutting speeds. For maximum cutting speed, fast melt removal is important.

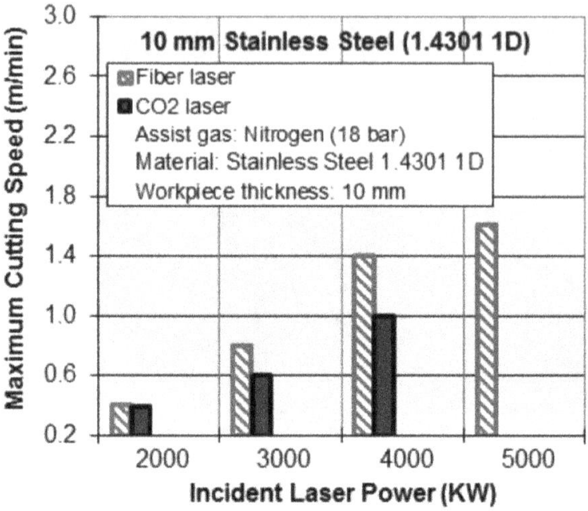

Figure 11.10. Cutting speed for fiber laser and CO_2 laser for various incident powers [33].

Application of mode-locked and -pulsed lasers in various domains such as harmonic generation, surgery, microscopy, and materials processing has been described. The development of newer applications is very feasible as lasers are developed in other wavelength ranges using rare-earth-doped materials in different glass hosts.

References

1. C. Winterfeldt, C. Spielmann, and G. Gerber, *Rev. Mod. Phys.*, **80**, (2008) 117.
2. J. Limpert, S. Hadrich, J. Rothhardt, M. Krebs, T. Eidam, T. Schreiber, and A. Tunnermann, *Laser Photon. Rev.*, **5**, (2011) 634.
3. P. Corkum, *Phys. Rev. Lett.*, **71**, (1993) 1994.
4. P. Corkum, N. H. Burnett, and M. Y. Ivanov, *Opt. Lett.*, **19**, (1994) 1870.
5. M. Lewenstein, P. Balcou, M. Y. Ivanov, A. L'Huillier, and P. B. Corkum, *Phys. Rev. A*, **49**, (1994) 2117.
6. S. Ghimire, *et al.*, *Nat. Phys.*, **7**, (2011) 138.
7. S. Ghimire and D. A. Reis, Stanford PULSE Institute, SLAC National Accelerator Laboratory Menlo Park, California, USA.
8. E. Lorek, E. W. Larsen, C. M. Hey, S. Carlström, D. Palecek, D. Zigmantas, and J. Mauritsson, *Rev. Sci. Inst.*, **85**, (2014) 123106.

9. A. Vogel and V. Venugopalan, Pulsed laser ablation of soft biological tissues. In *Optical-Thermal Response of Laser- Irradiated Tissue* (Dordrecht Heidelberg, London, New York: Springer, 2011), p. 551.
10. C. Latz, T. Asshauer, C. Rathjen, and A. Mirshahi, *Micromachines*, **12**, (2021) no. 2.
11. N. Tinne, G. Knoop, N. Kallweit, *et al.*, *J. Biomed. Opt.*, **19**, (2014), 048001.
12. T. Asshauer, C. Latz, A. Mirshahi, and C. Rathjen, *Adv. Opt. Technol.*, **10,** (2021) 393.
13. J. S. Chang, I. N. Chen, W. M. Chan, J. C. Ng, V. K. Chan, and A. K. Law, *J. Cataract Refract. Surg.*, **40**, (2014) 29.
14. J. A. Izatt, M. R. Hee, E. A. Swanson, *et al.*, *Arch. Ophthalmol.*, **112**, (1994) 1584.
15. J. Ma, Z. Qin, and G. Xie, *Appl. Phys. Rev.*, **6**, (2019) 021317.
16. J. Zou, F. Talbot, A. Tata, L. Ermini, K. Franjic, M. Ventura, J. Zheng, H. Ginsberg, M. Post, D. R. Ifa, D. Jaffray, R. J. D. Miller, and A. Zarrine-Afsar, *Anal. Chem.*, **87**, (2015) 12071.
17. H. Petersen, F. Tavakoli, S. Kruber, A. Münscher, A. Gliese, N.-O. Hansen, S. Uschold, D. Eggert, W. D. Robertson, T. Gosau, S. Sehner, M. Kwiatkowski, H. Schl€uter, U. Schumacher, R. Knecht, and R. J. D. Miller, *Lasers Surg. Med.*, **48**, (2016) 385.
18. R. Mouras, A. Mani, E. Bagnoli, T. Soulimane, C. Silien, and S. A. M. Tofail, In MICS 2018, OSA Technical Digest (Optical Society of America, 2018), Paper No. MT3C.7.
19. A. Vogel, *Appl. Phys. B*, **68**, (1999) 271.
20. A. K. Riau, Y. C. Liu, and N. C. Lwin, *et al.*, *Invest. Ophthalmol. Vis. Sci.*, **55** (2014) 3185.
21. A. F. Fercher, W. Drexler, C. K. Hitzenberger, and T. Lasser, *Rep. Prog. Phys.*, **66** (2003) 239.
22. J. S. Schuman, C. A. Puliafito, J. G. Fujimoto, and J. S. Duker, *Optical Coherence Tomography of Ocular Diseases* (New Jersey: Slack Inc., 2012).
23. A. Ustione and D. W. Piston, *J. Microscopy*, **243**, (2011) 221.
24. R. M. Williams, D. W. Piston, and W. W. Webb, *Faseb. J.*, **8**, (1994) 804.
25. W. Denk, J. H. Strickler, and W. W. Webb, *Science*, **248**, (1990) 73.
26. W. Denk, D. W. Piston, and W. W. Webb, Chapter 28: Multi-photon molecular excitation in laser-scanning microscopy. In J. B. Pawley (ed.), *Handbook of Biological Confocal Microscopy*, (3rd edn.) (New York: Springer Science+Business Media, LLC, 2006).
27. H. He, *et al. Adv. Photon. Nexus*, **1**, (2022) 026001.

28. A. C. Millard, P. W. Wiseman, D. N. Fittinghoff, K. R. Wilson, J. A. Squier, and M. Muller, *Appl. Opt.*, **36** (1999) 7393.
29. M. Mielke, D. Gaudiosi, K. Kim, M. Greenberg, X. Gu, R. Cline, X. Peng, M. Slovick, N. Allen, M. Manning, M. Ferrel, N. Prachayaamorn, and S. Sapers, *J. Laser Micro/Nanoeng.*, **5**, (2010) 53.
30. S. S. Wellershoff, *et al.*, *Appl. Phys. A*, **69** (Suppl.), (1999) S99.
31. F. Korte, *et al.*, *Appl. Phys. A*, **69** (Suppl.), (1999) S7.
32. P. P. Seriols and J. M. Fernández, Characterization of a fiber laser system for materials processing (2019). https://diposit.ub.edu/dspace/bitstream/2445/141640/1/PANIAGUA%20SERIOLS%20Pol.pdf
33. C. Wandera, *Fiber Lasers in Material Processing*, Chapter 18 in Fiber Laser Editor M. Paul EBOOK (PDF) ISBN 978-953-51-4207-2 ISBN 978-953-51-2257-9.
34. H. Hügel, *Opt. Lasers Eng.*, **34**, (2006) 213.
35. J. Canning, *Opt. Lasers Eng.*, **44**, (2006) 647.
36. C. Thomy, T. Seefeld, and F. Vollertsen, *Adv Mater. Res.*, **8**, (2005) 171.
37. V. Kancharla, In Proc. 25th Int. Congr. Appl. Lasers Elect. Opt., ICALEO 2006 (October 30–November 2), Scottsdale, Arizona, USA, Paper 1301, 2006, p. 579.
38. L. Quintino, A. Costa, R. Miranda, D. Yapp, V. Kumar, and C. J. Kong, *Mater. Design*, **28**, (2007) 1231.
39. W. M. Steen, *Laser Material Processing*, (3rd edn.) (London: Springer-Verlag, 2003), pp. 83–90, 111–122, 126, 163–165.

Index

www.ingramcontent.com/pod-product-compliance
Lightning Source LLC
LaVergne TN
LVHW010332220926
848484LV00004BB/884